Lecture Notes in Physics

Springer-Verlag Berlin Heidelberg GmbH

The Editorial Policy for Proceedings

The series Lecture Notes in Physics reports new developments in physical research and teaching – quickly, informally, and at a high level. The proceedings to be considered for publication in this series should be limited to only a few areas of research, and these should be closely related to each other. The contributions should be of a high standard and should avoid lengthy redraftings of papers already published or about to be published elsewhere. As a whole, the proceedings should aim for a balanced presentation of the theme of the conference including a description of the techniques used and enough motivation for a broad readership. It should not be assumed that the published proceedings must reflect the conference in its entirety. (A listing or abstracts of papers presented at the meeting but not included in the proceedings could be added as an appendix.)

When applying for publication in the series Lecture Notes in Physics the volume's editor(s) should submit sufficient material to enable the series editors and their referees to make a fairly accurate evaluation (e.g. a complete list of speakers and titles of papers to be presented and abstracts). If, based on this information, the proceedings are (tentatively) accepted, the volume's editor(s), whose name(s) will appear on the title pages, should select the papers suitable for publication and have them refereed (as for a journal) when appropriate. As a rule discussions will not be accepted. The series editors and Springer-Verlag will normally not interfere with the detailed editing except in fairly obvious cases or on technical matters.

Final acceptance is expressed by the series editor in charge, in consultation with Springer-Verlag only after receiving the complete manuscript. It might help to send a copy of the authors' manuscripts in advance to the editor in charge to discuss possible revisions with him. As a general rule, the series editor will confirm his tentative acceptance if the final manuscript corresponds to the original concept discussed, if the quality of the contribution meets the requirements of the series, and if the final size of the manuscript does not greatly exceed the number of pages originally agreed upon. The manuscript should be forwarded to Springer-Verlag shortly after the meeting. In cases of extreme delay (more than six months after the conference) the series editors will check once more the timeliness of the papers. Therefore, the volume's editor(s) should establish strict deadlines, or collect the articles during the conference and have them revised on the spot. If a delay is unavoidable, one should encourage the authors to update their contributions if appropriate. The editors of proceedings are strongly advised to inform contributors about these points at an early stage.

The final manuscript should contain a table of contents and an informative introduction accessible also to readers not particularly familiar with the topic of the conference. The contributions should be in English. The volume's editor(s) should check the contributions for the correct use of language. At Springer-Verlag only the prefaces will be checked by a copy-editor for language and style. Grave linguistic or technical shortcomings may lead to the rejection of contributions by the series editors. A conference report should not exceed a total of 500 pages. Keeping the size within this bound should be achieved by a stricter selection of articles and not by imposing an upper limit to the length of the individual papers. Editors receive jointly 30 complimentary copies of their book. They are entitled to purchase further copies of their book at a reduced rate. As a rule no reprints of individual contributions can be supplied. No royalty is paid on Lecture Notes in Physics volumes. Commitment to publish is made by letter of interest rather than by signing a formal contract. Springer-Verlag secures the copyright for each volume.

The Production Process

The books are hardbound, and the publisher will select quality paper appropriate to the needs of the author(s). Publication time is about ten weeks. More than twenty years of experience guarantee authors the best possible service. To reach the goal of rapid publication at a low price the technique of photographic reproduction from a camera-ready manuscript was chosen. This process shifts the main responsibility for the technical quality considerably from the publisher to the authors. We therefore urge all authors and editors of proceedings to observe very carefully the essentials for the preparation of camera-ready manuscripts, which we will supply on request. This applies especially to the quality of figures and halftones submitted for publication. In addition, it might be useful to look at some of the volumes already published. As a special service, we offer free of charge LATEX and TEX macro packages to format the text according to Springer-Verlag's quality requirements. We strongly recommend that you make use of this offer, since the result will be a book of considerably improved technical quality. To avoid mistakes and time-consuming correspondence during the production period the conference editors should request special instructions from the publisher well before the beginning of the conference. Manuscripts not meeting the technical standard of the series will have to be returned for improvement.

For further information please contact Springer-Verlag, Physics Editorial Department II, Tiergartenstrasse 17, D-69121 Heidelberg, Germany

Jürgen Parisi   Stefan C. Müller
Walter Zimmermann (Eds.)

A Perspective Look at Nonlinear Media

From Physics to Biology and Social Sciences

Springer

Editors

Jürgen Parisi
Fachbereich Physik
Abteilung Energie- und Halbleiterforschung
Universität Oldenburg
D-26111 Oldenburg, Germany

Stefan C. Müller
Institut für Experimentelle Physik
Abteilung Biophysik
Universität Magdeburg
D-39016 Magdeburg, Germany

Walter Zimmermann
Institut für Festkörperforschung
Forschungszentrum Jülich
D-52425 Jülich, Germany
and
Max-Planck-Institut für Physik Komplexer Systeme
Nöthnitzer Straße 38
D-01187 Dresden, Germany

Cataloging-in-Publication Data applied for.
Die Deutsche Bibliothek – CIP-Einheitsaufnahme
A perspective look at nonlinear media: from physics to biology and social sciences / Jürgen Parisi...
(ed.).

(Lecture notes in physics; Vol. 503)
ISBN 978-3-662-14190-8   ISBN 978-3-540-69681-0 (eBook)
DOI 10.1007/978-3-540-69681-0
ISSN 0075-8450
ISBN 978-3-662-14190-8

Typesetting: Camera-ready by the authors/editors
Cover design: *design & production* GmbH, Heidelberg
SPIN: 10644050   55/3144-543210 - Printed on acid-free paper

Preface

Concepts of nonlinear physics are appreciated in and applied to an increasing amount of research disciplines, ranging from physics to biology and social sciences. A considerable number of publications and conferences reflect that development. Two national meetings, sponsored by the WE-Heraeus-Stiftung Hanau, took place in the Physikzentrum Bad Honnef in 1995 and 1996. Both were organized by the present editors who are members of a German scientific network on nonlinear physics of complex systems[1]. It is the overall intention of these meetings to bring together lecturers and experienced and younger researchers from physics, chemistry, biology, biomedicine, and engineering. Recent developments and possible future directions should be highlighted and brought closer to a broader audience and, particularly, the younger generation of promising researchers. The contributions to the meeting in November 1995 built the framework of a recent volume of Springer's Lecture Notes in Physics[2]. In its preface, the main aspects of that spirit have been sketched already.

Following the event of 1995, we wanted in 1996 a further meeting on additional topics of nonlinear media in progress. The enthusiasm of the participants confirmed to the organizers that this enterprise was timely and served as an organizing center for the community. Lively discussions took place during and after the lectures, particularly also in front of the posters. They lasted up to the late evening in the cellar of the Physikzentrum Bad Honnef. Our prevailing observations underscored the stimulating atmosphere in that interdisciplinary field. With this volume, we intend to offer a selection of articles on such kind of nonlinear topics in progress, again ranging from physics and chemistry to biology and some application of social sciences, which have been partly presented at the Heraeus meeting in November 1996, to the attention of an even wider audience.

In view of the huge variety of topics addressed, it appears futile to put the articles into one specific order. However, their actual succession can be

---

[1] The program on pattern formation in dissipative continuous systems – experiment and theory in quantitative comparison is supported financially by the Deutsche Forschungsgemeinschaft (DFG).

[2] J. Parisi, S.C. Müller, and W. Zimmermann (eds.): Nonlinear Physics of Complex Systems – Current Status and Future Trends (Springer, Berlin, Heidelberg 1996)

characterized by certain keywords that are important to the field of nonlinear sciences. Elementary physical aspects of quantum optics and electron crystallization are followed by either cellular or flow patterns in fluids and also in granular media. Traffic flow exhibits related features and, therefore, a bridge can be built to the social sciences – reviewed here in terms of sociodynamical structures. We also find social behavior at the cellular level of biological systems, as exemplified in a study of optimized strategies in the life cycle of amoeba colonies. Furtheron, we are led to organizational aspects even at the molecular level in immunology as well as in the evolution of nucleic acids. A neurobiological work presents the control of brain structures via neuronal excitation, in close analogy to chemical wave fronts propagating in an excitable chemical medium. Those chemical patterns are looked at both in bulk solutions and on surfaces in heterogeneous systems. From regular structures, we turn to the more complex behavior in biology and physics, particularly in an article on hydrodynamical turbulence. The succession is completed by three contributions on low-dimensional dynamics in solid-state physics and concluded by an hypothesis about gravity.

Acknowledgments The editors would like to express their gratitude to all authors for taking a large amount of effort and time to prepare the broad spectrum of contributions printed below. Special thanks are due to Wolf Beiglböck from Springer-Verlag Heidelberg and Volker Schäfer from Heraeus-Stiftung Hanau for encouraging us to edit this volume – moreover, to Sabine Lehr, Rosi Klepac, Brigitte Reichel-Mayer, and Urda Beiglböck from Springer-Verlag Heidelberg, Jutta Hartmann from Heraeus-Stiftung Hanau, and Ilona Dwehus from University of Oldenburg for skillful assistance and valuable technical and secretarial support.

Oldenburg   Jürgen Parisi
Magdeburg   Stefan C. Müller
Jülich/Dresden   Walter Zimmermann
November 1997

Contents

Quantum Chaos in Rydberg Atoms

H. Held, J. Schlichter and H. Walther

Sektion Physik der Universität München and
Max-Planck-Institut für Quantenoptik
85748 Garching, Fed. Rep. of Germany

Abstract. Three experiments performed on rubidium Rydberg atoms in strong external fields are presented which probe different characteristic quantum features of the classically chaotic system. First, a new excitation scheme is described which allows for the observation of level statistics and long-periodic orbits in the case of static magnetic and electric fields. Second, the observation of wave packet evolution along quasi-Landau orbits is reported for the case of a magnetic field only and for parallel electric and magnetic fields. Finally, experiments on the microwave ionization of atomic Rydberg states are reviewed. Dynamical localization and its gradual destruction by microwave noise are observed.

1 Introduction

Up to now, the relation of quantum mechanics and classical chaos is still an open question. Research in the field of quantum chaos aims to resolve this difficulty. The method used is two-fold: First, quantum data are analyzed with respect to fingerprints of classical chaos, and second, underlying semiclassical theories which utilize asymptotic expansions in $\hbar$ are elaborated and tested. In this context it is remarkable that Rydberg atoms in strong external fields are ideal candidates for testing the behavior of quantal systems with a classically chaotic counterpart. The classical dynamics of Rydberg atoms exposed to a strong static magnetic (and, eventually, an additional electric field) displays a continuous transition from regular to chaotic behavior as the relative strength of the external perturbation is increased. The same holds in the case of Rydberg atoms exposed to a strong microwave field. Since in all these cases the Hamiltonian is a sum of power functions of position and momentum, one can introduce scaled variables which completely define the classical dynamics, and span the parameter space in which quantum and classical dynamics of the system are to be compared.

Up to now semiclassical methods provide the main tool in order to understand the correspondence between quantum mechanics on the one hand and classical mechanics on the other hand. Since spectra – or, more precisely, excitation cross sections – are the experimentally most easily accessible quantities, semiclassical theory has been elaborated most for the analysis of this kind of experimental data sets (Du and Delos 1988), (Dando et al. 1995).

While the case of integrable classical dynamics can be well understood via EBK (torus) quantization, the case of a globally chaotic or, generically, mixed regular/chaotic phase space still have to be elucidated. Although Gutzwiller's trace formula (Gutzwiller 1982) and related expressions could successfully establish the connection between sinusodial modulations in level densities or excitation cross sections on the one hand and the shortest periodic or closed orbits of the underlying classical dynamics on the other hand, the keys for exploring the fingerprints of nonlinearity on the fine-scale spectral properties seem to be hidden behind severe convergence problems of the corresponding asymptotic expansions. It is therefore of large interest to confront the existing theory with experimental studies of the fine-scale level structure displayed by the quantum system.

Up to now, experiments on Rydberg atoms exposed to strong static fields have focused mostly on stationary properties ("energy domain spectroscopy"). There is no tradition of time domain spectroscopy (testing the time evolutution of non-stationary states) comparable to that of energy domain spectroscopy. The experiments described in section 3 are among the first attempts to close this gap. We will focus on the dynamics of quantum wave packets excited by short laser pulses. The general idea of the experiments is to probe the evolution of these wave packets as the strength of the magnetic field is increased, i.e. as the structure of classical phase space undergoes a metamorphosis from globally regular to globally chaotic structure (Friedrich and Wintgen 1989).

In contrast to the conservative systems described so far, Rydberg atoms in a microwave field – an explicitly time dependent system – naturally allow for the study of energy exchange between the different degrees of freedom. A rough estimate of this exchange is the ionization probability of the atom, one of the quantities that can be measured directly. It contains some information on the dynamics, but in a strongly convoluted form (Delande and Buchleitner 1994). In section 4 we present a set of ionization experiments on Rydberg atoms exposed to coherent as well as stochastic microwave fields.

In the following three sections we therefore contribute to the whole range of Rydberg atom quantum chaology experiments from purely stationary to driven and stochastic systems.

2 Spectroscopy of Rydberg Atoms in Static Magnetic and Electric Fields

Since classical chaos is a long-time phenomenon it is natural to expect universal chaos-induced quantum features to show up in fine-scale properties on energy scale due to the time-energy uncertainty. Therefore, those features should become visible if high-resolution spectra are analyzed. By definition, the nearest-neighbor spacing distribution (NNS) of individual energy levels is the spectral property which refers to the fine-scale characteristics most. For classically chaotic systems, a well-known hypothesis (Bohigas et al. 1984) predicts the same spectral properties as for spectra which are generated by gaussian random matricies. Remarkably, this hypothesis explicitly includes low-dimensional systems. This establishes an interpretation of chaos as a generator of complexity. So far, there is no sane understanding of the underlying mechanism. Therefore, empirical NNS of classically chaotic systems are highly desirable. Until the Garching experiments, there had been no published experimental quantum NNS of a system with well-defined chaotic classical counterpart. We have taken such data in 2D- (pure magnetic field) and 3D-case (crossed magnetic and electric fields).

In the case of systems with generalized time-reversal symmetry – like Rydberg atoms in homogeneous magnetic and electric fields – random matrix theory predicts a Wigner-like NNS which shows level repulsion. Contrary, generic classically regular systems have been proved (Berry and Tabor 1977) to display Poissonian quantum spectra (level clustering). Our spectral resolution is sufficient to significantly distinguish between these two different cases. The high resolution is achieved by means of a new two-step excitation of rubidium. The former problem of simultaneously exciting from ten ground-state hyperfine levels and therefore superimposing ten Rydberg-series (Raithel et al. 1991) is successfully overcome (see Fig. 1) by this method.

A well defined structure of classical phase space facilitates the search for fingerprints of nonlinear classical motion in the experimentally observed spectrum of a quantum system. This is achieved by fixing the values of the scaled parameters

$$\varepsilon := EB^{-2/3}, \ f := FB^{-4/3}, \tag{1}$$

(E=energy, B=magnetic field, F=electric field[1]) during one experimental scan. For fixed ε and f, experimental NNS underly the following two restrictions: On the one hand, E has to be chosen high enough in order to induce proper level-mixing. On the other hand, E has to be low enough to keep the mean level spacing large compared to the experimental line width.

At first, we studied the most simplest, the 2D-case ($f = 0$, i.e. pure magnetic field). Fig. 3 shows that our experimental line width is small compared to the mean level spacing.

In Fig. 4, accumulated NNS for a chaotic ($\varepsilon = -0.24$) and a mixed situation ($\varepsilon = -0.315$) (Friedrich and Wintgen 1989), (Held et al. 1997) are shown. In the first case, the NNS is Wigner-like, while in the second we observe an intermediate distribution. This means that we obtain an experimental verification of the above hypothesis!

The same spectra were analyzed with respect to the underlying semiclassical theory, i.e. the spectra were Fourier-analyzed and the resonances related to the closed orbits of the classical system. For the first time, the theory was tested up to the Heisenberg time limit. The Heisenberg time defines an orbit's length resolving the mean level spacing and sets therefore the relevant

---

[1] If not explicitly mentioned, atomic units are used throughout the article.

1. One-Step-Scheme:

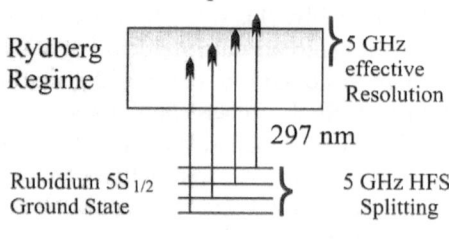

Rydberg Regime

5 GHz effective Resolution

297 nm

Rubidium 5S $_{1/2}$ Ground State

5 GHz HFS Splitting

2. Two-Step-Scheme:

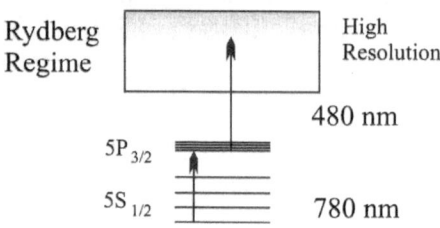

Rydberg Regime

High Resolution

480 nm

5P $_{3/2}$

5S $_{1/2}$

780 nm

Fig. 1. One-step versus two-step excitation scheme. In the old one-step scheme several Rydberg series are superimposed due to the hyperfine splitting of the ground state. For the sake of clearity, only the four Paschen-Back hyperfine levels of the [87]Rb isotope are indicated. In the real experiment six additional transitions from the [85]Rb isotope show up. With a two-step excitation only one of the hyperfine levels is selected and an unwanted superposition of Rydberg series is avoided. Fig. 2 demonstrates how the first excitation step selects one of the ten hyperfine transitions.

time scale for the desired explanation of the NNS. The theory was confirmed within the experimental limitations (Held et al. 1997).

In addition, we took data for the 3D, the crossed-fields case. Rydberg atoms in crossed electric and magnetic fields are one of the rare experimentally accessible three dimensional systems displaying classical chaos. Our NNS show significant deviations from a Poissonian distribution, consistent with a mixed classical phase space. We demonstrated that it will be possible to test the above hypothesis also in the 3D-case, in connection with an intricate analysis of the related 5D energy shell in phase space.

3 Observation of Quasi-Landau Wave Packets

3.1 Detection of wave packet motion

A technique suggested by Noordam et al. (Noordam et al. 1992) and implemented first by Christian et al. (Christian et al. 1993) probes the temporal

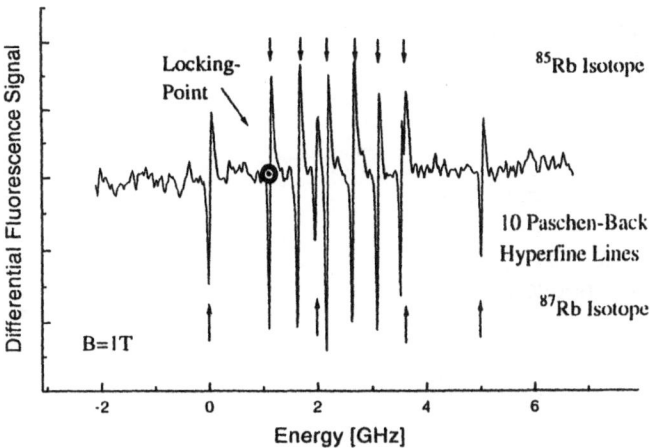

Fig. 2. Stabilization of the first excitation step. The differential spectrum shows ten possible Paschen-Back-HFS $5S_{1/2} \rightarrow 5P_{3/2}$ transitions. The 780 nm laser is locked on the first of the stronger ^{85}Rb transitions. The dominant contribution to the measured HFS splitting is caused by the $5S_{1/2}$ state. Therefore the energy width of the above pattern (≈ 5 GHz) equals the effective resolution of the one-step scheme.

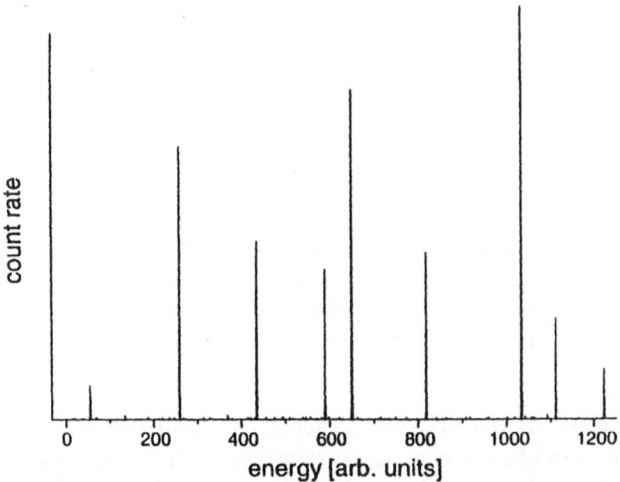

Fig. 3. Part of an expermimental spectrum for mixed phase space ($\varepsilon = -0.315$). The experimental line width of 30 MHz is small compared to the mean level spacing. This is an important prerequisite for experimental NNS.

evolution of an initially excited wave packet with a second, identical but temporarily delayed wave packet. A short ultraviolet pulse excites a wave packet, and after a time t_d a second, identical pulse interacts with the atom. Since both pulses are equivalent, the second pulse produces an identical wave packet

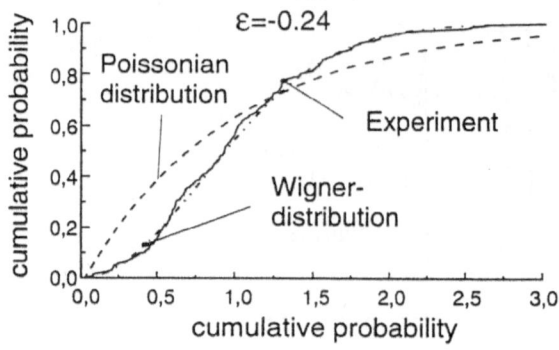

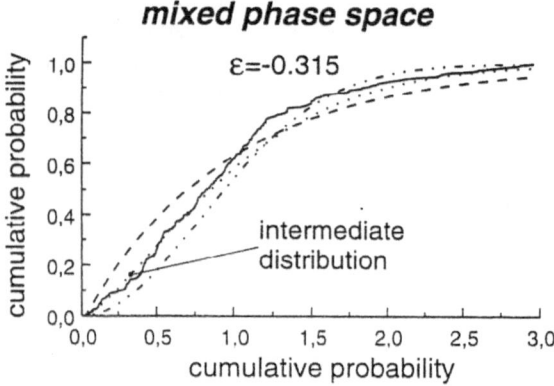

Fig. 4. Accumulated experimental NNS for chaotic and mixed phase space. In the chaotic case, the experimental curve is Wigner-like, in the mixed intermediate. Therefore the high-resolution spectra confirm the NNS-hypothesis outlined before. For each spectrum, between 230 and 270 levels have been analyzed.

(assuming perturbative pulses). If the first wave packet has returned to the excitation region near the core after a time t_d, it overlaps with the second one and interferes with it. In analogy to interference of two optical pulses the relative phase between the wave packets determines whether they interfere constructively or destructively. We change this phase by scanning the delay by a small amount and detect a modulation in the number of excited atoms ($n > 30$). This modulation gives a measure of the period of the orbit traced by the wave packet and giving rise to the recurrence in the Rydberg signal.

More quantitatively, we analyze the superposition of a time-evolved Rydberg wave packet with a second identical one. The wave packet after the first excitation may be written as $|wp\rangle$. After a time t_d, the second pulse produces an additional contribution to the Rydberg state identical to the first one, and the total Rydberg part of the wave function is $|wp\rangle + \hat{K}(t_d)|wp\rangle$, where

$$\hat{K}(t) = \sum_n |n\rangle e^{-iE_n t}\langle n| \tag{2}$$

(n, E_n : eigenenergies and -states[2])
is the time evolution operator. We detect excited atoms, for which the total probability $P_T(t_d)$ is given by

$$P_T(t_d) = 2\langle wp|wp\rangle + 2Re\left\{\langle wp|\hat{K}(t_d)|wp\rangle\right\}. \tag{3}$$

Utilizing (2) and the finite bandwidth of the wave packet, $\langle wp|\hat{K}(t_d)|wp\rangle$ can be expressed in the case of a variation Δt_d much smaller than the pulse length τ, i.e. $\Delta t_d/\tau \ll 1$, as follows:

$$\langle wp|\hat{K}(t_d + \Delta t_d)|wp\rangle \approx \langle wp|\left\{\sum_{n,\ |E_n - E_{\bar{n}}|<1/\tau} |n\rangle e^{-iE_n(t_d+\Delta t_d)}\langle n|\right\}|wp\rangle \tag{4}$$

$$\approx \langle wp|\left\{\sum_{n,\ |E_n - E_{\bar{n}}|<1/\tau} |n\rangle e^{-iE_n t_d - iE_{\bar{n}}\Delta t_d)}\langle n|\right\}|wp\rangle$$

$$\approx \langle wp|\hat{K}(t_d)|wp\rangle e^{-iE_{\bar{n}}\Delta t_d},$$

where $E_{\bar{n}}$ denotes the wave packet's central energy determined by the laser frequency[3].

Therefore by varying Δt_d we let the Rydberg probability $P_T(t_d + \Delta t_d)$ (see (3)) oscillate with an amplitude $|\langle wp|\hat{K}(t_d)|wp\rangle|$ around a mean value $\langle wp|wp\rangle$[4]. Both quantities are determined in the experiment. The latter provides a proper normalization while the former is the object of primary importance. Such a type of wave packet correlation functions has been object of primary interset in most calculations investigating the validity of the semiclassical propagator (Tomsovic and Heller 1993) in classically chaotic systems.

3.2 Experimental setup

The laser source has been described elsewhere (Yeazell et al. 1993) and is only briefly reviewed here. A synchronously-pumped and cavity-dumped dye laser produces pulses of 6 ps in duration with an energy of 10 nJ at a repetition rate of 2.5 MHz. This optical pulse is frequency doubled in a BBO crystal

---

[2] The continuum can be neglected here.

[3] In order to simplify the notation, the ground state energy was set to zero.

[4] The factor 2 of (3) is suppressed.

with 10% conversion efficiency resulting in an ultraviolet pulse with a width of 4 ps and a wavelength of 297 nm.

The UV beam is sent to a 50% beam splitter to produce pairs of identical UV pulses. The pulses are recombined at another 50% beam splitter forming a Mach-Zehnder interferometer. The first arm has a variable delay line which can produce delays over a range of 1500 ps. In the second arm, the UV beam passes through a 1 mm thick quartz window mounted on a galvanometer. When the window is tilted about an axis parallel to its diameter, the pulse is delayed by a small time ($\Delta t_d \approx 0.3$ ps). We call the window "phase plate" to indicate that the small time delay corresponds to a phase shift between the two pulses. Both UV pulses are then focused by a 30 cm lens onto a beam of rubidium atoms. A homogeneous magnetic field, parallel to the direction of the atomic beam, is applied to the interaction region using a superconducting magnet. The excited atoms reach a region where they are field ionized. The electrons are then guided into a multi-channel plate detector by the magnetic field lines, giving a signal proportional to the Rydberg probability P_T.

Any modulation of the signal caused by interfering wave packets will be at the frequency of the modulation produced by the phase plate. We filter out this modulation by means of a lock-in technique. To provide a reference signal a second, delay-independent interferometer is used (Marmet et al. 1994).

3.3 Experimental results

We tested the above destribed detection technique by probing radial wave packets (B=0 and E=0) for values of $n = 41$ up to $n = 83$. The known features of radial wave packets were clearly observed (see e.g. Fig. 5 in the case of $n = 51$). The data show a couple of recurrences, fractional revivals, and full revivals (Yeazell and Stroud 1991). However, an additional modulation not observed in earlier measurements appeared. It is caused by the hyperfine splitting of the Rb ground state (Raithel et al. 1994).

Fig. 6 shows the recurrence signal $|\langle wp|\hat{K}(t_d)|wp\rangle|$ for increasing values of B with the UV beam polarized perpendicular to the **B** field. Here, we see maxima produced by the interference between the first, time evolved and the second wave packet. Since both UV pulses are identical, the scans should be symmetric about $t_d = 0$. For small B fields, the scans are symmetric for delays up to 150 ps. For B=3.3 T, the symmetry is recognizable up to t_d=60 ps. We use the symmetry to determine a range of confidence for the experimental results. Outside the limit defined by the symmetry (indicated by a vertical line), the evolution is not repeatable from day to day. However, within this limit, good agreement is obtained. This loss of symmetry suggests a sensitivity to initial conditions. That is the experimental conditions, particularly the pulse characteristics, do not remain absolutely unchanged over the span of a data run ($\approx$30 minutes for 100 ps delay). Sensitivity to initial conditions is a defining characteristic of a chaotic system. Nevertheless, the observed sensitivity in the quantum experiment has to be interpreted with great care.

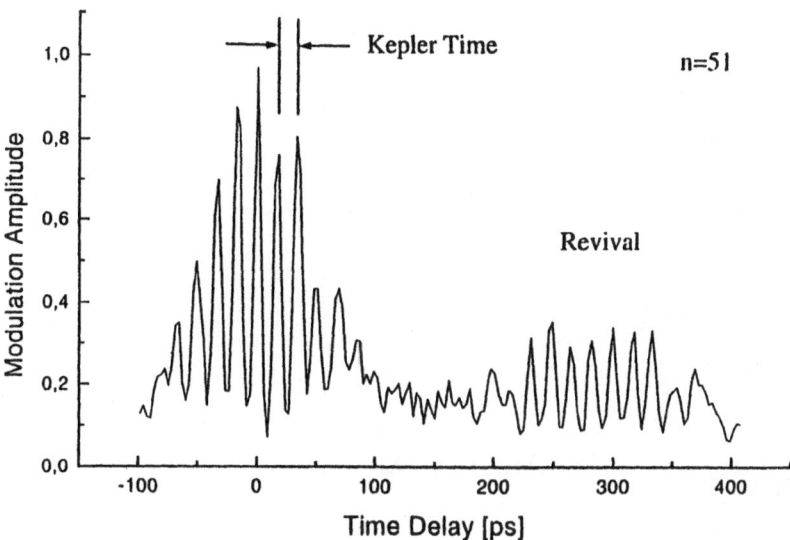

Fig. 5. Modulation amplitudes versus delay time for the Coulomb problem. The detection scheme is successfully tested: The quantum wave packet recurrence time coincides with the classical Kepler time.

For our experimental conditions, as the B field is increased beyond 2.2 T, the fraction of available phase space in which the classical trajectories are regular drops below 50% (Friedrich and Wintgen 1989). However, it should be noted that the region of phase space associated with the plane perpendicular to the B-field remains a region of regular motion.

We compare the results of the experiment with a semiclassical calculation. We related the recurrence amplitude $|\langle wp|\hat{K}(t_d)|wp\rangle|$ to classical orbits via a closed orbit expansion for the excitation cross section (Du and Delos 1988).

In Fig. 6 the results of the semiclassical calculations are superimposed over each measurement. All curves are normalized so that the peak at $t_d = 0$ has the same amplitude. For $B \leq 2.1$ T, the semiclassical theory reproduces the experimental results. Individual maxima in the modulation can be associated with successive transits in a single dominant orbit – the original quasi-Landau orbit – that lies in the plane perpendicular to **B**.

For $B > 2.1$T, the observation of the quasi-Landau orbit is less straightforward. The classical period becomes comparable to the laser pulse duration, therefore adjacent recurrences start to interfere. Additionally, a laser polarization perpendicular to the magnetic field leads to the excitation of superpositions of $l_z = 1$ and $l_z = -1$ states. Both manifolds are orthogonal, and exhibit identical spectra except for an energy shift of eB/m_e. For $t_d = (1 + 2k)(17.8 \text{ ps}/B \text{ [T]})$, with integers k, the Rydberg probabilities $P_T[l_z = \pm 1]$ of both manifolds oscillate with a phase difference π, hence cancel the modulation of the sum probability, and mask the wave packet recurrence.

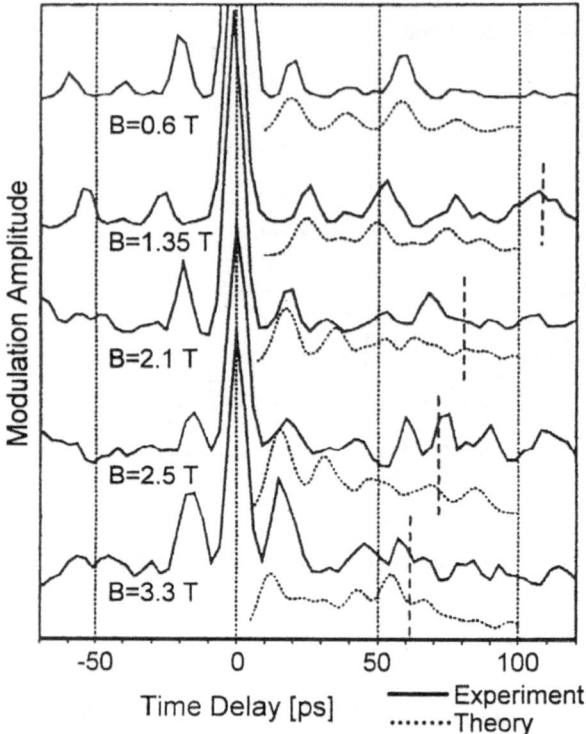

Fig. 6. Modulation amplitudes versus delay time for the indicated values of the magnetic field and central energy W=-36.1 cm^{-1}. The solid lines show the experimental results, whereas the dashed lines follow from a semiclassical calculation. The vertical long-dashed lines indicate the range over which the data is reproducible. This range is derived from the symmetry of the experimental results (Marmet et al. 1994).

We also excited wave packets with a laser polarization parallel to **B**, again at $E = -36.1$ cm^{-1} and fixed magnetic field stregth $B = 1.5$ T. Since in the case of parallel polarization only one Zeeman manifold contributes to the Rydberg signal, a masking of wave packet recurrences described above does not have to be taken into account in this case. The dominant quasi-Landau orbit observed in the case of the polarization perpendicular to **B** cannot be excited with parallel polarization. The recurrence signal from the remaining orbits turned out to be four times smaller. In order to modify the underlying classical dynamics we added an electric field of 700, 1400, 2100, and 2800 V/m parallel to **B**. We could identify the recurrences of four different orbits, one of them bifurcated from an orbit parallel to **B** between 700 and 1400 V/m.

4 Microwave Ionization

Up to now, in this paper experiments have been reported testing conservative systems either via energy domain or via time domain spectroscopy. We now focus on a driven system, Rydberg atoms exposed to a linearly polarized microwave field of frequency ω and amplitude F:

$$H = \frac{\mathbf{p}^2}{2} - \frac{1}{r} + Fz\cos(\omega t). \tag{5}$$

We introduce scaled variables corresponding to (5):

$$F_0 = n_0^4 F, \omega_0 = n_0^3\omega, \tag{6}$$

where n_0 denotes the hydrogen principal quatum number.

Once the driving field F exceeds a critical value, the classical electron may exhibit chaotic motion, absorb energy from the driving field and diffuse over increasing values of the classical action, until it reaches the continuum and thus will be ionized.

A simple model based on the quantum treatment of the periodically kicked rotor predicts the following:

For $\omega_0 < 1$ the scaled microwave amplitude $F_0(10\%)$ necessary to ionize 10% of the irradiated atoms with ω_0 decreases as $\omega_0^{-1/3}$ following the *classical* threshold. In contrary, for $\omega_0 > 1$, $F_0(10\%)$ follows the so-called "quantum delocalization border" which lies systematically *above* the classical threshold:

$$F_0(10\%) = \frac{\omega_0^{7/6}}{\sqrt{8\,n_0}}\sqrt{1 - \frac{n_0^2}{n_c^2}}, \tag{7}$$

where n_c defines the effective experimental continuum threshold, as states with a principal quantum number situated above n_c will be ionized by inevitable stray fields inside the apparatus (Casati et al. 1988).

The interpretation is as follows: Quantum dynamics may mimick the classically diffusive energy exchange between the electron and the field for a short interaction time of several field cycles before quantum mechanical interference "freezes" the energy transport in the quantum system.

4.1 Experimental setup

An atomic beam successively passes spatially separated regions where laser-excitation, microwave-interaction, and detection are performed. Since all atoms should interact with the microwave for exactly the same time, the experiment is set up in a pulsed way. One cycle of the pulsed experiment consists of the following steps:

With use of up to 10 mW of near-UV radiation obtained from an intracavity-doubled, rhodamine 6G cw ring-dye-laser operating with a typical bandwidth of 1 MHz, we can excite ground-state ^{85}Rb$(5s)\,^2S_{1/2}(F = 3)$ atoms directly to $^{85}\mathrm{Rb}(n_0 p)\,^2 P_{3/2}$ Rydberg states with $n_0 = 40$ to $n_0 = 135 \ldots 170$. The continuous laser radiation (polarized parallel to the atomic beam axis) is sent through an electro-optic switch, which is opened for 6 μs to excite a spatially confined bunch of Rydberg states, with a repetition time of 350 μs.

Within about 20 μs the laser-excited atoms move into the center of the microwave interaction region, which is machined in the dimensions of a standard "M-band" waveguide. The atomic beam crosses the waveguide perpendicular to its long side. The thermal velocity spread of the beam and the 6 μs duration of the laser pulse are small enough to be sure that all atoms experience the same microwave pulse. At the onset of the microwave pulse, all Rydberg atoms are already well within the microwave interaction region; similarly, at the end of the pulse all atoms are still within that region. Therefore, an influene of the microwave fringe field in the waveguide holes can be excluded. Additionally, the microwave frequency is chosen in such a way that – due to the resulting mode structure – all atoms experience exactly the same pulse.

In the noise measurements it is necessary to add noise to the coherent microwave in a well defined way. The noise is generated in the travelling-wave-tube amplifier the input of which was closed with a 50 Ω terminator. A small fraction of the microwave noise is added to the coherent radiation by a coaxial directional coupler.

Finally, the population distribution and the ionization probability of the atoms are analyzed by electric field ionization after the microwave interaction.

4.2 Interaction with coherent microwave radiation

Unscaled representation of the data. We now come to the presentation of our experimental results and start with the discussion of the ionization threshold fields obtained for a purely monochromatic microwave field. We performed ionization experiments at a microwave frequency of 8867 MHz, and for initial values of the principal quantum number ranging from $n_0 = 55$ to $n_0 = 95$. The interaction time with the microwave field was fixed at t = 5 μs.

In order to avoid complications which arise from the question how to scale field amplitude and frequency for non-hydrogenic atoms as rubidium we plot our results in Fig. 7 on a double logarithmic scale, in unscaled, laboratory units. As can be seen from the fits performed in the low- and high-n regime of both plots, the slope of $\log F(10\%)$ as a function of $\log n_0$ is significantly larger for $n_0 \in [70, 95]$ ($n_0 \in [69, 89]$) than for $n_0 \in [55, 70]$ ($n_0 \in [55, 69]$). In Fig. 7, the fits suggest dependences $F(10\%) \sim n_0^{-4.8 \pm 0.3}$ for $n_0 \in [55, 70]$ and $F(10\%) \sim n_0^{-1.2 \pm 0.1}$ for $n_0 \in [70, 95]$, respectively. This is in good agreement with both the power law $\omega_0^{-1/3}$ and (7) after transformation to unscaled parameters.

Scaled representation of rubidium data. Since in the experiments not hydrogen but rubidium is used, the scaling properties mentioned in (6) do

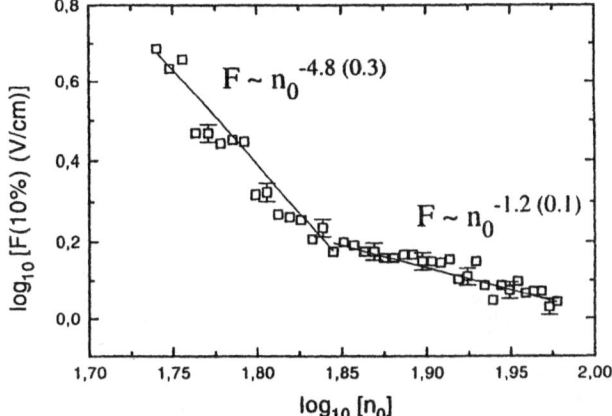

Fig. 7. Unscaled ionization threshold field $F(10\%)$, as a function of n_0, on a double logarithmic scale. Microwave interaction time: 5 μs. Initial states of the rubidium Rydberg atoms: $\ell = 1$, $m = 0$, $n_0 = 55 \ldots 95$. The straight lines are fitted to the low-n and to the high-n behaviour of the thresholds, respectively (Benson et al. 1995). The representative error bars indicate the uncertainty of the ionization threshold. A jump in the slope as predicted by dynamical localization theory is clearly visible.

no more apply. However, for the sake of comparison, we introduce an empirical scaling which no more relies on the explicit form of the Hamiltonian but rather on the spectral characteristics of the unperturbed rubidium atom. Whereas for hydrogen the Kepler frequency is the reference frequency the driving frequency has to be compared to, we chose in the case of the nonvanishing quantum defects δ_ℓ (with $\ell \leq 3$) of Rb the frequency of the dominant transition as a reference.

Starting from $n_0 P$ states, our reference frequency will therefore be the $n_0 P \to (n_0 - 1)D$ transition frequency which is about 0.3 times the spacing of nearby hydrogenic levels of rubidium. This leads to the definition

$$\omega_{0,pd} = \omega / \omega_{n_0 P \to (n_0-1)D}, \tag{8}$$

where we shall skip the subscript "pd" in the following. How to account for the atomic core in the normalization of the microwave amplitude is less evident. We refer here to the static field limit of the ionization threshold which scales as n_0^{*-4}, with n_0^* the effective principal quantum number $n_0^* = n_0 - \delta_\ell$ of the initial state $n_0 P$,

$$F_0 = F n_0^{*-4}, \tag{9}$$

which has the same form as the classical scaling introduced by (6).

Fig. 8 represents the unnormalized data from Fig. 7 in scaled units as just defined in eqs. (8) and (9). The abcissa is scaled according to (8), as well as in units of n_0^{*3}, the latter to allow for an easier comparison to the quantum

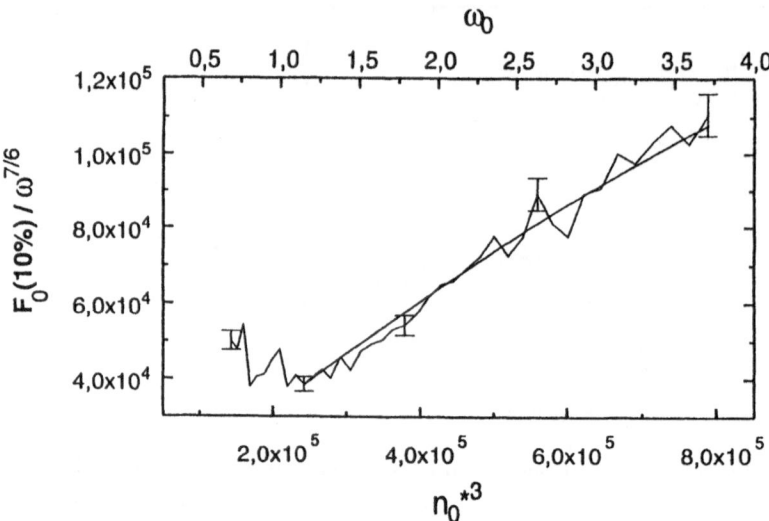

Fig. 8. Scaled ionization threshold field $F_0(10\%)$, as a function of n_0^{*3}, and of the scaled (see eq. (8)) frequency ω_0, respectively. We plot $F_0(10\%)/\omega^{7/6}$ vs. n_0^{*3} (Benson et al. 1995). The data show the predicted minimum at $\omega_0 \approx 1$.

delocalization border (7), where we replace the (numerically determined and not completely well defined (Chirikov 1991)) prefactor $1/\sqrt{8}$ by an adjustable parameter b, to obtain

$$\frac{F_0(10\%)}{\omega^{\frac{7}{6}}} = bn_0^{*3}\sqrt{1 - \frac{n_0^{*2}}{n_c^2}}. \tag{10}$$

$F_0(10\%)/\omega^{7/6}$ is exactly what is plotted in Fig. 8.

The data show a minimum at $\omega_0 \approx 1$, allow for fitting parameters b= 0.174 ± 0.001, $n_c \approx 150$, and therefore confirm eq. (7).

4.3 Influence of additional noise

We also studied the impact of additive noise on the ionization of Rydberg atoms in a monochromatic microwave field.

We show that the presence of a small amount of noise decreases the ionization thresholds in the regime of localization.

Fig. 9 displays $F_0(10\%)$ as a function of the scaled frequency in cases of additional amounts of noise. For the monochromatic case (solid line) we again observe an increase of the ionization threshold as a function of the scaled frequency, indicating dynamical localization. With an increasing noise level the resulting ionization field strength shows a smaller slope which finally gets negative. The influence of the additional noise increases when we increase

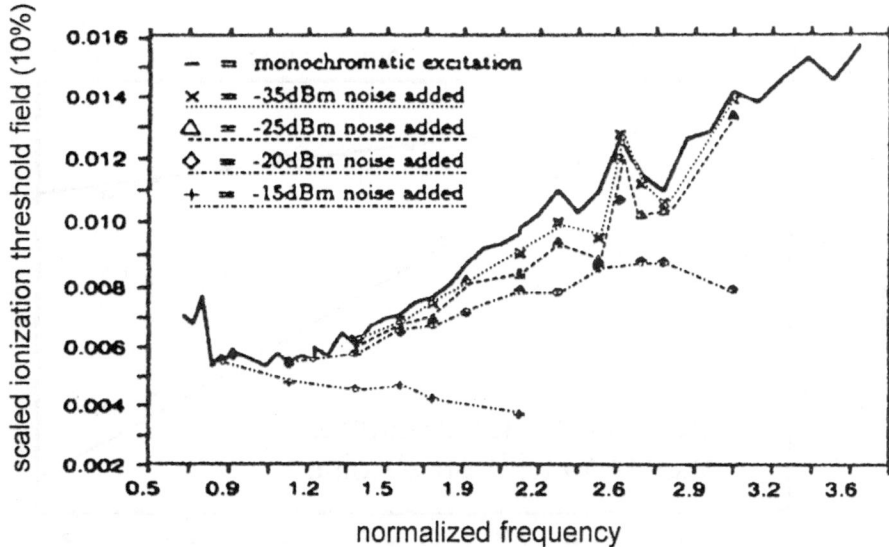

Fig. 9. Scaled ionization threshold field $F_0(10\%)$, as a function of normalized frequency ω_0 (Arndt et al. 1991). The solid line connects the values measured for monochromatic excitation. The dotted lines represent threshold fields observed in the presence of additive noise. Increasing amounts of noise seem to result in a decrease of dynamical localization.

the value of the initial principal quantum number n_0. We interpret our experimental observations in terms of destruction of the coherence properties of the quantum system due to the presence of additive noise; the system then approaches the classical diffusion process.

4.4 Time dependence of the ionization field strength

Finally, we performed an experimental study of the dependence of the ionization threshold on the microwave pulse length in the region $n_0 > 70$ where the phenomenon of localization is observed.

In Fig. 10 $F_0(10\%)$ is plotted as a function of time, with $n_0 = 85$. From this fully logarithmic plot it is apparent that in both cases the data are in agreement with a power-law behavoir $F_0(10\%) \propto \tau^{-\alpha}$ valid over three orders of magnitude. Least-square fits to the data results in $\alpha = 0.13 \pm 0.04$ (monochromatic case) and $\alpha = 0.48 \pm 0.04$ (noise case).

From our experimental observation we can extract the functional form of the quantum diffusion coefficient D_q for excitation by noise only. The system reaches a defined degree of ionization when the product $D_q \tau$ equals

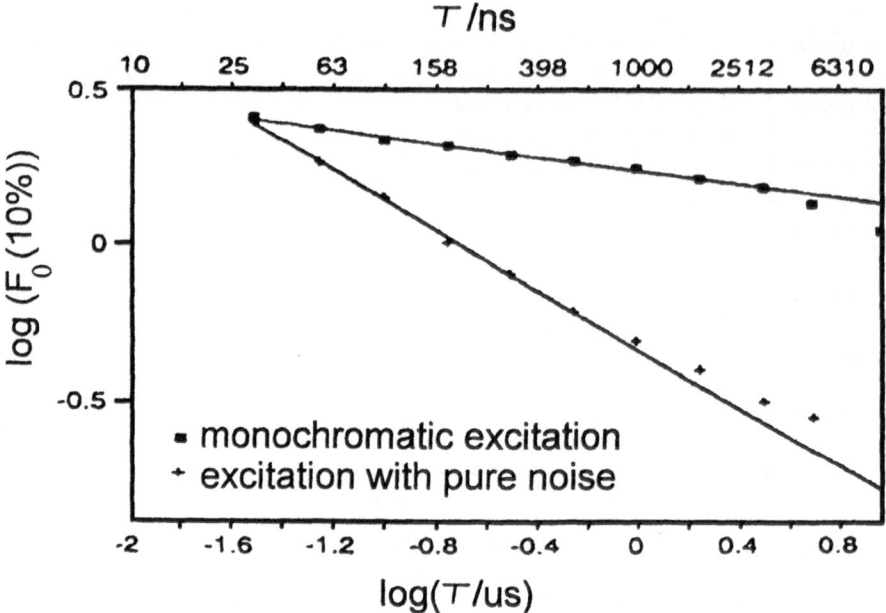

Fig. 10. Time dependence of the the scaled ionization threshold field $F_0(10\%)$ for coherent signal and broadband noise excitation (Arndt et al. 1991). In both cases $F_0(10\%)$ shows a power-law behavior over orders of magnitude.

a given constant value K which defines the amount of ionization. We tested the results for D_q according to $D_q \propto \epsilon^\kappa$, found for the quantum kicked rotor subjected to a broadband external noise signal. The theory predicts $\kappa = 2$. Our experimenatl result is $\kappa = 2.1 \pm 0.1$. It is worth noting that although the theoretical study was performed for the kicked rotor, the results are expected to be model independent. Moreover, the functional form of the diffusive constant coincides with the one obtained for the classical limit of D_q (Ott et al. 1984): $D_{cl} \propto \epsilon^2$.

In the case of a monochromatic microwave signal dynamical localization theory predicts that the ionization thresholds are time independent ($D_q = 0$). In contrast with this theoretical conclusion our experimental results again show a power-law decrease of the ionization threshold, here given by $\tau^{-0.13}$. Ref. (Buchleitner and Delande 1995) supports this dependence being caused by a weak coupling of the involved Floquet states to the continuum. However, the observed decay is much slower than in the case with only noise present.

5 Conclusion

We have tested spectral as well as temporal aspects of the dynamics of rubidium Rydberg atoms in static and in time-periodic external fields.

First, spectroscopy in static magnetic and electric fields has been performed. A new excitation scheme has been set up which gives us access to the fine-scale properties of the quantum spectrum and hence to the experimental study of the long time characteristics of the quantum dynamics. Closed-orbit theory is confirmed for times up to the Heisenberg limit. Experimental level statistics give support to the NNS conjecture of Wigner statistics in the chaotic case. In addition, the dominant closed orbits have been observed in real time for the case of parallel electric and magnetic fields, by the excitation of Rydberg wave packets.

Finally in a driven system, the Rydberg atom in a microwave field, the competition between classical diffusion and dynamical localization which is a quantum phenomenon has been studied in ionization experiments. A one dimensional model of the kicked rotor has been found to be qualitatively valid as far as dynanimcal localization and its destruction by noise are concerned. A first study of the time dependence of the ionization process revealed details so far neglected by dynamical localization theory.

All of these different kinds of experiments on Rydberg atoms exposed to external fields have demonstrated clear fingerprints of nonlinear classical dynamics on quantum mechanics. In that sense, signatures of quantum chaos have been observed.

References

Arndt M., Buchleitner A., Mantegna R. N., Walther H. (1991), Phys. Rev. Lett. **67**, 2435

Benson O., Buchleitner A., Raithel G., Anrdt M., Mantegna R. N., Walther H. (1995), Phys. Rev. A **51**, 4862

Berry M. V., Tabor (1977), Proc. R. Soc. London A **356**, 375

Bohigas O., Giannoni C., Schmit C. (1984), Phys. Rev. Lett. **52**, 1

Buchleitner A., Delande D. (1995), Chaos, Solitons & Fractals **5**, 1125

Casati G., Guarneri I., Shepelyansky D. L. (1988), IEEE J. Quantum Electron. **24**, 1420

Chirikov B. V. (1991): *Chaos and Quantum Physics*, Les Houches Session LII, edited by M.-J. Giannoni, A. Voros, and J. Zinn-Justin, North-Holland, Amsterdam

Christian J. F., Broers B., Hoogenraad J. H., van der Zande W. J., Noordam L. D. (1993), Opt. Comm. **103**, 79

Dando P., Monteiro T., Delande D., Taylor K. (1995), Phys. Rev. Lett. **74**, 1099

Delande D., Buchleitner A. (1994), Advances in Atomic, Molecular, and Optical Phys. **34**, 85

Du M. L., Delos J. B. (1988), Phys. Rev. A **38**, 1896

Friedrich H., Wintgen D. (1989), Phys. Rep. **183**, 37

Gutzwiller M. C. (1982), Physica (Amsterdam) **5D**, 183

Held H., Schlichter J., Raithel G., Walther H. (1997), to be published

Marmet L., Held H., Raithel G., Yeazell J. A., Walther H. (1994), Phys. Rev. Lett. **72**, 3779

Noordam L. D., Duncan D. I., Gallagher T. F. (1992), Phys. Rev. A **45**, 4734

Ott E., Anderson Jr. T. M., Hanson J. D. (1984), Phys. Rev. Lett. **53**, 2187

Raithel G., Fauth M., Walther H. (1991), Phys. Rev. A **44**, 1898

Raithel G., Held H., Marmet L., Walther H. (1994), J. Phys. B **27**, 2849

Tomsovic S., Heller E. J. (1993), Phys. Rev. E **47**, 282

Yeazell J. A., Raithel G., Marmet L., Held H., Walther H. (1993), Phys. Rev. Lett. **70**, 2884

Yeazell J. A.,Stroud Jr. C. R. (1991) Phys. Rev. A **43**, 5153

Crystallization of Electrons and Holes

P. Fulde

Max-Planck-Institut für Physik komplexer Systeme,
Nöthnitzer Straße 38,
01187 Dresden (Germany)

Abstract. The original concept of Wigner for electron crystallization is extended by considering electrons in orbitals close to nuclei. It is shown that in this case electron (or hole) crystallization can take place at high densities. This is in distinction to the homogenous electron gas considered by Wigner where only at very low densities electrons are forming a lattice. We show that Yb_4As_3 is an example for the case studied here. Because of the special anti-Th_3P_4 structure the material has, the crystallization of $4f$ holes can be modelled well by a band Jahn-Teller effect of strongly correlated electrons.

1 Introduction

Eugene Wigner was the first who posed the question what would happen when for an electron gas the Coulomb repulsion energy of the electrons becomes larger than their kinetic energy [1, 2]. For that purpose he considered a homogeneous gas of electrons with a uniformly smeared positive background to ensure charge neutrality (jellium model).

The corresponding Hamiltonians is

$$H = \sum_{\mathbf{p},\sigma} \epsilon(\mathbf{p}) a_{\mathbf{p}\sigma}^\dagger a_{\mathbf{p}\sigma} + \frac{1}{2\Omega} \sum_{\substack{\mathbf{pkq} \\ \sigma\sigma'}} v_{\mathbf{q}} c_{\mathbf{p}+\mathbf{q}\sigma}^\dagger c_{\mathbf{k}-\mathbf{q}\sigma'}^\dagger c_{\mathbf{k}\sigma'} c_{\mathbf{p}\sigma} \tag{1}$$

where $\epsilon(\mathbf{p}) = p^2/2m$ is the kinetic energy of an electron with momentum $\mathbf{p}$, Ω is the total volume and

$$v_{\mathbf{q}} = \frac{4\pi e^2}{q^2}(1 - \delta_{\mathbf{q}0}). \tag{2}$$

This form of $v_{\mathbf{q}}$ ensures that in the long wave-length limit the electron interactions cancel with those of the positive background. The following simple estimates for the kinetic and potential energies hold. Due to Pauli's principle the Fermi momentum k_F of the electrons is related to their density ρ through $k_F = (2\pi)(3\rho/(8\pi))^{1/3}$ so that the average kinetic energy per particle $\epsilon_{kin} = (3/5)(k_F^2/2m)$ is proportional to $\epsilon_{kin} \sim \rho^{2/3}$. On the other hand, the interaction energy per electron, ϵ_{int}, which for a homogeneous electron gas equals the exchange energy is proportional to $\epsilon_{int} \sim \rho^{1/3}$. Therefore, at sufficient low densities ρ the interactions of the electrons dominate their kinetic energy and the conditions which Wigner envisaged are fulfilled. What will be the ground state in that case? The one **Wigner** proposed is that the electrons will form a lattice, often referred to as Wigner lattice. In that configuration the electrons minimize their Coulomb repulsion because they stay apart from each other as far as possible. The kinetic energy reduces to the zero-point fluctuations of the electrons around their equilibrium lattice positions. Various estimates have been made as regards the critical density below which the formation of a Wigner lattice takes place. For that purpose it is customary to associate a characteristic length r_0 with a given density ρ and to use for it the mean radius r_0 per particle, i.e., $(4\pi/3)r_0^3 = \rho^{-1}$. When expressed in units of Bohr's radius $a_B = 0.529\text{Å}$ the relation $r_0 = r_s a_B$ holds. It is believed that electron crystallization takes place for $r_s \geq 70$ although there is considerable uncertainty in this estimate. Since the lattice parameter is large in this case, the exchange is negligible, the reason being that it falls off exponentially. A review discussing the various energy contributions is found in Ref. [3]. The excitations of a Wigner lattice are phonon-like with the electrons vibrating around the lattice sites. In accordance with that picture the low-temperature specific heat is Debye like, i.e., $C(T) = aT^3$. A lattice structure has been observed for electrons on the surface of ^4He [4]. Other examples where lattice formations of the electrons have been suggested to occur are inversion layers in semiconductors [5], or low-carrier systems like CeN or other rare-earth pnictides [6]. However, up to now the possible existence of Wigner lattices in solids was shown only indirectly by their influence on other physical properties of the systems like transport.

A rather different case of electron localization was envisaged by Verwey in an attempt to explain the enormous changes with temperature which take place in the electric conductivity of magnetite Fe_3O_4 [7]. Because oxygen has a valency of $2-$, the three Fe ions must have valencies $2 \times Fe^{3+} + 1 \times Fe^{2+}$.

Magnetite has the spinel structure with two types of cation sites, i.e., type A and B. One of the two Fe^{3+} ions is occupying an A site while the remaining Fe^{3+} and the Fe^{2+} occupy B sites. With an average valency of $+2.5$ of the cations on the B sites the $Fe^{2+} - Fe^{3+}$ ions effectively attract each other while there is a repulsion between pairs of ions $Fe^{2+} - Fe^{2+}$ and $Fe^{3+} - Fe^{3+}$. Below the Verwey transition these short-range interactions lead to charge ordering (or crystallization) of the extra electrons provided by Fe^{2+}. According to Verwey this ordering takes place along rods of which there exist two interpenetrating families. According to his theory one of the two families is occupied by Fe^{2+} while the other one is occupied by Fe^{3+}. However, this proposed structure of the ordered state is still a subject of controversy. At finite temperatures the free energy $F = U - TS$ is minimized, implying that with increasing T an increasing amount of disorder is favoured. The Verwey transition has been viewed as a classical example of an order-disorder phase transition.

A third scenario for electron localization was suggested by Mott [8] and Hubbard [9] (Mott-Hubbard metal-insulator transition). Mott considered a chain of atoms with one electron per site (e.g., hydrogen or sodium-like atoms). He predicted a transition from delocalized to localized electron behaviour when the interatomic distance becomes sufficiently large. Hubbard considered a lattice of sites with one orbital each and found electron localization when there is one electron per site and the on-site Coulomb repulsion of electrons is sufficiently larger than their kinetic energy. The Hamiltonian he used is of the form

$$H = \sum_{i,j,\sigma} t_{ij} a_{i\sigma}^\dagger a_{j\sigma} + U \sum_i n_{i\uparrow} n_{i\downarrow} \tag{3}$$

where the index i refers to different lattice sites and $n_{i\sigma} = a_{i\sigma}^\dagger a_{i\sigma}$. Usually, only nearest-neighbour matrix elements $t_{ij} = -t$ are taken into account. The Hubbard Hamiltonian (3) as it is customarily called, was first written down independently by Gutzwiller [10], Hubbard [9] and Kanamori [4] and has been a subject of countless investigations since.

It is noticed that the physical origin of electron localization or crystallization is quite different in the three scenarios just described, despite the fact that the Coulomb repulsion plays a major role in all three of them. In the Wigner-lattice case it is the long-range part of the interaction which is of importance while in the Mott-Hubbard case it is the on-site interaction which is considered only. In both cases though, it is the competition between the electron-electron repulsion and the kinetic energy which leads to electron localization if the former is dominant. The Verwey theory of the transition in Fe_3O_4 is based on the assumption of an order-disorder phase transition and does not account at all for the kinetic energy of the electrons. Here it is the competition between a short-range electron repulsion and the entropy which determines the transition.

The above considerations suggest still another way of obtaining electron crystallization. Assume, that instead of a homogeneous electron gas we are dealing with itinerant electrons tied closely to atomic sites. In that case the kinetic energy may be quite small even at high electron densities, provided the overlap of wave functions at neighbouring sites is small. As a consequence, electron crystallization should already take place at high densities. The valence electrons with the strongest attachment of their wave function to atomic sites are the $4f$ electrons of rare-earth ions. It is therefore a challenging task to identify specific rare-earth systems in which electron crystallization or lattice formation takes place at easily accessible temperatures. Such a system is Yb_4As_3 which we discuss in the following.

2 Yb_4As_3 - an Unusual Semimetal

The intermetallic compound Yb_4As_3 has the anti-thorium phosphite structure (see Fig. 1a). The Yb ions can be thought of as occupying four families of chains (or rods) which are directed along the diagonals of a cube. This does not imply that we are dealing here with a truly one-dimensional structure since the distance between Yb ions in a chain is larger than between Yb ions belonging to different chains. However, thinking of them in terms of occupying four families of interpenetrating chains is very useful as we shall see below (see Fig 2). The unit cell of Yb_4As_3 is cubic and contains 16 Yb and 12 As ions. Therefore one also refers to the system as a body-centered cubic rod structure.

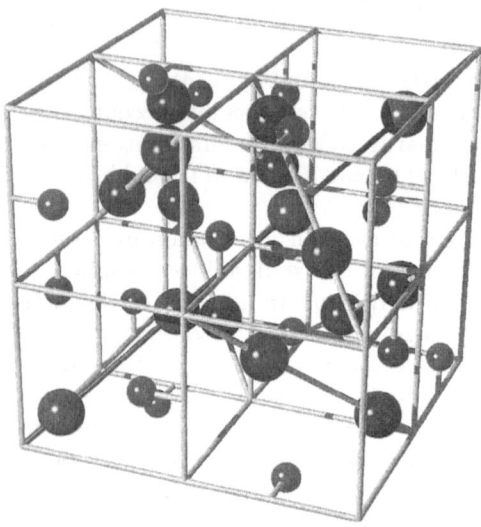

Fig. 1. Structure of Yb_4As_3. The Yb ions (large spheres) form chains with directions along the diagonals of a cube. The As ions are represented by the small spheres.

Counting valence electrons one notices that with As^{3-} the Yb ions must have different valencies, i.e., $Yb_3^{2+}Yb^{3+}As_3^{3-}$. While Yb^{2+} has a filled $4f$ shell Yb^{3+} has one $4f$ hole. At high temperatures this hole is shared between four Yb ions and the system is metallic. This is in accordance with measurements of the Hall constant which yields a carrier concentration of approximately $1/4$ per Yb ion [12]. At a temperature $T = 290K$ the system undergoes a weak first-order phase transition which we interpret as being due to hole crystallization. Due to their Coulomb repulsion holes avoid nearest neighbour configurations and ensemble in chains of one direction, e.g., [1,1,1] only (remember that nearest-neighbour Yb ions belong to different chains). While at high temperatures each chain contains on average $1/4$ holes per site, at low temperatures three of the four families of chains are unoccupied by holes while the remaining one has nearly one hole per site. Configurations of Yb ions with two $4f$ holes on a site (i.e., Yb^{4+} configurations) can be excluded for energetic reasons, implying that $4f$ holes must be treated as being strongly correlated.

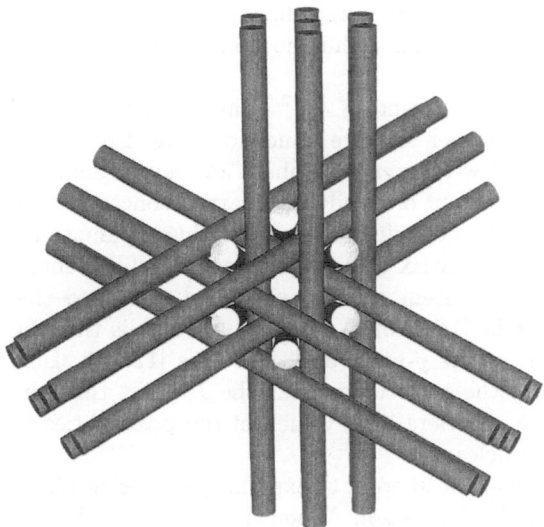

Fig. 2. Schematic representation of the rod structure which is obtained when only the Yb ions of Yb_4As_3 are considered.

Those configurations are only virtually excited and lead to a spin-spin interaction between holes on nearest-neighbour sites or neighbouring sites in a chain. The Hamiltonian which may be used to describe the formation of a hole lattice superimposed on the underlying anti-Th_3P_4 structure is

$$H = -\sum_{ij\sigma} t_{ij}\hat{f}_{i\sigma}^{+}\hat{f}_{j\sigma} + \sum_{<ij>} q^2 \frac{e^{-\lambda_D/r_{ij}}}{r_{ij}} n_i n_j + \sum_{<ij>} J_{ij}\underline{S}_i\underline{S}_j. \qquad (4)$$

The first term describes the kinetic energy of the $4f$ holes. The operator $\hat{f}_{i\sigma}^{+} = f_{i\sigma}^{+}(1 - n_{i-\sigma})$ with $n_{i-\sigma} = f_{i-\sigma}^{+}f_{i-\sigma}$ creates a $4f$ hole on site i with effective-spin index σ provided that there is not already one with spin-σ. The effective spin index refers to the crystal-field ground state doublet of the $J = 7/2$ multiplet of the $4f^{13}$ configuration. The unconditioned f-hole creation operator is $f_{i\sigma}^{+}$. The second term represents the Coulomb repulsion of holes with effective charge q and a screening length λ_D. The latter depends on the charge-carrier concentration. The density operators are $n_i = \sum_\sigma n_{i\sigma}$. Finally, the last term in Eq. (4) describes the spin-spin coupling of holes due to indirect exchange. This term is important because it contributes to finding the energy of the ground-state configuration. For example, compare the Coulomb repulsion in the following two configurations: (a) one family of chains filled with holes and the other three being empty and (b) ordered Yb ions in all four families of chains according to $Yb^{3+} - Yb^{2+} - Yb^{2+} - Yb^{2+} - Yb^{3+}...$ Calculations show [13] that the latter has a slightly lower repulsion energy than the former. Inclusion of the spin-spin coupling changes this picture since it favours the former configuration (the latter has no holes on next-nearest neighbour sites while the former has).

The strong electron correlations preventing $4f^{12}$ configurations are responsible that chains with one hole per site are nonconducting. Despite of this Yb_4As_3 is a semimetal in the low-temperature phase and not a semiconductor. From the measured Hall constant one deduces a carrier concentration of $10^{-3}/Yb$ ion. For this semimetallic behaviour several possible explanations suggest themselves. Due to hopping matrix elements between the filled short chains and empty long chains the system can be self-doped, provided the different parameter values are right [14 - 16]. Another possibility is that an overlap is taking place between an As p-band and $4f$ states [12, 17]. Due to a very small effective mass of the p electrons near the Γ point this can also result in a low concentration carriers [17]. Which of the possibilities is realized in Yb_4As_3 is presently unclear.

The Hamiltonian (4) is not sufficient in order to explain the different physical features of Yb_4As_3 for the following reason. When the holes ensemble on one family of chains then due to the smaller size of Yb^{3+} ions as compared with Yb^{2+} ones the length of those chains shrinks slightly while that of the chains being emptied increases correspondingly so that the volume of a unit cell remains unchanged. Thus a lattice distortion is associated with the ordering of holes which can not be described by the Hamiltonian (4) because it does not contain lattice degrees of freedom. This suggest treating the hole ordering phenomenon in terms of a band Jahn-Teller effect of strongly correlated electrons [14].

3   Band Jahn-Teller Description

We want to set up a model which is able to explain semiquantitatively the most important of the unusual properties of Yb_4As_3. For that purpose we describe the four families of chains by four degenerate quasi-one dimensional bands associated with them. Our model is based on a band Jahn-Teller effect of correlated electrons $(CBJT)$. It lifts the fourfold degeneracy and results in a shrinkage of the crystal, e.g., in [1, 1, 1] direction and an associated expansion in the three other directions. The effective Hamiltonian is of the form

$$H = -\sum_{\mu}\sum_{<ij>\sigma} t_\mu f^+_{i\mu\sigma} f_{j\mu\sigma} + U\sum_{\mu}\sum_{i} n_{i\mu\uparrow} n_{i\mu\downarrow} \qquad (5)$$
$$+ \epsilon_\Gamma \sum_{\mu}\sum_{i\sigma} \Delta_\mu n_{i\mu\sigma} + N_L c_\Gamma \epsilon_\Gamma^2.$$

The subscript $\mu = 1, ...4$ refers to the four different chain directions and $< ij >$ denotes neighbouring sites on a chain. The first term describes the kinetic energy of holes moving in a chain. Hopping between chains is neglected here. In the distorted phase the t_μ may differ for the elongated and the contracted chains and an ansatz of the form

$$t_\mu = t_+ \delta_{\mu 1} + t_- (1 - \delta_{\mu 1}) \tag{6}$$
$$t_+ = t\, e^{\lambda \epsilon_\Gamma}, \qquad\qquad t_- = t\, e^{-\lambda \epsilon_\Gamma / 3}$$

seems appropriate with ϵ_Γ defined below. From LDA calculations [18] one can extract a band width of $4t = 0.2eV$. The second term takes the on-site Coulomb repulsions of holes into account. The third term is due to a volume-conserving coupling of the f bands to the trigonal strain ϵ_Γ with $\Gamma = \Gamma_5$. This coupling is described by a deformation potential of the form

$$\Delta_\mu = \frac{\Delta}{3}(4\delta_{\mu 1} - 1). \tag{7}$$

The basis for this is the Coulomb repulsion of holes on neighbouring sites. It is treated here as an effective attraction between holes on next-nearest neighbour sites, i.e., nearest sites on a chain. Here the tacit assumption has been made that due to the last term in Eq. (4) the ground state is one with the holes positioned in one type of chains only. The last term in Eq. (5) is the elastic energy associated with the lattice deformation. Here N_L is the number of sites in a chain and c_Γ is the bulk elastic constant for which we choose $c_\Gamma / \Omega = 4 \cdot 10^{11}$ erg/cm^3 with Ω denoting the volume of a unit cell. Note that the lattice constant is $a_0 = 8.789 \mathring{A}$.

The Hamiltonian (5) is suitable for describing a band Jahn-Teller transition. It lifts the four-fold degeneracy of the f bands by the spontaneous appearance of a trigonal strain which lowers the symmetry. In order to demonstrate this we consider first the case of $U = 0$. This neglects the effect of the strong electron correlations on the CBJT-transition which prevent occupancies of sites by two holes. Near the transition, where each of the chains is still nearly quarter-filled this is not a too bad approximation. However, it is not acceptable for low temperatures where in the distorted phase all holes are ensembled in one family of chains so that they contain nearly one hole per site. The condition for a CBJT transition is $\Delta^2 / (t\epsilon_\Gamma) > 3$ [14]. Choosing $\Delta = 5eV$ leads to a transition temperature of $T_c \simeq 250K$ which is close to the experimental value of $T_c^{exp} \simeq 300K$. The Grüneisen parameter $\Omega_G = \Delta / (4t)$ is 25 in this case which seems reasonable when compared with other mixed valence systems. The temperature dependence of the spontaneous strain as well as the associated occupation numbers n_μ of the chains are shown in Fig. 3. As a consequence of the trigonal strain the fourfold degenerate one-dimensional f bands split into three (upper) and one (lower). The energy difference of their centers is $(4/3) \mid \epsilon_\Gamma \Delta \mid$ with the equilibrium strain given by $\epsilon_\Gamma = -\Delta / (2c_\Gamma) \approx 0.02$. It is noted that band refilling by a CBJT transition and hole crystallization are equivalent pictures, in particular when with the help of large U values double occupancies of sites by holes are excluded. This brings us to the effects of finite values of U.

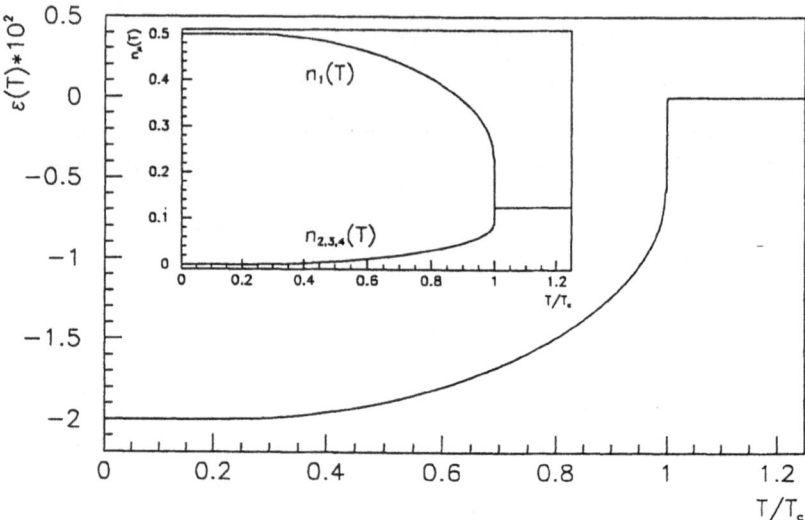

Fig. 3. Temperature dependence of the trigonal strain $\epsilon_\Gamma(T)$ below T_c. The insert shows the T dependence of the occupation numbers of the four one-dimensional f-bands (from Ref. [14])

Since the work of Lieb and Wu [19] it is well known that the one-dimensional Hubbard model can be solved exactly. The same solution can be adapted to the present Hamiltonian because the chains are not directly coupled. Only an indirect coupling exists via the strain ϵ_Γ which enters t_μ and the chemical potential. The Lieb-Wu type of equations based on the Bethe ansatz must therefore be solved self-consistently [16]. While for $U = \infty$ analytic solutions can be obtained one must resort to numerical solutions when U is arbitrary. For $U = \infty$ and $U = 0$ one finds for the ground-state energy per site [16]

$$\frac{E_s}{N_L} = -\frac{2q}{\pi}[sin(n_1\frac{\pi}{q})e^{\lambda\epsilon_\Gamma} + 3sin(n_2\frac{\pi}{q})e^{-\lambda\epsilon_\Gamma/3}]$$
$$- \epsilon_\Gamma(n_1 - n_2) + c_\Gamma\epsilon_\Gamma^2 \tag{8}$$

where $q = 1$ for $U = \infty$ and $q = 2$ for $U = 0$. Furthermore, n_1, n_2 are the hole concentrations in the shortened chain and in the other three chains, respectively. Here and in the following the energy scale is set by taking $t = 1$. In order to find the equilibrium strain ϵ_Γ we must minimize E_s with respect to ϵ_Γ under the condition that the chemical potential is the same in all chains.

At $T = 0$ this implies that $\delta E_1/\delta n_1 = \delta E_2/\delta n_2$, and therefore we must know E_1 and E_2 separately. One finds

$$\frac{E_i}{N_L} = -\frac{2q}{\pi}t_i sin(n_i \frac{\pi}{q}) - \epsilon_\Gamma \Delta_i n_i + \frac{c_\Gamma}{4}\epsilon_\Gamma^2 \qquad (9)$$

which leads to the constraint

$$\epsilon_\Gamma = 1.5[cos(n_2\frac{\pi}{q})e^{-\lambda\epsilon_\Gamma/3} - cos(n_1\frac{\pi}{q})e^{\lambda\epsilon_\Gamma}]. \qquad (10)$$

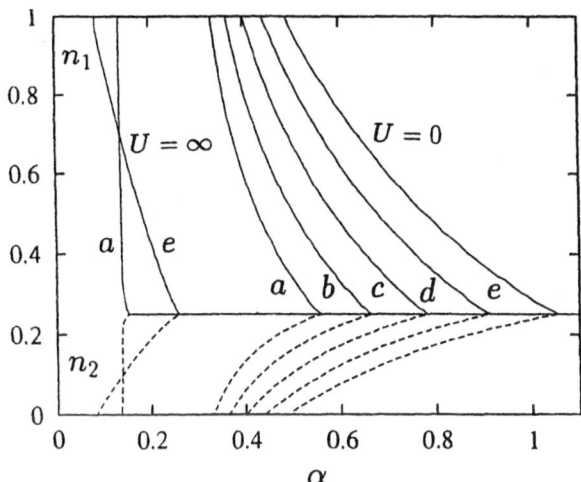

Fig. 4. Hole concentration n_1 (solid lines) and n_2 (dashed lines) as function of c_Γ ($= \alpha\Delta^2$) for various values of λ and $U = 0$ as well as $U = \infty$. Curves $a - e$ correspond to $\lambda/\Delta = 0$, 0.05, 0.10, 0.15, and 0.2 respectively (from Ref. [16])

After minimization of E_s/N_L with respect to ϵ_Γ (or n_1) the hole concentrations in the different chains are obtained. Without the CBJT transition, i.e., for $\epsilon_\Gamma = 0$ it is $n_1 = n_2$. But when c_Γ is sufficiently small so that $\Delta^2/(c_\Gamma t) \geq 3$ we have $n_1 > n_2$. Thereby two different cases must be distinguished. For the largest part of the parameter space we find $n_1 = 1$ and $n_2 = 0$, i.e., all holes collect in the short chains while the long chains do not contain holes. This is the same situation as shown in Fig. 3. However, depending on the size of λ, U and c_Γ there is also a parameter regime for which $n_1 \neq n_2 \neq 0, \epsilon_\Gamma \neq 0$. This is demonstrated in Fig. 4. When $n_1 \neq n_2 \neq 0$ holds the system is a conductor rather than an insulator and the situation corresponds to self-doping, but here without interchain hopping as we want to stress. The above estimate of $\alpha \simeq 0.25$ suggests that Yb_4As_3 could well be in the regime $n_1 \neq 0$ for a reasonable interaction strength U and bandwidth parameter λ. A general finding is that the region with a CBJT distorted ground state is reduced by the local Coulomb repulsion U. But it is never suppressed completely even in the limit $U = \infty$. This is different when a mean-field (MF) approximation is made to the Hamiltonian (5). In that case there exists a critical value U_c depending on c_Γ above which a CBJT transition does not take place. This limitation is an artifact of the MF approximation.

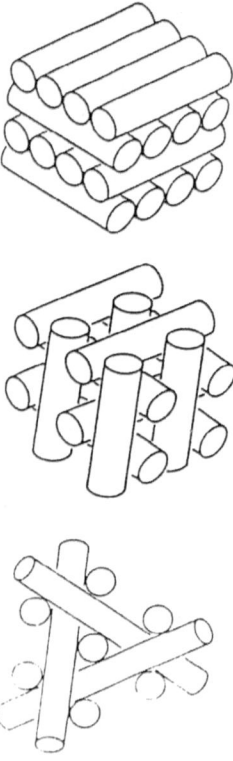

Fig. 5. Two-, three- and four-rod structure corresponding to the B sites of magnetite, to A 15 compounds and to Yb_4As_3, respectively (from Ref. [20])

The above considerations demonstrate that a CBJT description of the hole crystallization in Yb_4As_3 is indeed a suitable scheme. Of course, this is possible only due to the rod like structure of the Yb sublattice. The chain structure leads to another important effect. A chain with one electron per site and strong correlations is equivalent to a Heisenberg chain, the spin excitation of which are known to give raise to a low temperature specific heat $C = \gamma T$. The γ coefficient depends on the size of the exchange coupling and in the case of Yb_4As_3 is of the order of 0.2 Joule/$(mol \cdot K^2)$. Materials with such a large linear specific heat coefficient are usually called heavy-fermion systems because it corresponds to the one of free electrons but with a huge effective mass. The one to one correspondence between magnon like magnetic excitations and those of an electron gas is solely due here to the one-dimensional structure formed by the Yb^{3+} ion. The description of a structural phase transition in conjunction with an electronic charge redistribution in terms of a band Jahn-Teller phase transition has been suggested before by Labbé and Friedel [21] in order to explain the martensitic phase transition in A 15 compounds such as V_3Si, Nb_3Sn etc. There we are dealing with a structure consisting of three families of interpenetrating chains of transition metal ions (e.g., V or Nb). The chains are along the sides of a cube. Although electron correlations on transition metal sites are not as strong as on rare-earth sites it is tempting to think also here of an electron crystallization taking place. This would require a reconsideration of the theory of those materials. It would imply that there exist rod structures with two (e.g., magnetite), three (e.g., A 15 compounds) and four (e.g., Yb_4As_3) families of intersecting chains (rods) (see Fig. 5) in which electron crystallization is taking place. Clearly, it seems that these are not the only ones in which this phenomenon is taking place.

Acknowledgement

I want to thank Dr. P. Thalmeier and Dr. B. Schmidt for a long standing fruitful collaboration. I also thank Prof. H.G. von Schnering for stimulating discussions.

References

[1] E.P. Wigner, Phys. Rev. **46**, 1002 (1934)

[2] E.P. Wigner, Trans. Faraday Soc. **34**, 678 (1938)

[3] C.M. Care and N.H. March, Adv. Phys. **24**, 101 (1975)

[4] R.S. Crandall, Phys. Rev. A **8**, 2136 (1973)

[5] J. Durkan, R.J. Elliott and N.H. March, Rev. Mod. Phys. **40**, 812 (1968)

[6] T. Kasuya, J. Alloys and Compounds, **192**, 217 (1993) and earlier work cited there

[7] E.J.W. Verwey and P.W. Haaymann, Physica **8**, 979 (1941)

[8] N.F. Mott, Phil. Mag. **6**, 287 (1961)

[9] J. Hubbard, Proc. R. Soc. London A **276**, 238 (1963), *ibid*, A **281**, 401 (1964)

[10] M.C. Gutzwiller, Phys. Rev. Lett. **10**, 159 (1963)

[11] J. Kanamori, Prog. Theor. Phys. **30**, 275 (1963)

[12] A. Ochiai, T. Suzuki and T. Kasuya, J. Phys. Soc. Jpn. **59**, 4129 (1990)

[13] H.G. von Schnering, private communication

[14] P. Fulde, P. Thalmeier and B. Schmidt, Europhys. Lett. **31**, 323 (1995)

[15] S. Blawid, Hoang Anh Tuan and P. Fulde, Phys. Rev. B **54**, 7711 (1996)

[16] Y.M. Li, N. d'Ambrumenil and P. Fulde, Phys. Rev. Lett. **78**, 3386 (1997)

[17] H. Eschrig, private communication

[18] K. Takegahara and Y. Kaneta, J. Phys. Soc. Jpn. **60**, 4009 (1991)

[19] E.H. Lieb and F.Y. Wu, Phys. Rev. Lett. **20**, 1445 (1968)
[20] M. O'Keeffe and K.S. Andersson, Acta Crystallogr. A **33**, 914 (1977)
[21] J. Labbé and J. Friedel, J. Physica **27**, 153 and 303 (1966)

Natural Patterns in Nonequilibrium Systems

M. Bestehorn and R. Friedrich

Institut für Theoretische Physik und Synergetik
Universität Stuttgart
Pfaffenwaldring 57/4, 70550 Stuttgart, Germany

Abstract. We discuss the theoretical description of the formation of natural cellular patterns by means of rotationally invariant order parameter equations. Starting from the hydrodynamic basic equations it turns out that the order parameter equations have nonlocal character. We present a suitable approximation scheme of the nonlocal terms by local ones which is based on a systematic gradient expansion. A truncation of this expansion leads to model equations which are widely used in the theoretical treatment of pattern formation in complex systems. Its application to spiral turbulence is studied in detail.

1 Introduction

There are numerous examples of systems far from equilibrium which exhibit an instability leading to the formation of well-ordered patterns (Haken 1983, Manneville 1990, Busse and Kramer 1990, Newell et al. 1993, Cross and Hohenberg 1993). In spatially extended systems various planforms are observed like roll structures, square and hexagonal patterns. Well-known examples include hydrodynamic systems, like Rayleigh-Bénard convection with its various modifications (Busse 1978, Kramer and Pesch 1995, Friedrich et al. 1990), the Faraday-instability (Edwards and Fauve 1993), chemical instabilities leading to Turing structures (Turing 1952, Castets et al. 1990, Ouyang and Swinney 1991), but also pattern formation in the transverse field of lasers (Arecchi 1991).

All these systems share the common property that they have, for a certain range of control parameters, a stable stationary state which is, for large aspect ratio systems, independent on the horizontal coordinates $\mathbf{x} = (x, y)$. As one or several control parameters are changed this state undergoes an instability leading to the formation of cellular patterns.

Although the basic mechanisms leading to pattern formation in the above mentioned systems are completely different, it turned out during the last 25 years that a common and unified mathematical description can be achieved close to instability (Haken 1987, Cross and Hohenberg 1993). The state variables depend on order parameters for which a closed system of evolution equations, namely the order parameter equations, can be obtained.

2 Order Parameter Equations

We shall investigate evolution laws as, for instance, the hydrodynamic basic equations of the form

$$\dot{\mathbf{q}}(\mathbf{r}, t) = L(\nabla, \sigma)\mathbf{q}(\mathbf{r}, t) + \Gamma : \mathbf{q}(\mathbf{r}, t) : \mathbf{q}(\mathbf{r}, t) \quad . \tag{1}$$

Thus we restrict our attention to systems with a quadratic nonlinearity. The state vector $\mathbf{q}(\mathbf{r}, t)$ describes the deviation form a basic state, which is time independent and homogeneous in the horizontal directions (x, y), σ denotes one or a set of control parameters.

The stability of the basic state $\mathbf{q} = 0$ is investigated by a normal mode analysis neglecting the nonlinearities. Due to translational symmetry in the horizontal plane the normal modes take the form $(\mathbf{x} = (x, y))$:

$$\Phi_{j,\mathbf{k}}(\mathbf{x}, z) = \Phi_j(\mathbf{k}, z)e^{i\mathbf{k}\cdot\mathbf{x}} \quad . \tag{2}$$

The corresponding eigenvalue problem reads:

$$\lambda_j(\mathbf{k})\Phi_j(\mathbf{k}, z)e^{i\mathbf{k}\cdot\mathbf{x}} = L(\nabla, \sigma)\Phi_j(\mathbf{k}, z)e^{i\mathbf{k}\cdot\mathbf{x}} \quad . \tag{3}$$

The discrete index j specifies the mode structure in vertical direction (see Fig.1). The continuous wave vector $\mathbf{k}$ defines orientation as well as the wave length of the plane waves.

In order to deal with the nonlinear properties the state vector $\mathbf{q}(\mathbf{r}, t)$ is expanded into the complete set of normal modes:

$$\mathbf{q}(\mathbf{r}, t) = \sum_j \int d^2k \xi_j(\mathbf{k}, t)\Phi_j(\mathbf{k}, z)e^{i\mathbf{k}\cdot\mathbf{x}} \quad . \tag{4}$$

Inserting this expansion into the evolution equation (1) one can derive a set of differential equations for the mode amplitudes $\xi_j(\mathbf{k}, t)$:

$$\dot{\xi}_j(\mathbf{k}, t) = \lambda_j(\mathbf{k})\xi_j(\mathbf{k}, t) \tag{5}$$

$$\sum_{j',j''} \int d^2k' \int d^2k'' \delta(\mathbf{k} - \mathbf{k}' - \mathbf{k}'')\Gamma_{j;j',j''}(\mathbf{k}; \mathbf{k}', \mathbf{k}'')\xi_{j'}(\mathbf{k}', t)\xi_{j''}(\mathbf{k}'', t) \quad .$$

Here, the matrix elements $\Gamma_{j;j',j''}(\mathbf{k}; \mathbf{k}', \mathbf{k}'')$ are introduced according to:

$$\Gamma_{j;j',j''}(\mathbf{k}; \mathbf{k}', \mathbf{k}'') = <\Phi^\dagger_{j,\mathbf{k}}(\mathbf{x}, z)|\Gamma : \Phi_{j',\mathbf{k}'}(\mathbf{x}, z) : \Phi_{j'',\mathbf{k}''}(\mathbf{x}, z) > \quad . \tag{6}$$

The brackets denote a suitably defined scalar product and $\Phi^\dagger$ is the solution of the adjoint problem (3). Due to (3) the linear part of (5) is diagonal. The eigenvalues $\lambda_j(\mathbf{k})$ are continuous functions of the wave vector $\mathbf{k}$. For rotationally invariant systems they only depend on the modulus $|\mathbf{k}|$. There are $j = 1, ...\infty$ different bands of modes (Fig.1). If the basic state is stable, all bands have negative growth rates $\text{Re}[\lambda_j(\mathbf{k})] < 0$. An instability arises if

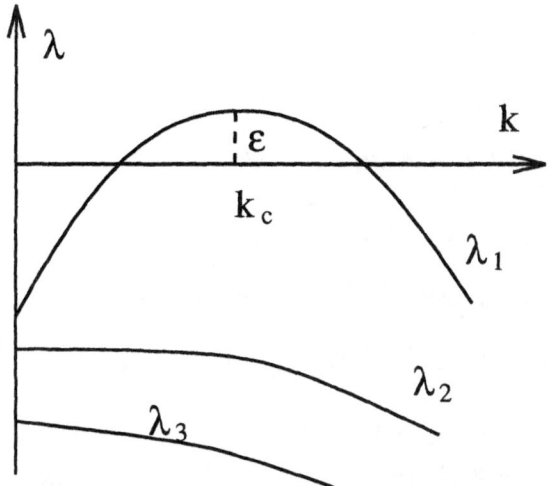

Fig. 1. The systems under consideration posses eigenvalues that are grouped in several (usually infinitely many) bands. The above band may cross the k-axis for certain values of the control parameter and mark the onset of an instability with typical wave number k_c. Modes that belong to this band are named order parameters. All other bands stay below the k-axis for all parameter values. They belong to linearly damped modes which can be eliminated.

one (or several) bands have modes with vanishing or even positive growth rates, at least in a region around a certain critical wave vector k_c. The bands formed of modes with negative growth rates are denoted as stable bands.

We denote the mode amplitudes of the critical and the stable bands with $\xi_u(\mathbf{k}, t)$, $\xi_s(\mathbf{k}, t)$ respectively. The amplitudes $\xi_u(\mathbf{k}, t)$ define the order parameters. An adiabatic elimination of the modes (Haken 1983, 1987) of the stable bands leads to a closed set of evolution equations for the order parameters. In lowest order one obtains:

$$\xi_s(\mathbf{k}, t) = \sum_{u', u''} \int d^2\mathbf{k}' \int d^2\mathbf{k}'' \delta(\mathbf{k} - \mathbf{k}' - \mathbf{k}'') \times \tag{7}$$

$$\times \frac{\Gamma_{s;u',u''}(\mathbf{k}; \mathbf{k}', \mathbf{k}'')}{i(\omega_{u'}(\mathbf{k}') + \omega_{u''}(\mathbf{k}'')) - \lambda_s(\mathbf{k})} \xi_{u'}(\mathbf{k}', t) \xi_{u''}(\mathbf{k}'', t) \quad .$$

Here, $\omega_u(\mathbf{k})$ denotes the imaginary part of the eigenvalue $\lambda_u(\mathbf{k})$.

Inserting the resulting expression for the stable modes into the equations for the unstable modes yields the following order parameter equations up to cubic order:

$$\dot{\xi}_u(\mathbf{k},t) = \lambda_u(\mathbf{k})\xi_u(\mathbf{k},t) + \sum_{u',u''} \int d^2\mathbf{k}' \int d^2\mathbf{k}'' \delta(\mathbf{k}-\mathbf{k}'-\mathbf{k}'') \times$$

$$\times \Gamma^2_{u;u',u''}(\mathbf{k};\mathbf{k}',\mathbf{k}'')\xi_{u'}(\mathbf{k}',t)\xi_{u''}(\mathbf{k}'',t)$$

$$+ \sum_{u',u'',u'''} \int d^2\mathbf{k}' \int d^2\mathbf{k}'' \int d^2\mathbf{k}''' \delta(\mathbf{k}-\mathbf{k}'-\mathbf{k}''-\mathbf{k}''') \times$$

$$\times \Gamma^3_{u;u',u'',u'''}(\mathbf{k};\mathbf{k}',\mathbf{k}'',\mathbf{k}''')\xi_{u'}(\mathbf{k}',t)\xi_{u''}(\mathbf{k}'',t)\xi_{u'''}(\mathbf{k}''',t) \quad . \quad (8)$$

The mode coupling coefficients $\Gamma^3_{u;u',u'',u'''}(\mathbf{k};\mathbf{k}',\mathbf{k}'',\mathbf{k}''')$ take the form

$$\Gamma^3_{u;u',u'',u'''}(\mathbf{k};\mathbf{k}',\mathbf{k}'',\mathbf{k}''') = \sum_{s\neq u} \int d\mathbf{k}_s \delta(\mathbf{k}_s - \mathbf{k}' - \mathbf{k}'') \times$$

$$\times [\Gamma_{u;u',s}(\mathbf{k};\mathbf{k}',\mathbf{k}_s) + \Gamma_{u;s,u'}(\mathbf{k};\mathbf{k}_s,\mathbf{k}')] \times$$

$$\times \frac{\Gamma_{s,u'',u'''}(\mathbf{k}_s;\mathbf{k}'',\mathbf{k}''')}{i[\omega_{u''}(\mathbf{k}'') + \omega_{u'''}(\mathbf{k}''') - \omega_s(\mathbf{k}_s)] - \lambda_s(\mathbf{k}_s)} \quad . \quad (9)$$

It is convenient to transform the order parameter equation from $\mathbf{k}$-space into real space. We define order parameter fields $\Psi_u(\mathbf{x},t)$ by the Fourier transforms of the amplitudes $\xi_u(\mathbf{k},t)$:

$$\Psi_u(\mathbf{x},t) = \int d\mathbf{k}\xi_u(\mathbf{k},t)e^{i\mathbf{k}\cdot\mathbf{x}}. \quad (10)$$

These fields then obey the following equation:

$$\dot{\Psi}_u(\mathbf{x},t) = \lambda_u(-i\nabla)\Psi_u(\mathbf{x},t) \quad (11)$$

$$+ \sum_{u',u''} \int d^2\mathbf{x}'d^2\mathbf{x}'' \Gamma^2_{u;u',u''}(\mathbf{x}-\mathbf{x}',\mathbf{x}-\mathbf{x}'')\Psi_{u'}(\mathbf{x}',t)\Psi_{u''}(\mathbf{x}'',t)$$

$$+ \sum_{u',u'',u'''} \int d^2\mathbf{x}'d^2\mathbf{x}''d^2\mathbf{x}''' \Gamma^3_{u;u',u'',u'''}(\mathbf{x}-\mathbf{x}',\mathbf{x}-\mathbf{x}'',\mathbf{x}-\mathbf{x}''') \times$$

$$\times \Psi_{u'}(\mathbf{x}',t)\Psi_{u''}(\mathbf{x}'',t)\Psi_{u'''}(\mathbf{x}''',t).$$

The kernels $\Gamma^2_{u;u',u''}(\mathbf{x}-\mathbf{x}',\mathbf{x}-\mathbf{x}'')$ are obtained from the Fourier transforms of the mode coupling coefficients $\Gamma^2_{u;u',u''}(\mathbf{k};\mathbf{k}',\mathbf{k}'')$ etc. It is evident how higher order terms have to be included.

3   Approximation of the Mode Coupling Coefficients

It is desirable to obtain suitable approximations for the linear operators $\lambda_u(i\nabla)$ as well as for the kernels $\Gamma^2_{u;u',u''}(\mathbf{x}-\mathbf{x}',\mathbf{x}-\mathbf{x}'')$ etc. of the non-linear terms.

Let us start with the linear terms. Due to rotational symmetry in the horizontal plane the eigenvalues in (3) only depend on k^2. If there is one band of unstable modes with real eigenvalues, $\lambda_u(k^2)$ can be expanded at k_c with respect to $k^2 - k_c^2$.

$$\lambda_u(k^2) = \left.\frac{\partial \lambda_u}{\partial \sigma}\right|_{k_c \sigma_c} \sigma_c \varepsilon \; + \; \frac{1}{2} \left.\frac{\partial^2 \lambda_u}{\partial (k^2)^2}\right|_{k_c \sigma_c} (k^2 - k_c^2)^2 \quad . \tag{12}$$

Furthermore, we have introduced a reduced control parameter $\varepsilon = \sigma/\sigma_c - 1$ which measures the deviation from threshold (cf. Fig.1).

Transforming the expression (12) to real space leads, after appropriate scaling of time and space to the linear operator of the Swift-Hohenberg equation (Swift and Hohenberg 1977):

$$\varepsilon - (1 + \Delta)^2 \quad . \tag{13}$$

Here and for the following Δ denotes the Laplacian with respect to the horizontal coordinates $\mathbf{x}$.

An approximation of the nonlinear mode coupling coefficients turns out to be more involved. An important step consists in investigating the dependence of the mode coupling coefficients Γ^2 and Γ^3 in k-space on the various k-vectors. This dependence is to some extent determined by the underlying symmetries of the system.

For the sake of simplicity we assume from now on the case of one order parameter field $\xi(\mathbf{k}, t)$ or $\Psi(\mathbf{x}, t)$, we therefore may drop the indices u and denote $\mathbf{k}'$ by $\mathbf{k}_1$ etc.. The extension to more order parameter fields is evident.

Translational symmetry is responsible for the selection rules of the k-vectors:

$$\mathbf{k} = \mathbf{k}_1 + \mathbf{k}_2 \quad \text{for } \Gamma^2, \qquad \mathbf{k} = \mathbf{k}_1 + \mathbf{k}_2 + \mathbf{k}_3 \quad \text{for } \Gamma^3. \tag{14}$$

Rotational symmetry implies that the mode coupling coefficients only depend on the absolute values of the wave vectors, the scalar products as well as the vertical components of the cross products between the various wave vectors involved in the mode coupling:

$$k_i^2, \quad \mathbf{k}_i \cdot \mathbf{k}_j, \quad \mathbf{e}_z \cdot [\mathbf{k}_i \times \mathbf{k}_j] \tag{15}$$

with $\mathbf{e}_z$ the unit vector in vertical direction. For systems which are invariant under reflections with respect to arbitrary planes perpendicular to the fluid layer, the dependency on the cross products drops out (in the following, we shall concentrate on this case).

These symmetry considerations allow us to find suitable representations of the mode coupling coefficients. Let us first have a look at the quadratic mode coupling term Γ^2. This term depends on

$$\Gamma^2 = \Gamma^2(k^2, k_1^2, k_2^2). \tag{16}$$

since the scalar products as well as all other cross products can be expressed according to (Fig.2a):

$$\mathbf{k}_1 \cdot \mathbf{k}_2 = \frac{1}{2}\left(k^2 - k_1^2 - k_2^2\right),$$

$$\mathbf{k} \cdot \mathbf{k}_1 = k_1^2 + \mathbf{k}_1 \cdot \mathbf{k}_2, \qquad \mathbf{k} \cdot \mathbf{k}_2 = k_2^2 + \mathbf{k}_1 \cdot \mathbf{k}_2. \qquad (17)$$

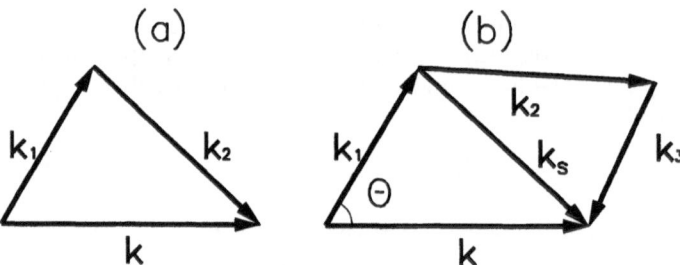

Fig. 2. Wave vector selection in Fourier space for quadratic (a) and cubic (b) non-linear terms.

(a) An arbitrary triangle is fixed by its three sides, leading to expressions in real space that include three Laplace operators

(b) An arbitrary square needs for a unique description its four sides and one diagonal, for instance $|\mathbf{k}_s| = |\mathbf{k}_2 + \mathbf{k}_3|$. This leads to cubic expressions including five Laplacians.

As already done for the linear part we may again expand (16) with respect to the band width and obtain

$$\Gamma^2(k^2, k_1^2, k_2^2) = \sum_{ijl} A_{ijl}(k_c^2 - k^2)^i(k_c^2 - k_1^2)^j(k_c^2 - k_2^2)^l, \qquad (18)$$

which may be transformed back to real space. Before doing this we shall turn to the mode coupling terms of third order. For the following we shall assume that they arise solely due to elimination of stable modes, i.e. the nonlinearity of the system under consideration is quadratic. Then the mode coupling term takes the form:

$$\Gamma^3 = \Gamma^3(k^2, k_1^2, k_2^2, k_3^2, (\mathbf{k}_2 + \mathbf{k}_3)^2). \qquad (19)$$

As before in (18), we may expand Γ^3 with respect to k_i^2 at k_c^2. But now we need an additional variable to define the arbitrary square of four-mode-coupling drawn in Fig.2b. We may chose the diagonal $(\mathbf{k}_2 + \mathbf{k}_3)^2$ and obtain formally:

$$\Gamma^3 = \sum_{n;ijlm} B_{n;ijlm}(-1)^n(k_c^2 - k^2)^i(k_c^2 - k_1^2)^j(k_c^2 - k_2^2)^l(k_c^2 - k_3^2)^m(\mathbf{k}_2 + \mathbf{k}_3)^{2n}$$

$$(20)$$

Now, we may transform to real space according to (10). The order parameter equation for the field $\Psi(\mathbf{x}, t)$ then reads with the linear part given in (13):

$$\dot{\Psi}(\mathbf{x}, t) = [\varepsilon - \tilde{\Delta}^2]\Psi(\mathbf{x}, t) + \sum_{ijl} A_{ijl}\tilde{\Delta}^i\Psi(\mathbf{x}, t)\tilde{\Delta}^j\Psi(\mathbf{x}, t)\tilde{\Delta}^l\Psi(\mathbf{x}, t)$$

$$+ \sum_{n;ijlm} B_{n;ijlm}\tilde{\Delta}^i\{\tilde{\Delta}^j\Psi(\mathbf{x}, t)\Delta^n[\tilde{\Delta}^l\Psi(\mathbf{x}, t)\tilde{\Delta}^m\Psi(\mathbf{x}, t)]\} \quad (21)$$

with the abbreviation

$$\tilde{\Delta} = 1 + \Delta \quad . \tag{22}$$

Note that the spatial coordinates are scaled so that the critical wave length is 2π, i.e. $k_c = 1$.

We note that the expansions in (21) can be grouped into two quite different subjects. One is the expansion with respect to the band width and is systematic in the distance from threshold. This is sufficient up to quadratic order. Terms of cubic and higher order in Ψ need for additional variables which reflect the behavior of the mode coupling with respect to one ore more coupling angles between the plane waves. There is no convergence criterion in any small parameter for this second expansion. In the spirit of a Fourier series, one must take "enough" terms to approximate the angular dependence sufficiently well.

A truncation of the formal Taylor expansions (21) leads to an evolution equation with local nonlinear interaction terms, i.e. terms involving only the field $\Psi(\mathbf{x}, t)$ as well as its spatial derivatives. The simplest approximation using only the term $B_{0;000}$ leads to the Swift-Hohenberg equation.

4 Order Parameters of Pseudo Scalar Type

It is well established that the convective instability in systems with low Prandtl numbers is not only governed by an order parameter which belongs to the convective roll structure. As first noticed by Siggia and Zippelius 1981, there are effects of large scale horizontal drift motions due to the fact that these motions are only weakly damped. A secondary spatially slowly varying field arises which has to be considered as a further order parameter. This situation has been modeled by introducing as a second order parameter field the stream function of the vortical velocity field. However, the stream function is a pseudo scalar which changes sign under reflection at an arbitrary plane vertical to the fluid layer. Furthermore, we assume that the two fields are coupled by quadratic terms. For this case, the set of order parameter equations read:

$$\dot{\Psi}(\mathbf{x}, t) = \lambda(\Delta)\Psi(\mathbf{x}, t) + \sum_{n;ijlm} B_{n;ijlm}\tilde{\Delta}^i\{\tilde{\Delta}^j\Psi(\mathbf{x}, t)\Delta^n[\tilde{\Delta}^l\Psi(\mathbf{x}, t)\tilde{\Delta}^m\Psi(\mathbf{x}, t)]\}$$

$$+ \sum_{i;jl} C_{i;jl}\tilde{\Delta}^i\mathbf{e}_3 \cdot [\nabla\tilde{\Delta}^j\Psi(\mathbf{x}, t) \times \nabla\Delta^l\Phi(\mathbf{x}, t)], \tag{23}$$

$$\tau(\Delta)\dot{\Phi}(\mathbf{x},t) = \gamma(\Delta)\Phi(\mathbf{x},t) + \sum_{n;ijlm} D_{i;jl}\tilde{\Delta}^i\mathbf{e}_3 \cdot [\nabla\tilde{\Delta}^j\Psi(\mathbf{x},t) \times \nabla\Delta^l\Psi(\mathbf{x},t)] \quad .$$

$$(24)$$

Here, $\gamma(-k^2)/\tau(-k^2)$ denotes the linear growth rates of the normal modes of the pseudo scalar field $\Phi(\mathbf{x},t)$.

5 Spiral Turbulence

The class of systems described by (23,24) has become highly important by the observation that they are able to describe the so-called spiral turbulence (Bestehorn et al. 1993) which has been investigated experimentally by Morris et al. 1993 and recently in direct numerical simulations of the three dimensional hydrodynamic equations by Pesch 1996. But even a drastically simplified system of two coupled order parameter fields can describe the basic features of spiral turbulence.

Here we discuss that system first written down by Manneville 1983a,b. It can be regarded as a simplified case of (23,24) in the same sense as the Swift-Hohenberg equation is a simplified case of (21) (here and in the following we suppress the argument t):

$$\dot{\Psi}(\mathbf{x}) = \left[\varepsilon - (1+\Delta)^2\right]\Psi(\mathbf{x}) - \Psi^3(\mathbf{x}) - g(\nabla\Psi(\mathbf{x}) \times \nabla\Phi(\mathbf{x}))\mathbf{e}_z , \quad (25)$$

$$\Delta\dot{\Phi}(\mathbf{x}) = -Pr\Delta\Phi(\mathbf{x}) + (\nabla\Psi(\mathbf{x}) \times \nabla\Delta\Psi(\mathbf{x}))\mathbf{e}_z \quad . \quad (26)$$

Here, g is a constant which was estimated by Manneville from the hydrodynamic equations to $g \approx 100$. The Prandtl number Pr plays a crucial role, both in experiments as well as in theory. For large Pr the mean flow is strongly damped and unimportant, the model yields the same results as the Swift-Hohenberg equation. For very small Pr phase turbulence close to instability renders the patterns time dependent and rather disordered. If $Pr \approx 1$ (compressed CO_2), spontaneous formation of spirals sets in and the structure is organized by targets and rotating spirals of different sizes and orientation. Fig.3 shows a numerical solution of (25,26).

For $Pr \approx 1$ one may be tempted to eliminate the meanflow by putting

$$\dot{\Phi} = 0$$

and write down a closed order parameter equation of third order for Ψ alone. However, as we shall show now, nonlocal terms remain which seem to be the main reason for spiral formation. From (26) we find:

$$\Phi(\mathbf{x}) = \frac{1}{Pr} \int d^2\mathbf{x}' \, G(\mathbf{x} - \mathbf{x}')(\nabla\Psi(\mathbf{x}') \times \nabla\Delta\Psi(\mathbf{x}'))\mathbf{e}_z \quad (27)$$

with the Green's function

$$\Delta G(\mathbf{x}) = \delta(\mathbf{x}) \quad (28)$$

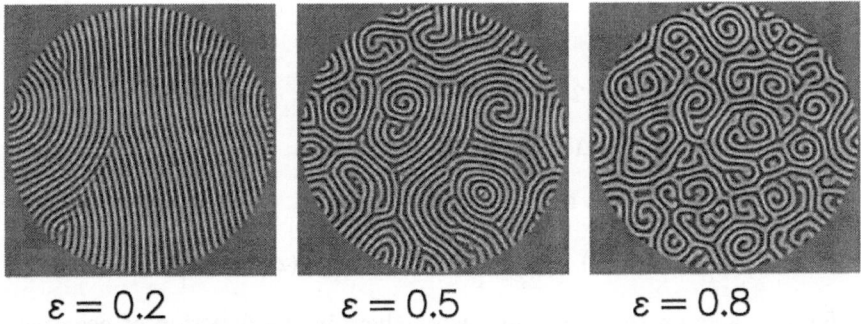

$$\varepsilon = 0.2 \qquad \varepsilon = 0.5 \qquad \varepsilon = 0.8$$

Fig. 3. Numerical solutions of (25,26) for different ε and $Pr = 1$. The structure on the left is stable whereas the two other states are organized by spirals and show turbulent behavior .

as 2D-potential of a point charge. Inserting this into (25) we obtain the Swift-Hohenberg equation plus additional nonlocal cubic terms. It is more clear to treat the following steps in Fourier space. Therefore we write the additional terms (here denoted by Γ_g^3) in the form of (8):

$$\Gamma_g^3(\mathbf{k}; \mathbf{k}_1, \mathbf{k}_2, \mathbf{k}_3) = \frac{g}{2Pr}\left([\mathbf{k}_1 \times (\mathbf{k}_2 + \mathbf{k}_3)] \cdot (\mathbf{k}_2 \times \mathbf{k}_3)\right)\frac{k_3^2 - k_2^2}{(\mathbf{k}_2 + \mathbf{k}_3)^2} \quad . \quad (29)$$

The scalar products of two cross products can be written as products of two scalar products. After some manipulations we find

$$\Gamma_g^3 = \frac{g}{4Pr}(k_3^2 - k_2^2)(\mathbf{k}_1 \cdot (\mathbf{k}_2 - \mathbf{k}_3)) + \frac{g}{4Pr}\frac{(k_3^2 - k_2^2)^2}{(\mathbf{k}_2 + \mathbf{k}_3)^2}(\mathbf{k}_1 \cdot (\mathbf{k}_2 + \mathbf{k}_3)). \quad (30)$$

The first term can be cast into the local form of (21) and yields in real space:

$$N_{loc}(\Psi) = \frac{g}{4Pr}\left[\tilde{\Delta}\Psi\Delta\Psi^2 - \Psi\Delta(\Psi\tilde{\Delta}\Psi) + \Psi^2\tilde{\Delta}^2\Psi - \Psi(\tilde{\Delta}\Psi)^2\right]. \quad (31)$$

The second expression is nonlocal due to the singularity at $\mathbf{k}_2 + \mathbf{k}_3 = 0$. It reads:

$$\frac{g}{2Pr}(\nabla\Psi(\mathbf{x})) \cdot \nabla\left[\int d^2\mathbf{x}'\, G(\mathbf{x} - \mathbf{x}')\left(\Psi(\mathbf{x}')\tilde{\Delta}^2\Psi(\mathbf{x}') - (\tilde{\Delta}\Psi(\mathbf{x}'))^2\right)\right]. \quad (32)$$

If we introduce the "potential" $U(\mathbf{x})$ as

$$U(\mathbf{x}) = \frac{g}{2Pr}\int d^2\mathbf{x}'\, G(\mathbf{x} - \mathbf{x}')\left(\Psi(\mathbf{x}')\tilde{\Delta}^2\Psi(\mathbf{x}') - (\tilde{\Delta}\Psi(\mathbf{x}'))^2\right) \quad (33)$$

we may write the complete order parameter equation in the form

$$\dot{\Psi}(\mathbf{x}) + \mathbf{F}(\mathbf{x}) \cdot \nabla\Psi(\mathbf{x}) = \left[\varepsilon - (1 + \Delta)^2\right]\Psi(\mathbf{x}) - \Psi^3(\mathbf{x}) + N_{loc}(\Psi(\mathbf{x})) \quad (34)$$

with the "force" field

$$\mathbf{F}(\mathbf{x}) = -\nabla U(\mathbf{x}) \tag{35}$$

and U as a solution of the 2D-Poisson equation

$$\Delta U(\mathbf{x}) = \rho(\Psi(\mathbf{x}) : \Psi(\mathbf{x})) \tag{36}$$

where the "charge" follows from (33):

$$\rho(\Psi(\mathbf{x}) : \Psi(\mathbf{x})) = \frac{g}{2Pr}\left[\Psi(\mathbf{x})\tilde{\Delta}^2\Psi(\mathbf{x}) - (\tilde{\Delta}\Psi(\mathbf{x}))^2\right]. \tag{37}$$

The force F acts on the rolls and pushes them in the direction of negative gradient of U, as is expected for an overdamped motion. In turn, negative charges are formed in the core of a spiral or in the center of a target pattern. This negative charge produces a maximum in the potential and drives the concentric rolls outside the center (see Fig.4). In the case of a spiral this has the effect of rotation.

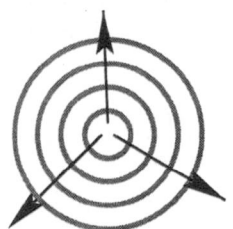 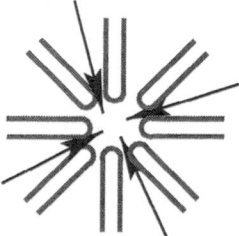

Fig. 4. Concentric rolls or spirals produce a negative charge (37) and via the Poisson equation a local maximum in U. The rolls are driven down hill and the rings travel outward, spirals rotate (left).

A positive charge accounts for a minimum in U and the rolls form a star around the positive sink (right).

In regions where rolls meet at grain boundaries, positive charges are produced by the expression (37), giving rise to a minimum of U. The rolls travel to that grain boundary and vanish there in a sink. Fig.5 shows the structure of Fig.3, middle frame, together with the potential calculated from (37) and (36). An impression of the dynamics is given if one identifies the black regions with hills, the white regions with valleys of U and let the rolls travel downwards in the so obtained landscape.

We summarize the mechanism of spiral formation suggested from that picture: rolls bend and are the origin of negative charges, according to (37), which produce local maxima in the potential. The force let the rolls travel away from the negative charge, leading to a stronger bending and finally to the formation of a spiral or a target. Once formed, these spirals are sources

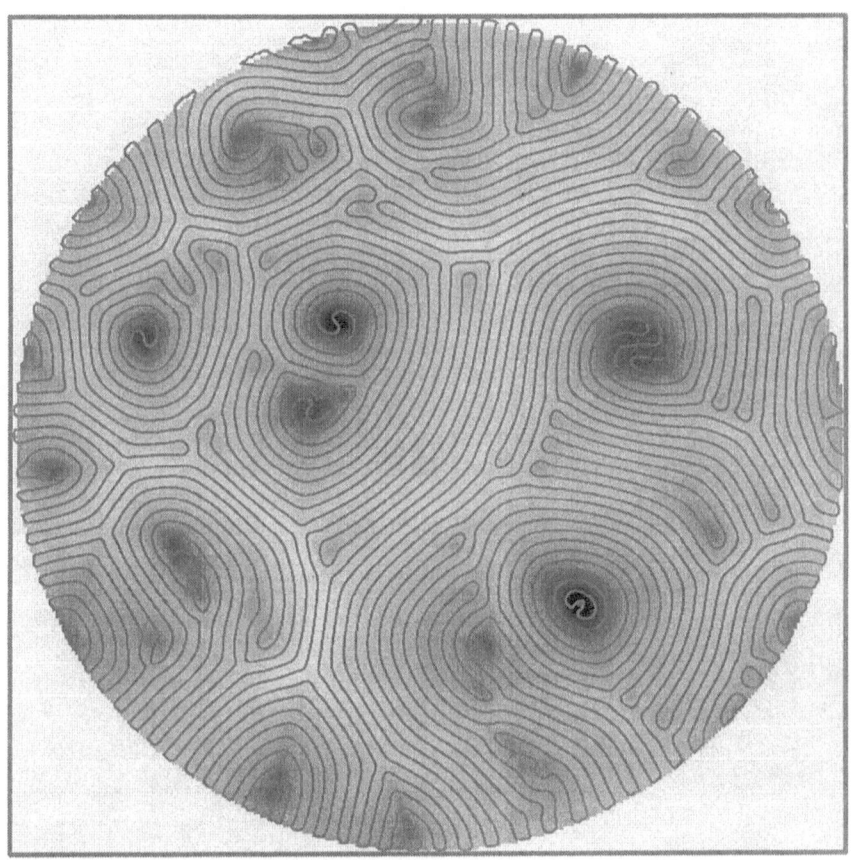

Fig. 5. Spiral turbulence at $\varepsilon = 0.5$. Dark regions denote maxima of U, clear regions minimal.

for the force field and, permanently emitting rolls, rotate. The rolls vanish in sinks of the force field which are located at grain boundaries or near side walls.

This mechanism becomes even more evident, if we fix some point charges from the outside, i.e. compute the force field from given and immobile delta-shaped sources and sinks, according to

$$\mathrm{div}\mathbf{F}(\mathbf{x}) = -\sum_{\ell} Q_{\ell}\delta(\mathbf{x} - \mathbf{x}_{\ell}) \quad . \tag{38}$$

Fig.6 shows time series obtained by numerical integration of (34) with $\mathbf{F}$ from (38) and $N_{loc} = 0$. The case of a positive charge or a sink in the force field leads to a formation of a star. However this structure is highly unstable, leading to a rather large Lyapunov potential of the variational part of (34) and therefore is never formed without external fixing.

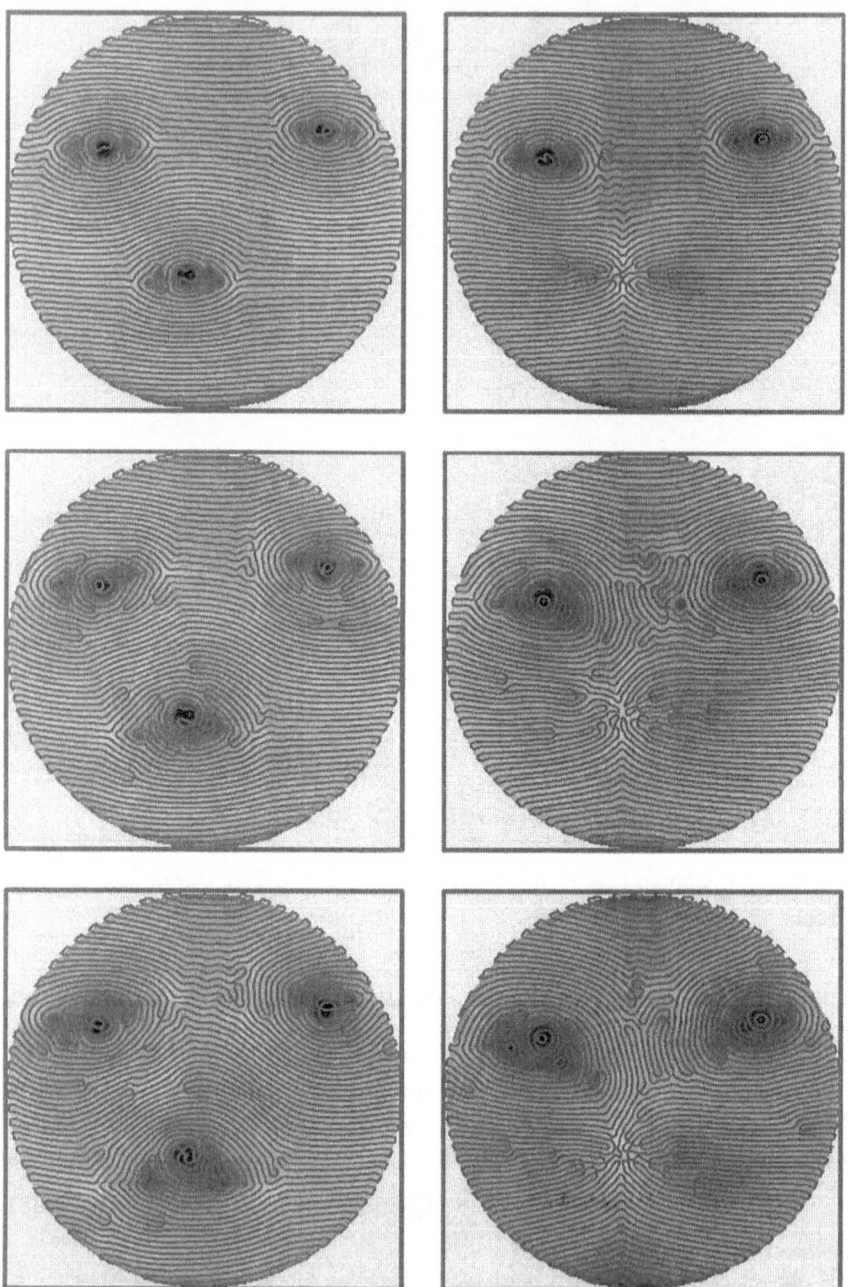

Fig. 6. Externally fixed charges lead to the formation of spirals for $Q_\ell < 0$ and stars ($Q_\ell > 0$). The left row, top to bottom, shows a time series for three sources, the right row that for two sources in the upper part and one sink in the lower part of the layer. The sink is energetically unfavorable and not found spontaneously.

It remains to examine the influence of the local term $N_{loc}(\Psi)$ caused by the coupling of the mean flow to the roll structure. In Fig.7 we show the region of stable rolls for (34) if we neglect the force field. It can be seen that the stability region is shifted to values of smaller wave vectors and the stable band shrinks considerably for the value of ε used in our computations.

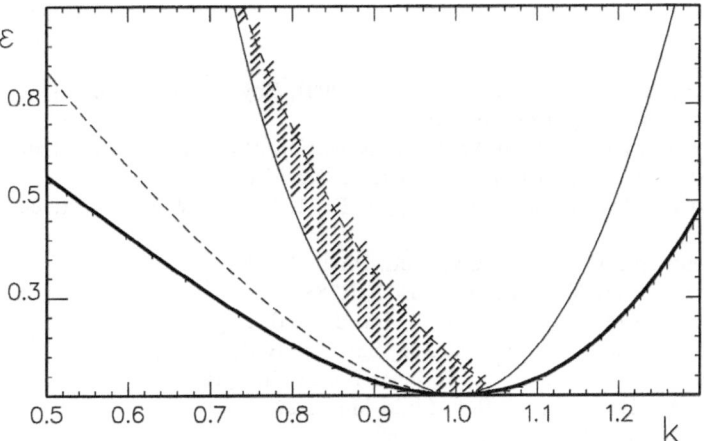

Fig. 7. In contrast to the Swift-Hohenberg equation, where rolls with almost all wave vectors $|k| > 1$ can be stable, the inclusion of the expression $N_{loc}(\Psi)$ from (31) inclines the stability region to larger wave lengths and narrows the band considerably. Rolls are only stable in the shaded area. The stability region is bounded by the Eckhaus instability (small k) and by the cross roll instability (large k).

6 Conclusion

In the present paper we have given a detailed derivation of order parameter equations describing the evolution of patterns in nonequilibrium systems. Starting from a general set of equations of motion, we argued that the reduction of the many degrees of freedom to a view "relevant" ones, namely the order parameters, close to an instability leads to a nonlocal order parameter equation in form of an integro-differential equation. We have shown that the nonlocal terms may be approximated by a suitable local gradient expansion. This expansion takes into account that close to threshold the order parameter field is excited only in a finite band around a typical critical wave number. This allows us to perform a systematic expansion of the nonlocal terms in powers of the band width. A secondary expansion takes into account the nonlinear interaction between plane waves of different directions.

The dynamics of spirals was studied in detail. We showed that two terms are important for their formation and stabilization. One is local and accounts for a certain wave length selection, the other has nonlocal character and can be expressed in terms of a force field that is derived from a potential. The charges that determine this potential origin from deformations and defects of the pattern itself and lead to the self-organized formation of spirals.

References

Arecchi F.T.(1991), Physica D 51, 450

Bestehorn M., Fantz M., Friedrich R., Haken H. (1993), Phys. Lett. **A174**, 48

Busse F.H. (1978), Rep. Prog. Phys. **41**, 1929

Busse F.H., Kramer L. (eds.) (1990):*Nonlinear Evolution of Spatiotemporal Structures in Dissipative Continuous Systems* Plenum, New York

Castets V., Dulos E., Boissonade J., De Kepper P. (1990), Phys. Rev. Lett. **64**, 2953

Cross M.C., Hohenberg P.C. (1993), Rev. Mod. Phys. **65**, 851

Edwards W.S, Fauve S. (1993), Phys. Rev. **E47**, R788

Friedrich R., Bestehorn M., Haken H. (1990), Int. J. Mod. Phys. **B4**, 365

Haken H. (1983):*Synergetics. An Introduction*, Springer Berlin, 3rd print

Haken H. (1987):*Advanced Synergetics*, Springer Berlin, 2nd print

Kramer L., Pesch W. (1995), Annu. Rev. Fluid Mech. **27**, 515

Manneville P. (1983a), J. Phys. (Paris) **44**, 759

Manneville P. (1983b), J. Phys. (Paris) **44**, L-903

Manneville P. (1990):*Dissipative Structures and Weak Turbulence*, Academic, San Diego, Ca

Morris S.W., Bodenschatz E., Cannell D.S., Ahlers G. (1993), Phys. Rev. Lett. **71**, 13

Newell A.C., Passot T., Lega J. (1993), Annu. Rev. Fluid Mech. **25**, 399

Ouyang Q., Swinney H.L. (1991), Nature **352**, 610

Pesch W. (1996), Chaos **6**, 348

Siggia D., Zippelius A. (1981), Phys. Rev. Lett. **47**, 835

Swift J., Hohenberg P.C. (1977), Phys. Rev. **A15**, 319

Turing A.M. (1952), Philos. Trans. R. Soc. London **327B**, 37

Linear Aspects of the Faraday Instability

Hanns Walter Müller

Universität des Saarlandes, Postfach 15 11 50, D-66041 Saarbrücken, Germany

Abstract. An introduction to the theoretical description of parametrically driven surface waves (Faraday waves) is presented. The focus is on the latest developments in the solution of the linear stability problem. Numerical techniques as well as analytical approaches are discussed and compared. The damping within the convective bulk is compared to dissipation in the viscous boundary layers. Depending on their relative weight, the surface waves resonate either subharmonically or harmonically with respect to the external drive.

1 Introduction

The observation of standing surface patterns at the free surface of a fluid layer under vertical vibration dates back to Faraday (1831). The current interest in this system is due to its great variability and a richness of observed patterns. Simply by changing the excitation frequency Ω the aspect ratio (container size/wavelength) can be tuned in a wide range. At low aspect ratios (low excitation frequencies) a small number of dynamical degrees of freedom produces interesting chaotic dynamics (Ciliberto and Gollub (1985), Simonelly and Gollub (1989), Gollub (1991)). On the other hand, at high aspect ratios the observation of convective structures becomes independent of the geometry and a variety of well orderd patterns resembling two-dimensional crystals (including quasi-periodic ones) have been observed (Tufillaro et al. (1989), Douady (1990), Fauve et al. (1992), Christiansen et al. (1992), Edwards and Fauve (1994), Müller (1993) , Kumar and Bajaj (1995), Kiyashko et al. (1996), Binks and van de Water (1997)). In the present publication we focus on the case of large aspect ratio systems.

Another advantage of the Faraday system is the small dynamical relaxation time: Usually it takes a couple of seconds until a stationary well ordered pattern can be observed. This is unlike convection experiments where the typical time constant is of the order of minutes or more.

Faraday already noticed that the surface waves resonate with twice of the forcing period. Rayleigh (1883) suspected that this phenomenon results from a parametric excitation. But only in the 50th, it has been theoretically proved by Benjamin and Ursell (1954), that the stability problem for an inviscid fluid can be reduced to a parametrically driven pendulum (Mathieu oscillator). The predicted dispersion relation is in close agreement with measurements on low-viscosity fluids. In order to explain the finite acceleration amplitude for the onset of surface waves, damping must be introduced into the formulation.

Traditionally this is done on a phenomenological level by evaluating the rate of viscous energy dissipation (Landau and Lifshitz (1987)). This approach ignores the formation of boundary layers close to the borders of the container, where additional dissipation occurs.

The purpose of this article is to give an introduction into the theoretical description of Faraday waves and to review recent developments in the treatment of the linear stability problem since Benjamin and Ursell (1954). In Section 2 we introduce the basic evolution equations of the nonlinear problem. We then linearize and provide the eigenvalue system which governs the onset of the Faraday instability. Numerical as well as analytic approaches will be shown in Section 3. Besides the usual subharmonic surface response, the theoretical approach also predicts a harmonic resonance. We discuss in detail the mechanisms giving rise to a bicritical situation, where both instabilities compete. In the last section we discuss possible implications of the bicriticality with respect to nonlinear pattern formation.

2 Evolution equations and basic state

In this section the basic hydrodynamic equations of motion and the boundary conditions are presented for a viscous fluid (Kumar and Tuckerman (1994), Beyer and Friedrich (1995)). We consider a layer of an incompressible fluid of uniform density ρ, kinematic viscosity ν subjected to a vertical sinusoidal vibration with frequency 2Ω. In the frame of reference which moves with the oscillating container one observes a modulated graviational acceleration $g(t) = g_0 + a \cos 2\Omega t$ in vertical direction $\mathbf{e}_z$. The time evolution is described by the mass-continuity and the Navier-Stokes equation

$$\nabla \cdot \mathbf{v} = 0 \tag{1}$$

$$\partial_t \mathbf{v} + (\mathbf{v} \cdot \nabla)\mathbf{v} = -\frac{\nabla p}{\rho} - g(t)\mathbf{e}_z + \nu \nabla^2 \mathbf{v}, \tag{2}$$

where $\mathbf{v} = \mathbf{v}(\mathbf{r}, t) = (u, v, w)$ is the velocity field and $p(\mathbf{r}, t)$ is the pressure. The horizontal coordinates are denoted by $\mathbf{r}_\perp = (x, y)$. The equations of motion are suppelemented by boundary conditions. For simplicity we consider a fluid layer which extends to infinity in lateral direction. In vertical direction it is limited by a top free surface at $z = 0$ and a rigid bottom at $z = -h$. Deviations from the flat surface are described by the surface elevation $z = \eta(\mathbf{r}_\perp, t)$. The kinematic surface condition reads

$$\partial_t \eta + (u|_\eta \partial_x + v|_\eta \partial_y)\eta = w|_\eta, \tag{3}$$

where $|_\eta$ abbreviates evaluation at $z = \eta$. The surface normal vector in outwards direction is $\mathbf{n}(\mathbf{r}_\perp, t) = (-\partial_x \eta, -\partial_y \eta, 1)/\sqrt{1 + (\nabla_\perp \eta)^2}$. At the liquid air interface the force per unit area is balanced by the surface tension

$$(\sigma^{air} - \sigma) \cdot \mathbf{n} = \alpha(\nabla \cdot \mathbf{n}). \tag{4}$$

Here $\sigma_{ij} = -p\delta_{ij} + \sigma'_{ij}$ is the stress tensor of the fluid, $\sigma'_{ij} = \rho\nu(\nabla_i v_j + \nabla_j v_i)$ denotes the viscous stress tensor and α is the coefficient of surface tension. The stress tensor of the covering air can be approximated by the constant atmospheric pressure, $\sigma^{air}_{ij} \simeq -p_{at}\delta_{ij}$, because the dynamic viscosity of air $\rho^{air}\nu^{air}$ is at least one order of magnitude smaller than that of a fluid. At the bottom of the container the no-slip condition reads

$$\mathbf{v} = 0 \quad \text{at } z = -h. \tag{5}$$

For sufficiently small drive amplitude a the fluid remains at rest and the surface is flat, $\eta = 0$. In this situation the basic state of the system is described by the pressure field $p_B(\mathbf{r}, t) = p_{at} - \rho g(t)z$, which balances the modulated gravity acceleration. After introduction of the deviation $\pi(\mathbf{r}, t) = (p - p_B)$ we obtain the evolution equations

$$\nabla \cdot \mathbf{v} = 0 \tag{6}$$

$$\partial_t \mathbf{v} + (\mathbf{v} \cdot \nabla)\mathbf{v} = -\frac{\nabla\pi}{\rho} + \nu\nabla^2\mathbf{v}. \tag{7}$$

for the bulk of the fluid. At the surface we get

$$\pi|_\eta - \mathbf{n} \cdot \sigma'|_\eta \cdot \mathbf{n} = \rho g(t)\eta + \alpha(\nabla \cdot \mathbf{n}) \tag{8}$$

$$\mathbf{t} \cdot \sigma'|_\eta \cdot \mathbf{n} = 0, \tag{9}$$

where $\mathbf{t}$ denotes any tangential vector perpendicular to $\mathbf{n}$. In Eq. (8) the parametric character of the drive becomes evident: The modulated gravity acceleration $g(t)$ couples multiplicatively to the surface elevation η. The Eqs. (3,5,6,7, 8,9) represent the full set of nonlinear hydrodynamic evolution equations.

3 Linear stability

3.1 Linear evolution equations

In order to investigate the stability of the rest state it is appropriate to linearize the evolution equations around the basic solution $\mathbf{v} = \pi = \eta = 0$. Then, all surface evaluations $|_\eta$ can be taken at $z = 0$. Accordingly, e.g. the pressure in Eq. (8) $\pi|_\eta$ is substituted by its leading order Taylor expansion $\pi|_0$.

In the absence of lateral boundary conditions the hydrodynamic fields can be expanded in plane waves with wave number $\mathbf{k} = (k_x, k_y)$ according to

$$\eta(\mathbf{r}_\perp, t) = \eta_k(t)e^{i\mathbf{k}\cdot\mathbf{r}_\perp} \tag{10}$$

$$w(\mathbf{r}, t) = w_k(z, t)e^{i\mathbf{k}\cdot\mathbf{r}_\perp}. \tag{11}$$

Similar expressions apply for the other velocity components. The pressure can be eliminated by operating twice with curl upon Eq. (7) and with $\nabla_\perp^2$ upon Eq. (8). One finally arrives at the following system of equations (Kumar and Tuckerman (1994), Kumar (1996)):

$$[\partial_t - \nu(\partial_{zz} - k^2)] (\partial_{zz} - k^2)w_k = 0 \tag{12}$$

in the bulk of the fluid,

$$[\partial_t - \nu(\partial_{zz} - k^2) + 2\nu k^2] \partial_z w_k|_0 = -(g(t) + \alpha/\rho k^2)k^2\eta_k. \tag{13}$$

$$\nu(\partial_{zz} + k^2)w_k|_0 = 0 \tag{14}$$

$$\partial_t\eta_k = w_k|_0 \tag{15}$$

at the free surface and

$$w_k|_{-h} = 0 \tag{16}$$

$$\partial_z w_k|_{-h} = 0 \tag{17}$$

at the bottom. The Eqs. (12,13,14,15, 16,17) constitute the linear eigenvalue problem which determines the onset of the Faraday instability.

3.2 Ideal fluids

For inviscid fluids ($\nu = 0$) the order of the spatial derivative in in Eq. (12) reduces by two, simultaneously the boundary conditions (14) and (17) drop out. With the solution

$$w_k(z,t) = A(t)\sinh k(z + h) \tag{18}$$

the velocity field can be eliminated and an ordinary differential equation for the surface elevation (Benjamin and Ursell (1954)) results

$$\ddot{\eta}_k + [\omega_0^2(k) + a\,k\,\tanh kh\,\cos 2\Omega t]\,\eta_k = 0. \tag{19}$$

with

$$\omega_0^2(k) = \tanh(kh)(g_0 k + \frac{\alpha}{\rho}k^3). \tag{20}$$

For vanishing drive, ($a = 0$), Eq. (19) describes undamped free surface waves $\eta_k(t) \propto e^{i\omega t}$ with a dispersion $\omega(k) = \omega_0(k)$. For finite modulation amplitude $a > 0$ Eq.(19) is the evolution equation of a parametrically driven pendulum (Mathieu oscillator). The rest state of the oscillator corresponds to the undeformed flat surface of the hydrodynamic problem. It is well known, that the resting pendulum can be destabilized by a gravity modulation. The stability criterion yields a series of resonance tongues in the parameter plane a vs. ω_0/Ω (see e.g. Nayfeh and Mook (1979)). By virtue of the dispersion relation $\omega_0(k)$ Mathieu's stability chart can be transformed into the marginal stability diagram $a(k)$ for the onset of Faraday waves (see. Fig. 1). The tongues touch the abscissa whenever the wave vector k fulfills the resonance condition

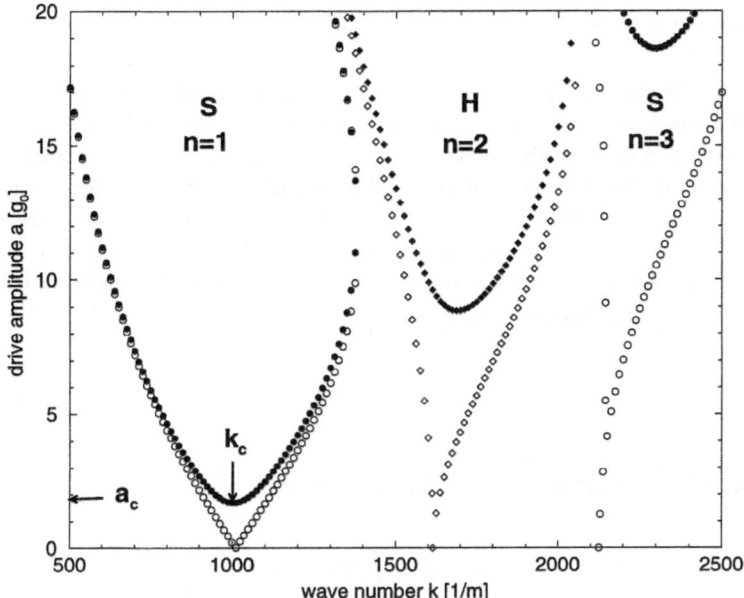

Fig. 1. The neutral stability chart for the onset of Faraday waves. Below (above) the resonance tongues the flat surface is stable (unstable). Tongues labeled by odd (even) n correspond to a subharmonic (harmonic) time dependence of the surface waves. Open symbols: for an inviscid fluid ($\nu = 0$) resulting from the undamped Mathieu oscillator (19). Solid symbols: for a viscous fluid ($\nu = 10mm^2/s$) resulting from the damped Mathieu oscillator (23). Parameters are $\rho = 0.934g/cm^3$, $\alpha = 20.1 \star 10^{-3} N/m$, $h = 1mm$, $\Omega/\pi = 50Hz$.

$\omega_0(k) = n\Omega$ (n integer). The tongues $a_S(k)$ ($a_H(k)$) labeled by odd (even) n are called subharmonic (harmonic), because the spectrum of $\eta_k(t)$ consists of odd (even) multiples of the basic frequency Ω.

Note that the reduction of the hydrodynamic problem to the parametric oscillator was only possible because the vertical and temporal dependence of the velocity field w separate (see Eq. (18)). As can be seen from Eq. (12), the separation is an immediate consequence of the inviscid approximation. For viscous fluids a more complicated spatio-temporal dependence arises. Nevertheless, in order to account for viscous dissipation it is tempting to introduce a damping term into Eq. (19). In the next Section we show how this can be justified by semi-quantitative arguments.

3.3 Phenomenological treatment of the damping

For small viscous dissipation $\varepsilon = \nu k^2/\Omega$ and for small velocity amplitudes an oscillating flow field can be considered to be irrotational (free of vorticity) except within a layer of thickness $\delta = \sqrt{2\nu/\Omega}$ along the system boundaries. If the dimension of the system (here the depth h of the layer) is large compared to the wavelength and also to the thickness of the vortical boundary layer, the latter can be ignored and the whole flow field behaves like an inviscid one. A damping coefficient can be estimated (Landau and Lifshitz (1987)) by the ratio $\gamma = |\langle \dot{E}\rangle/2\langle E\rangle|$, where

$$E = \rho \int \mathbf{v}^2 dV \tag{21}$$

is the total mechanical energy in the system and

$$\dot{E} = -2\rho\nu \int (\nabla_i v_j)^2 dV \tag{22}$$

is the rate of viscous energy dissipation. The integration extends over the fluid volume, while $\langle ... \rangle$ denotes time averaging. On using the inviscid flow field $w(z,t) \propto e^{i\mathbf{k}\cdot\mathbf{r}_\perp} \cos \Omega t \sinh k(z+h)$ (see Eq. (18)) the evaluation gives $\gamma = 2\nu k^2$. To account for viscous dissipation in the Mathieu equation a linear damping term is added giving

$$\ddot{\eta}_k + 2\gamma\dot{\eta}_k + \left[\omega_0^2 + ak \tanh kh \cos 2\Omega t\right]\eta_k = 0. \tag{23}$$

Fig. 1 compares the neutral stability curves resulting from the damped Mathieu oscillator (solid symbols) with the inviscid case (open symbols). In a systematic perturbation theory with the expansion parameter ε, one can show that the minima of the sequence of resonance tongues (see Fig.1) scale with $\varepsilon^{(1/n)}$. Thus for small ε it is always the subharmonic tongue $n = 1$ which provides the lowest minimum. This defines the critical threshold a_c and the critical wave number k_c for the onset of the Faraday instability. For the first tongue $n = 1$ the perturbation expansion yields at leading order

$$a_c = 8\frac{\Omega^2}{k_c}\varepsilon \coth k_c h = 8\nu\Omega k_c \coth k_c h, \tag{24}$$

where k_c solves the dispersion relation

$$\omega_0(k_c) = \Omega. \tag{25}$$

3.4 Numerical treatment of the stability problem

A consistent treatment of viscous dissipation (see Kumar and Tuckerman (1994), Kumar (1996)) requires the consideration of the full hydrodynamic eigenvalue problem. After temporal Fourier transform $\eta_k(t) = \int \bar{\eta}_k e^{i\omega t} d\omega$ (similar expression for w_k) the Navier-Stokes equation (12) can be solved analytically in the form

$$\bar{w}_{k,n}(z,\omega) = A e^{kz} + B e^{-kz} + C e^{qz} + D e^{-qz}, \tag{26}$$

where $q = \sqrt{k^2 + i\omega/\nu}$. By virtue of the boundary conditions (14,15,16,17) the coefficients A, B, C, D can be expressed in terms of $\bar{\eta}_k$. The remaining boundary condition (13) yields a tridiagonal system of the form

$$A(k,\omega)\,\bar{\eta}_k(\omega) + \frac{ak \tanh kh}{2}\left[\bar{\eta}_k(\omega + 2\Omega) + \bar{\eta}_k(\omega - 2\Omega)\right] = 0 \tag{27}$$

where

$$
\begin{aligned}
0 = A(k,\omega) = {}& \omega_0^2(k) + \\
& \nu^2 \frac{q(q^4 + 2q^2 + 5)\coth qh - (k^4 + 6k^2q^2 + q^4)\tanh kh}{q \coth qh - k \coth kh} - \\
& \nu^2 \frac{4qk^2(q^2 + k^2)\tanh kh/(\sinh kh \sinh qh)}{q \coth qh - k \coth kh}.
\end{aligned}
\tag{28}
$$

Without mechanical drive, $a = 0$, the equation $A(k,\omega) = 0$ is the dispersion relation for *free* damped surface waves. In the limit $\nu \to 0$ this expression reduces to the inviscid result $\omega(k) = \omega_0(k)$. If a finite drive, $a > 0$, is imposed, the Floquet theorem states that the solution can be represented by a discrete Fourier sum

$$\eta_k(t) = \sum_n \bar{\eta}_{k,n} e^{i(2n+\beta)\Omega t}, \tag{29}$$

where $\beta = 0$ correspond to the harmonic (H) response and $\beta = 1$ to the subharmonic (S). These two cases are to be investigated separately. The tridiagonal system (27) with ω replaced by $(2n+\beta)\Omega$ defines an infinite dimensional eigenvalue problem for the forcing amplitude a (Hill's infinite determinant). Any desired accuracy can be achieved by an approproiate high-dimensional cut-off. For given parameters $\nu, \alpha, \rho, h, \Omega$ one can follow the eigenvalues a as a function of the wave number k. This procedure defines the neutral stability curves $a_S(k)$ and $a_H(k)$ as shown in Fig .1. A subsequent minimization with respect to k provides the critical threshold. The outcomes of the numerical stability theory are in good agreement with experimental measurements (Kumar and Tuckerman (1994), Christiansen et al. (1992), Bechhoefer et al. (1995)).

As discussed above, for weak dissipation $\varepsilon \ll 1$ it is usually the first subharmonic tongue ($n = 1$) which defines the threshold a_c. For a typical experimental situation Fig.2 compares the numerically computed onset amplitude

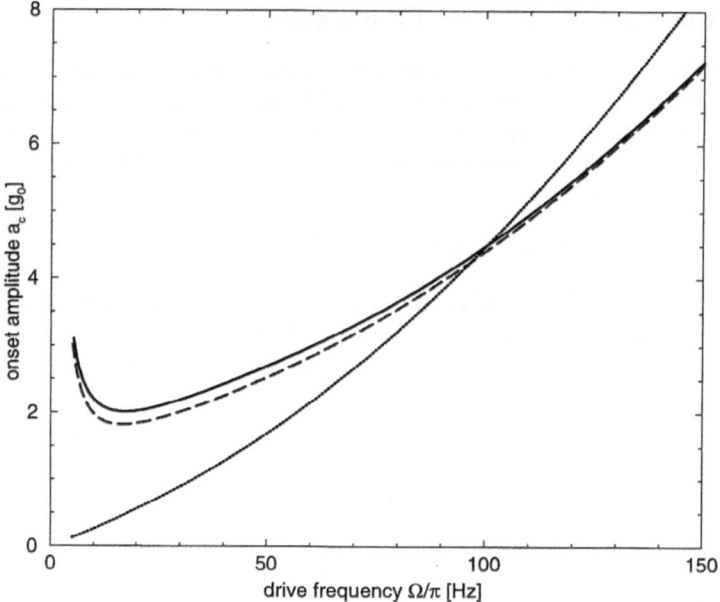

Fig. 2. The critical amplitude a_c for the onset of subharmonic Faraday waves. Solid line: exact numerical solution of the full hydrodynamic eigenvalue problem. Dotted line: stability boundary resulting from the heuristric approach (Eq. (24)). Dashed line: stability boundary according to the systematic perturbation expansion Eq. (36). Parameters are $\rho = 0.934 g/cm^3$, $\nu = 10 mm/s$, $\alpha = 20.1 \star 10^{-3} N/m$, $h = 1 mm$, $\Omega/\pi = 50 Hz$.

a_c (solid line) with the phenomenological result (Eq. (24), dotted line). Evidently, the heuristic damping coefficient γ over-estimates the dissipation at higer drive frequencies, while it under-estimates at small Ω. The first short-coming is due to the fact that formula (22) (which is used to evaluate γ) implicitly assumes that the velocity field vanishes all along the integration boundaries. Since this assumption fails at the free surface, the actual velocity gradients are smaller there, giving rise to a reduced effective damping.

At small Ω the heuristic treatment also fails as it predicts the damping coefficient $\gamma = 2\nu k^2$ to vanish in the long wavelength limit. This appears to be unphysical, as additional dissipation comes into play when the wavelength of the pattern compares to the depth of the layer (i.e. $kh = O(1)$). In this case, which we denote as the "shallow water limit", the neglect of the the viscous boundary layers is no longer legal. Rather the dissipation re-increases towards small k (i.e. smaller Ω). Since the subharmonic tongue $n = 1$ in the marginal stability diagram is the first who experiences this extra dissipation, it is shifted upwards and the neighboring harmonic tongue $n = 2$ competes in providing the absolute minimum. Eventually, this leads to a bicritical situation. In order to develop a deeper understanding of the bicriticality it is useful to get an analytic description of the stability problem.

3.5 Analytic stability analysis in the limit of weak dissipation

In the limit of weak dissipation $\varepsilon \ll 1$ the linear eigenvalue problem can be solved perturbatively (Müller et al. (1997)). The additional assumption $h/\delta \gtrsim 3$ is necessary to approximate the hyperbolic functions in (28) by $\coth qh \simeq 1$ and $1/\sinh qh \simeq 0$. Otherwise the expansion in ε becomes non-analytic. The two approximations made are not very restrictive, they apply to almost all recent experiments. Note that no restiction upon the relation between h and k is made.

By expanding Eq.(28) in powers of $k/q = O(\sqrt{\varepsilon})$ one obtains

$$A(k,\omega) = \omega_0^2 - \omega^2 + i\omega\Omega\varepsilon(3 + \coth^2 kh) + 2\Omega^2 \frac{\varepsilon^{1/2}\sqrt{\varepsilon + i\omega/\Omega}^3}{\sinh 2kh} +$$
$$\varepsilon^{3/2}\Omega^2\sqrt{\varepsilon + i\omega/\Omega}\left(-6\tanh kh + \coth kh + \coth^3 kh\right) + O(\varepsilon^2) \quad (30)$$

This formulation is particularly useful for a comparison with the phenomenological approach (Sec. 3.3). The re-transformation of Eq. (27) into real space yields a Mathieu oscillator with non-local contributions

$$0 = \ddot{\eta}_k(t) + \nu k^2(3 + \coth^2 kh)\dot{\eta}_k(t) + \left[\omega_0^2 + a\,k\,\tanh kh\,\cos(2\Omega t)\right]\eta_k(t) +$$
$$\frac{2\sqrt{\nu}k}{\sqrt{\pi}\sinh 2kh}\int_{-\infty}^{t} G(t-\tau)(\nu k^2 + \partial_\tau)^2\eta_k(\tau)\,d\tau +$$
$$\nu^{3/2}k^3\frac{-6\tanh kh + \coth kh + \coth^3 kh}{\sqrt{\pi}}\int_{-\infty}^{t} G(t-\tau)(\nu k^2 + \partial_\tau)\eta_k(\tau)\,d\tau,$$
$$(31)$$

where $G(t) = \exp(-\nu k^2 t)/\sqrt{t}$ is the kernel. Besides a conventional damping $\propto \dot{\eta}_k$, which is related to the dissipation in the bulk, the two integrals also contribute to the damping: The moving surface emits velocity waves into the interior of the fluid, where history dependent dissipation occurs. The first memory integral scales like $O(\nu^{1/2})$ and is the leading dissipative contribution, if the wavelength $2\pi/k$ compares with the depth of the layer (shallow water limit $kh = O(1)$). This contritution is associated with the damping in the bottom boundary layer, because it dies out for $h \to \infty$. The second integral (c.f. Beyer and Friedrich (1995)) remains in the infinite depth limit; it is related to the free surface boundary layer and provides an $O(\nu^{3/2})$-correction. The various scaling of the bottom and surface dissipation arise due to the different character of boundary conditions: The no-slip condition at the bottom makes the tangential velocity die out rapidly within the boundary layer. The resulting velocity gradient is large and so is the viscous dissipation. This is unlike the surface boundary layer, where the boundary condition enforces some combinations of velocity gradients to vanish. There is no reason why the velocity gradients itselves should be large, thus dissipation beneath the surface is weaker.

In order to achieve analytic expressions for the threshold of the Faraday instability we expand the square roots in Eq.(30) giving

$$A(k,\omega) = -\omega^2 + \Omega^2\, X(k,\omega) + \omega_0^2, \tag{32}$$

where all viscous contributions up to $O(\varepsilon^{3/2})$ are collected in

$$
\begin{aligned}
X(k,\omega) = \Re(X) + i\,\Im(X) = \\
-\varepsilon^{1/2}\frac{\sqrt{2}|\omega/\Omega|^{3/2}}{\sinh 2kh} + \varepsilon^{3/2}\frac{|\omega/\Omega|^{1/2}}{2\sqrt{2}}(-15\tanh kh + 5\coth kh + 2\coth^3 kh) + \\
i\,\mathrm{sgn}(\omega)\left[\varepsilon^{1/2}\frac{\sqrt{2}|\omega/\Omega|^{3/2}}{\sinh 2kh} + \varepsilon|\omega/\Omega|(3 + \coth^2 kh) + \right. \\
\left. \varepsilon^{3/2}\frac{|\omega/\Omega|^{1/2}}{2\sqrt{2}}(-15\tanh kh + 5\coth kh + 2\coth^3 kh)\right].
\end{aligned}
\tag{33}
$$

The subharmonic resonance can be investigated by introducing

$$\bar\eta_k(\omega) = \alpha_1\delta(\omega - \Omega) + \alpha_2\delta(\omega + \Omega) + \bar\eta_k^{(1)} + \ldots \tag{34}$$

in Eq. (27). On assuming a small detuning $\Omega^2 - \omega_0^2 = O(X)$, the solvability condition for $\bar\eta_k^{(1)} = O(X)$ yields the neutral curve $a_S(k)$ by

$$\left[X(k,\Omega) + (\omega_0^2 - \Omega^2)\right]\left[X(k,-\Omega) + (\omega_0^2 - \Omega^2)\right] = \left(\frac{a_S\, k\tanh kh}{2\Omega^2}\right)^2. \tag{35}$$

The minimum of a_S with respect to ω_0^2 (see Footnote) defines the onset of the subharmonic response

$$a_{Sc} \simeq \frac{2\Omega^2}{k_{Sc}}\,\Im\left[X(k_{Sc}),\Omega)\right]\coth k_{Sc}h. \tag{36}$$

The critical wave number k_{Sc} solves the viscous dispersion relation

$$\omega_0^2(k_{Sc}) \simeq \Omega^2 - \Re\left[X(k_{Sc}),\Omega)\right]. \tag{37}$$

Fig. 2 shows the analytic approximation according to Eq. (36) by a dashed line. Favourable agreement with the exact numerical solution (solid line) is achieved. The shortcomings of the traditional approach (dotted line) are cured. Eq. (36) deserves a simple and reliable substitute for the numerical solution of the stability problem. The three contributions in $\Im(X)$ are related respectively to damping in the bottom boundary layer, the bulk, and the surface. They dominate at low, intermediate, and higher drive frequencies. In the latter case Eqs. (36,37) are well approximated by the infinite depth limit $kh \gg 1$

$$a_{Sc} \simeq \frac{2\Omega^2}{k_{Sc}}(4\varepsilon - 2\sqrt{2}\,\varepsilon^{3/2}) \tag{38}$$

$$\omega_0^2(k_{Sc}) \simeq \Omega^2 + 2\sqrt{2}\,\varepsilon^{3/2}. \tag{39}$$

In the opposite shallow water limit ($kh = O(1)$), which occurs at small Ω, we observe in Fig.2 a sharp re-increase of the threshold amplitude indicating that the dissipation in the bottom boundary layer begins to dominate. The subharmonic tongue $n = 1$ in the neutral chart is the first who experiences this small-k-damping and thus it is pushed up. If this effect is strong enough the neighboring harmonic tongue $n = 2$ provides the abolute minimum and thus defines a_c. This inverted situation is depicted in Fig.3. At a drive frequency where the minimum values of both tongues coincide a bicriticality occurs: A subharmonic mode with wave number $k = k_S$ competes with a harmonic mode at $k = k_H > k_S$. In order to predict this bicriticality a perturbation expansion for the harmonic tongue must also be performed. One obtains

$$a_{Hc} \simeq \frac{4\Omega^2}{k_{Hc}} \sqrt{\Im[X(k_{Hc}, 2\Omega)]} \coth k_{Hc} h \tag{40}$$

with the wave number k_{Hc} determined by

$$\omega_0^2(k_{Hc}) \simeq 4\Omega^2 + \frac{2}{3}\Im[X(k_{Hc}, 2\Omega)] - \Re[X(k_{Hc}, 2\Omega)]. \tag{41}$$

Fig. 4 compares the S- and H-thresholds for a fluid layer of water (dashed line) and another of silicone oil (solid line). As can be seen in the inset, the S- and H-boundaries intersect at rather low frequencies, when the shallow water condition $kh = O(1)$ is met. The bicriticality has been observed experimentally by Müller et al. (1997). Fig.5a presents a photograph of a surface pattern, which is in harmonic resonance with the external drive. The excitation frequency of the vessel is $9Hz$ and so is the frequency of the surface wave. To confirm the observation we show in Fig. 5b a pattern which is driven at $\Omega/\pi = 10Hz$. Inspite of the higher frequency the observed wavelength is considerably larger, indicating that the surface is oscillating subharmonically with $5Hz$.

3.6 The Faraday instability in the limit of strong dissipation

So far we have considered the case of weak dissipation $\varepsilon \ll 1$, which applies to almost all Faraday experiments undertaken in the past. However, an analytic

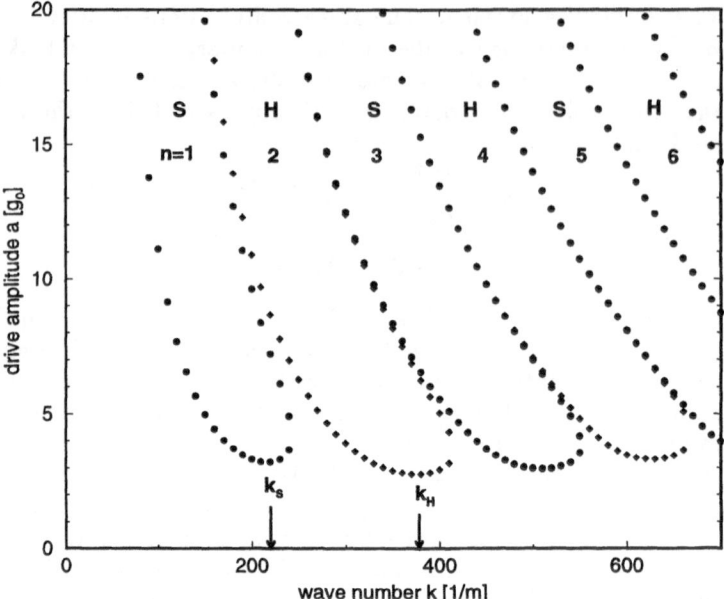

Fig. 3. At a drive frequency of $\Omega/\pi = 5Hz$ the subharmonic tongue is pushed up due to the damping in the bottom boundary layer, which dominates at small k. The neighboring harmonic tongue provides the absolute minimum at $k_c = k_H$ and thus defines the onset amplitude a_c. For $\Omega \simeq 7Hz$ the minima of the two tongues coincide giving rise to a bicritical situation. Fluid parameters as in Fig.2

treatment of the opposite limit $\bar{\varepsilon} = \omega/(\nu k^2) \ll 1$ is also possible (Cerda and Tirapegui (1997)). On expanding Eq. (28) in powers of $\bar{\varepsilon}$ one obtains

$$A(k, \omega) = i\frac{\omega^2}{\bar{\varepsilon}}G(k\,h) + \omega_0^2 - \omega^2 F(k\,h) + O(\bar{\varepsilon}), \qquad (42)$$

where the functions $G(x)$ and $F(x)$ stand for

$$G(x) = \frac{2\tanh x(\cosh 2x + 2s^2 + 1)}{\sinh 2x - 2x} \qquad (43)$$

$$F(x) = \tanh x\frac{3\cosh^2 x\,(\sinh 2x - 2x - 4x^3/3) + s^2\,(\sinh 2x - 2x)}{(\sinh 2x - 2x)^2}. \qquad (44)$$

After transformation of Eq. (27) into real space this gives an overdamped Mathieu oscillator. By using a WKB-technique one can show that – inspite of the dominating damping – the rest state can be destabilized by a sufficient

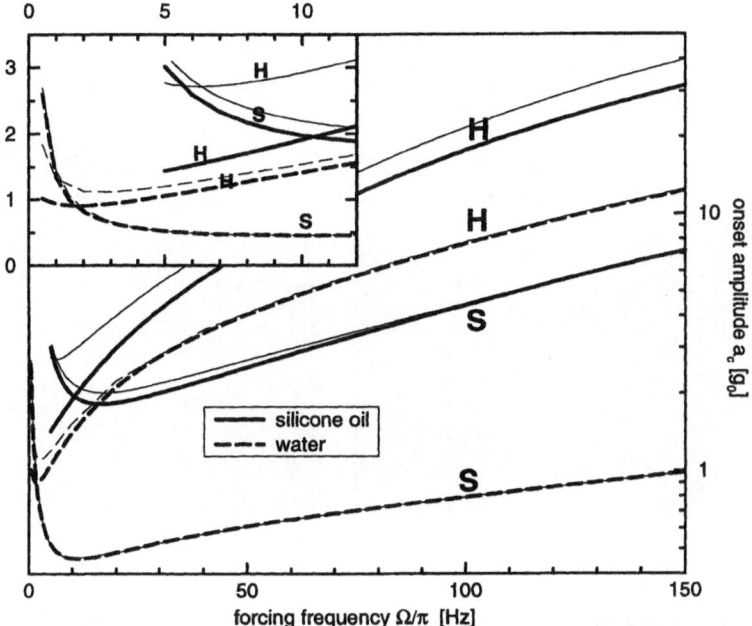

Fig. 4. The critical onset amplitude for the subharmonic (S) and harmonic (H) Faraday instability as a function of the forcing frequency Ω/π. Thick lines correspond to the analytic result Eq. (36), thin lines are obtained numerically as described in Sec. 3.4. Parameters for water (dashed): $\rho = 1.0 g/cm^3$, $\sigma = 72.4 * 10^{-3} N/m$, $\nu = 1 mm^2/s$, $h = 1 mm$; for silicone oil (solid): $\rho = 0.934 g/cm^3$, $\sigma = 20.1 * 10^{-3} N/m$, $\nu = 10 mm^2/s$, $h = 1 mm$.

gravity modulation. Depending on the excitation frequency Ω and the depth h, any of the harmonic or subharmonic tongues ($n = 1, 2, 3...$) in the neutral stability chart can provide the absolute minimum and thus determine a_c. In the strong damping approximation the critical wavelength $2\pi/k_c$ is nearly independnent of Ω but scales with the filling depth h. This is unlike the weak dissipation limit where k_c is determined by the dispersion.

The condition $\bar\varepsilon \ll 1$ applies when the viscous boundary layer extends over the whole fluid depth. This case is called the lubrication limit. Clearly, the strong damping gives rise to enormous critical accelerations for the onset of surface waves. Experiments in this parameter region have not been performed yet.

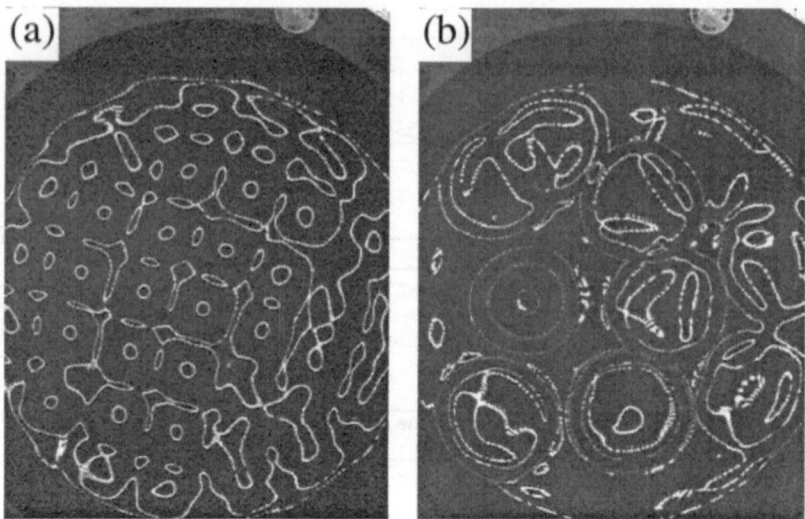

Fig. 5. Surface patterns as observed in the experiment. (a): Driving force and surface oscillate synchronously at $\Omega/\pi = 9Hz$ (harmonic response). (b) Subharmonic surface oscillation with $5Hz$ at a drive frequency $\Omega/\pi = 10Hz$. Parameters for the experiment: $\rho = 0.929g/cm^3$, $\nu = 8.9mm^2/s$, $\alpha = 0.0198N/m$, $h \simeq 0.8mm$.

4   Discussion

Nonlinear perturbation expansions for weak dissipation and small supercritical drive have demonstrated (Milner (1991), Zhang and Vinals (1996)) that subharmonic Faraday waves appear via a forwards pitchfork bifurcation, described by a Landau-equation of cubic order. Depending upon the nonlinear cross coupling between Fourier modes of different lateral orientation a variety of spatially periodic and quasi-periodic patterns has been predicted by Zhang and Vinals (1996), Chen and Vinals (1997) and experimentally verified by Binks and van de Water (1997). The absence of quadratic nonlinearities in the Landau-equations results from the subharmonic temporal symmetry of the critical modes: Since a subharmonic spectrum consists of odd multiples of the basic frequency Ω (see Eq. (29) with $\beta = 1$) any second order nonlinearity produces "even" Fourier spectra, which do not resonate with odd ones. The subharmonic symmetry in the Faraday experiment is similar to the Oberbeck-Boussinesq symmetry in convection, where quadratic nonlinearities in the amplitude equations are also forbidden.

In case of the *harmonic* Faraday resonace the situation is different: The temporal Fourier spectrum at onset of the instability is even. As the square of an even spectrum reproduces an even spectrum, quadratic coupling in the amplitude equation is possible. The corresponding generic nonlinear pattern consists of subcritcal hexagons. At some higher drive amplitude a secondary transition to another ordered pattern can be expected. However it is by far not clear, whether this will be a line structure as in Rayleigh-Bénard convection.

Presumably even more interesting than the harmonic Faraday instability is the parameter space nearby the bicriticality. Mediated by quadratic nonlinearities is a cross coupling between two subharmonic modes and a harmonic one. Since two different length scales $2\pi/k_S$ and $2\pi/h_H$ interact complicated convective structures are likely to appear.

Unfortunately in typical experimental setups the bicriticality appears at drive frequencies $< 10Hz$. This value is below the specifications of most commercial shaker systems. Clearly, small excitation frequencies generate patterns with a long wavelength (of the order of cm). Vessels of large diameters (with considerable weight) are therefore necessary to get rid of geometrical influences on pattern formation. Alternatively the shallow water limit can be achieved by reducing the filling depth h. Then, the bicriticality occurs at frequencies which are better accessible, but the quickly rising threshold amplitude requires powerful shakers. Neither theoretical nor experimental investigations of the harmonic Faraday instability or the bicriticality have been undertaken. This will be a challenge for future work on this subject.

References

T. B. Benjamin und F. Ursell, Proc. R. Soc. Lond. A **225** 505 (1954).

J. Bechhoefer, V. Ego, S. Manneville, B. Johnson, J. Fluid Mech. **288**, 325 (1995).

J. Beyer and R. Friedrich, Phys. Rev. E**51**, 1162 (1995).

D. Binks, W. van de Water, Phys. Rev. Lett. **78**, 4043 (1997).

E. Cerda, E. Tirapegui, Phys. Rev. Lett. 78, 859 (1997).

P. Chen and J. Vinals, preprint (1997).

S. Ciliberto und J. P. Gollub, J. Fluid Mech. **158**, 381 (1985).

B. Christiansen, P. Alstrom und M. T. Levinsen, Phys. Rev. Lett. **68**, 2157 (1992); J. Fluid Mech. **291**, 323 (1995).

S. Douady, J. Fluid Mech. **221**, 383 (1990).

W. S. Edwards und S. Fauve, J. Fluid Mech. **278**, 123 (1994).

M. Faraday, Philos. Trans. R. Soc. London **52**, 319 (1831).

S. Fauve, K. Kumar, C. Laroche, D. Beyens, Y. Garrabos, Phys. Rev. Lett. **68**, 3160 (1992).

J. P.Gollub, Physica D **51**, 501 (1991).

S. V. Kiyashko, L. N. Korzinov, M. I. Rabinovich, L. S. Tsimring, Phys. Rev. E **54**, 5037 (1996).

K. Kumar, L. Tuckerman, J. Fluid Mech. **279**, 49 (1994).

K. Kumar, K. Bajaj, Phys. Rev. E **52** R4606 (1995).

K. Kumar, Proc. R. Soc. London A **452**, 1113 (1996).

L. Landau and E. M. Lifshitz, Fluid Mechanics, 2nd edn. Pergamon Press (1987).

S. T. Milner, J. Fluid Mech. **225**,81 (1991).

H. W. Müller, Phys. Rev. Lett. **71**, 3287 (1993).

H. W. Müller, H. Wittmer, C. Wagner, J. Albers, K. Knorr, Phys. Rev. Lett. **78**, 2357 (1997).

A. H. Nayfeh and D. T. Mook, *Nonlinear oscillations*, Wiley (1979).

Lord Rayleigh, Phil. Mag. **16**, 50 (1883).

F. Simonelly, J. P. Gollub, J. Fluid Mech. **199**, 471, (1989).

N. B. Tufillaro, R. Ramshankar, J. P. Gollub, Phys. Rev. Lett. **62**, 422 (1989).

W. Zhang, J. Vinals, Phys. Rev. E **53**, R4286 (1996).

Footnote: Up to $O(X)$ minimization with respect to k or ω_0^2 is equivalent.

On Curved Cellular Flames

Andreas G. Class

Institut für Angewandte Thermo- und Fluiddynamik
Forschungszentrum Karlsruhe, 76021 Karlsruhe, Germany

Abstract. Laminar premixed flames exhibit cellular instabilities for sufficiently small Lewis-numbers. In the present article we study a modified Kuramoto-Sivashinsky-equation describing curved, cellular flames near the instability threshold. Pattern formation is strongly influenced by the sign of flame curvature. For a flame that is convex with respect to the burnt gases localized patterns are observed. Concave flames display a family of patterns that we call the cell reproduction cascade (CRC). CRC-patterns on large domains are unstable to a spontanous symmetry breaking instability.

1 Introduction

It is well known that flat or weakly curved premixed flames may exhibit a cellular instability. Sivashinsky (1977) was the first to derive a long wave theory near the cellular stability threshold. Weakly nonlinear evolution about a planar flame front is described by a Kuramoto-Sivashinsky equation. Later the theory was generalized by multiple authors among which Sivashinsky, Law and Joulin (1982) accounts for stagnation point flow, Margolis and Sivashinsky (1984) accounts for gravity, Matalon and Erneux (1984) includes the effect of general flow fields and Buckmaster and Ludford (1982) consider heat losses.

Here, we model curved flames which are stabilized by heat losses to a flame holder. Flame curvature is due to a curved shape of the screen. The motivation for the analysis of curved flames is to study flame stretch which has an important impact on the behavior of flames. For the screen-stabilized flame stretch is positive if the flame is concave as viewed from the burnt gases while the flame is compressed if it is convex. In contrast to the theory of Sivashinsky et. al. (1982) we may control flame stability and flame stretch independently also giving us access to negative flame stretch. We will specifically study flames of parabolic shape. For this special configuration flame stretch is a constant and weakly nonlinear evolution around the curved flame is described by a modified Kuramoto-Sivashinsky equation (MKS) which is similar to the equation found by Sivashinsky et. al. (1982) except that we have two independent parameters while in their theory these two parameters take identical values.

We analyze the stability of solutions of the MKS equation for curved parabolic flames on infinite transverse domains and show that for positive flame stretch any such solution is unstable to a drift instability that we call the exponential drift instability (EDI). This instability renders any localized pattern unstable on large domains. For nonlocalized patterns the EDI is related to symmetry breaking.

We evaluate the typical behavior of uniformly stretched or compressed flames by numerical computations where we identify phenomena as localization, time periodicity and chaos.

2 Formulation

Before analyzing MKS equation that describes flame dynamics we describe the geometry and the underlying flame model but we skip the lengthy derivation that is similar to the one in Class (1995).

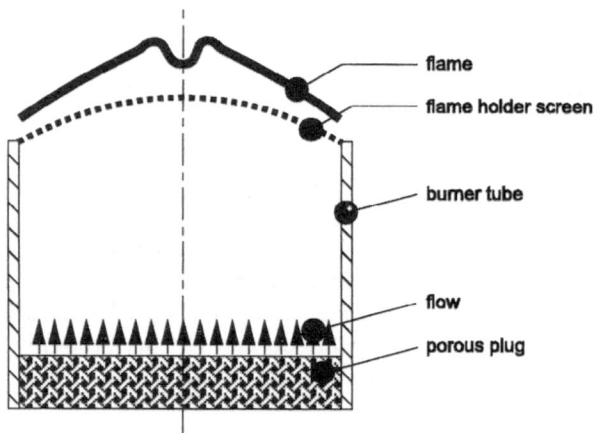

Fig. 1. Burner geometry

The model geometry is shown in Fig. 1. The mixture of fuel and oxidant enters the combustion chamber from a porous plug. At the porous plug the flow is purely axial and the flow speed, the temperature and the mixture composition is uniform. The flow velocity is smaller than the adiabatic flame speed. Above the porous plug there is a screen which has a negligible effect on the flow and mixture composition. The flame is located above the screen. The heat released in the combustion is transported upstream by heat conduction. Since the screen is located within the preheat zone it takes the temperature of the mixture which is higher than the temperature of the environment. Therefore, a fraction of the reaction heat is radiated to the environment by the screen thus reducing preheating of the oncoming gases. This mechanism reduces the reaction temperature and lowers the flame speed such that stable combustion conditions are realized.

Our model is similar to Sivashinsky (1977) assuming that density is constant but we account for heat losses. In the asymptotic limit of thin flames the reaction zone shrinks to a front which is located at $x = F(y, z, t)$ and the governing equations are the heat equation and a mass conservation equation for the limiting reactant which is given in suitable non dimensional notation as

$$\frac{\partial T}{\partial t} + u\frac{\partial T}{\partial x} = \nabla^2 T + \sigma Q - r \ , \tag{1}$$

$$\frac{\partial C}{\partial t} + u\frac{\partial C}{\partial x} = \frac{1}{\text{Le}}\nabla^2 C - Q \ , \tag{2}$$

$$Q = \exp\left(\frac{1}{2}N(T-1)\right)\sqrt{1 + (\nabla_\perp F)^2}\delta\left(x - F\right) \ , \tag{3}$$

$$r = q\left(T^4 - T_0^4\right)\delta\left(x - F_s\right) \ . \tag{4}$$

subject to the boundary conditions

$$\begin{aligned}
T(x, y, z, t) &= T_0 \text{ for } x \to -\infty \ , \\
C(x, y, z, t) &= 1 \text{ for } x \to -\infty \ , \\
\tfrac{\partial}{\partial x}T = C &= 0 \quad \text{ for } x \to +\infty \ , \\
T, C \text{ are finite} \quad &\text{ for } y, z \to \pm\infty
\end{aligned} \tag{5}$$

and initial conditions describing a state which is close to a planar burning flame.

Here T is scaled absolute temperature, C is concentration and (x, y, z, t) are the independent variables of space and time. The axial flow velocity is u which is a constant smaller than the adiabatic flame speed which is unity. The parameter σ measures the heat released by the combustion reaction and is defined as $\sigma = (T_b - T_0)/T_b$ where T_b is the adiabatic combustion temperature and T_0 is the temperature at which the mixture is supplied to the combustion chamber. The Lewis-number Le in the mass conservation equation is the ratio of thermal diffusivity to mass diffusivity and has a value which is for gas mixtures typically close to unity. In the model the combustion reaction is represented by the heat release σQ and the fuel consumption $-Q$

The Dirac delta function $\delta(x - F(y, z, t))$ concentrates the reaction term at the front and the square root accounts for the geometric shape of the flame. The exponential models the temperature dependence of the reaction, where N is the activation energy of the reaction which is typically large and therefore justifies the thin flame approximation. The sink r in the energy balance (1) is the radiative heat transfer to the environment which is active at the screen located at $x = F_s(y, z)$. The constant q is a nondimensional coefficient of radiation.

From flames propagating in free space ($u \to 1, r \to 0$) it is known that cellular patterns develope spontaneously from planar flames if the reduced Lewis number defined as $\lambda = (\text{Le} - 1)\,\sigma N$ is lower than a critical value $\lambda_c = -2$. For a weakly unstable situation where $\varepsilon = (\lambda - \lambda_c)/\lambda_c \ll 1$ long wave patterns with a critical wave length of $\mathcal{O}\left(\varepsilon^{-1/2}\right)$ develope on a slow $\mathcal{O}\left(\varepsilon^{-2}\right)$ time scale.

For uniform $u = 1 + \varepsilon^2 u^{(2)}$ i.e. close to unity and a sufficiently small radiation coefficient $q = \varepsilon^2\beta$ the evolution of the flame front position $F = \phi^{(0)} + \varepsilon\phi^{(1)}(\eta, \zeta, \tau) + \dots$, is described by a weakly nonlinear long wave theory in terms of rescaled space $\eta = \sqrt{\varepsilon}y$, $\zeta = \sqrt{\varepsilon}z$ and time $\tau = \varepsilon^2 t$. The standoff distance from the screen is at leading order constant

$$\phi^{(0)} = -\log(-\Theta_0 + \sqrt[4]{\Theta_0^4 - \frac{2}{\beta}u^{(2)}}).\tag{6}$$

Note, that flames may only be stabilized by the flame holder if the flow speed is smaller than the adiabatic flame speed $u^{(2)} < 0$.

Weakly non planar flame holder screens are represented by the shape function $F_s = \varepsilon\phi_s^{(1)}(\eta, \zeta) + \dots$.

Following a derivation that is similar to Class (1995), we derive an evolution equation for the flame front

$$\frac{\partial\phi^{(1)}}{\partial\tau} + 4\,\nabla_\perp^4\phi^{(1)} + \nabla_\perp^2\phi^{(1)} + \frac{1}{2}\left(\nabla_\perp\phi^{(1)}\right)^2 - u^{(2)}\,\phi^{(1)} = \rho,\tag{7}$$

which is a modified Kuramoto-Sivashinsky equation. The new terms in the MKS-equation as compared to the Kuramoto-Sivashinsky equation are the non differentiated term which acts to stabilize the flame and the forcing term which takes account of non planar flame holder screens

$$\rho = -u^{(2)}\phi_s^{(1)}.\tag{8}$$

Other effects such as spatially dependent flow and enthalpy at the burner outlet have been considered in Class (1995) where it was shown that these effect may be modeled by adding similar forcing terms ρ.

Sivashinsky et. al. (1982) has outlined the analysis for the stagnation point burner. Due to the stagnation point flow field the undifferentiated term $-u^{(2)}\,\phi^{(1)}$ is replaced by $(\nabla_\perp \mathbf{v})\,\phi^{(1)} = \alpha\,\phi^{(1)}$ where the flow vector field is $\mathbf{v} = \frac{1}{2}\alpha \cdot (\eta, \zeta)$ and an additional convective term $\mathbf{v}.\nabla_\perp\phi^{(1)}$ appears. Below we will show that we recover a similar convective term by considering curved flames.

2.1 Parabolic Curved Flames

We proposed curved-screen-stabilized flames as a simple configuration to study stretched flames. Flame stretch is defined following Williams (1975) as $K = \left(\frac{1}{A}\frac{dA}{dt}\right)$ where A is a surface element of the flame that travels with the flame and that is deformed by the flow tangential to the flame. In Matalon (1983) the general formula for K as a function of flame shape, flame evolution and flow is given. For purely axial flow and a weakly curved flame, stretch is given by

$$K \approx \nabla_\perp^2 \phi^{(1)} . \tag{9}$$

In this section we develop the necessary screen shape $\Phi_s^{(1)}$ that is required to set up a flow field exhibiting a constant flame stretch. Moreover, we emphasize that for real burners it is in general not possible to set up an ideal uniform axial velocity profile at the burner outlet. Due to viscous effects inside the burner there is typically a drop of the velocity profile near the burner rim. Similar arguments hold for the temperature and enthalpy profiles. This suggests that we study specific simple forcing profiles, with the most simple shape given by the parabolic form

$$\rho = a_0 + a_2(\eta^2 + \zeta^2) . \tag{10}$$

It may be verified easily that the parabolic shape

$$\phi^{(1)} = \phi_b^{(1)} = c_0 + c_2 \left(\eta^2 + \zeta^2 \right) \tag{11}$$

is now a basic solution to (7) so that flame stretch is given by

$$K = 4c_2 , \tag{12}$$

i.e. positive (negative) c_2 corresponds to positive (negative) flame stretch. The coefficients c_0 and c_2 are defined by the relations

$$a_0 = 4c_2 - u^{(2)}c_0 , \tag{13}$$

$$a_2 = c_2 \left(2c_2 - u^{(2)} \right) \tag{14}$$

or

$$c_2 = \frac{1}{4} \left(u^{(2)} \pm \sqrt{((u^{(2)})^2 + 8a_2)} \right) . \tag{15}$$

Note that the planar basic state for a planar flame holder screen corresponds to the plus sign in (15). For this case positive (negative) a_2 also correspond to positive (negative) c_2. Thus the flame bends in the same direction as the screen.

For convenience (7) is converted into a homogeneous equation by subtracting the curved basic solution from $\phi^{(1)}$. The deviation of the flame front from the curved basic state is denoted as

$$\psi \equiv \phi^{(1)} - \phi_b^{(1)} \tag{16}$$

and the evolution equation for ψ is:

$$\frac{\partial \psi}{\partial \tau} + 4\nabla_\perp^4 \psi + \nabla_\perp^2 \psi + \frac{1}{2} \left(\nabla_\perp \psi \right)^2 + \mathbf{v}_{\perp,\mathrm{eff}} . \nabla_\perp \psi - u^{(2)} \psi = 0 \tag{17}$$

where the effective velocity field is

$$\mathbf{v}_{\perp,\text{eff}} \equiv \nabla_\perp \phi_b^{(1)} \; . \tag{18}$$

Since quite general $\nabla_\perp \Phi_b^{(1)}$ may be realized by appropriate choices of the screen shape and the boundary conditions at the burner almost any $\mathbf{v}_{\perp,\text{eff}}$ corresponds to a physical realizable situation.

In particular, we study effective velocity profiles of parabolic curved flames

$$\mathbf{v}_{\perp,\text{eff}} = \frac{1}{2}\alpha \cdot (\eta, \zeta) \tag{19}$$

where the constant α is given by

$$\alpha = -2c_2. \tag{20}$$

Note, that for stagnation point flow and planar flames (Class (1995)) we obtain the same velocity field along the flame. Therefore we use the term effective stagnation point flow if $\alpha > 0$ and effective inverse stagnation point flow if $\alpha < 0$.

3 Pattern Formation

3.1 Linear Stability

Linear stability of a parabolic curved flame on an infinite domain is analyzed using the approach proposed by Sivashinsky et. al. (1982), where a coordinate transformation to stretching or compressing coordinates

$$\begin{aligned}
\widehat{\eta} &= \eta \exp\left(-\tfrac{1}{2}\alpha\tau\right) \; , \\
\widehat{\zeta} &= \zeta \exp\left(-\tfrac{1}{2}\alpha\tau\right) \; , \\
\tau &\to \tau \; .
\end{aligned} \tag{21}$$

cancels the convective term from linearized (17). In transformed coordinates we express the solution as a time dependent amplitude multiplied by harmonic functions.

$$\Psi = A(\tau) \exp\left(ik\widehat{\eta} + il\widehat{\zeta}\right) \; . \tag{22}$$

where k and l are wave numbers.

Substituting this ansatz into (17) yields an ordinary differential equation for the amplitude A

$$\frac{1}{A}\frac{dA}{d\tau} = l^{*2} - 4l^{*4} + u^{(2)} \tag{23}$$

with the effective wave number $l^* = \sqrt{(k^2 + l^2)\exp\left(-\tfrac{1}{2}\alpha\tau\right)}$.

For $-u^{(2)} < \frac{1}{16}$ there are modes that are temporarily amplified. The condition for amplification is:

$$\frac{1}{8}\left(1 - \sqrt{1 + 16u^{(2)}}\right) < l^{*2} < \frac{1}{8}\left(1 + \sqrt{1 + 16u^{(2)}}\right) . \tag{24}$$

Since the effective wave number is time dependent we find that the wave length of perturbations is either compressed or stretched in time for the cases $\alpha < 0$ and $\alpha > 0$ respectively. Therefore, any perturbation may be amplified for a finite time only until its wave number leaves the region of amplification (24). For large times the perturbation will finally decay to zero which means the system is stable with respect to any small perturbation. Since the duration of amplification grows as $|\alpha|$ decays, even small perturbations may become very large so that nonlinear effects may no longer be neglected. Thus we expect to find instability for sufficiently small $|\alpha|$ and $-u^{(2)} \ll \frac{1}{16}$. Our result is a generalization of Sivashinsky et. al. (1982) since we allow for both flame stretch and flame compression and also since the parameter $u^{(2)}$ that controls stability is independent of the parameter α that controls flame stretch.

3.2 Instability by Nonlinear Mode Interaction

In Sivashinsky et. al. (1982) it was proposed that instability of stretched flames is due to a continuous generation of short wave modes by nonlinear mode interaction. In this section we elaborate such a mechanism of nonlinear mode interaction for a stretched flame, i.e. $\alpha > 0$.

We assume an initial perturbation in form of a pure cosine mode

$$\psi = A(\tau)\cos(k(\tau)\hat{\eta}) \tag{25}$$

where $k(\tau) = k_0 \exp\left(-\frac{1}{2}\alpha\tau\right)$ accounts for the stretching effect of the flow field. $A(0)$ is assumed to be $\mathcal{O}(1)$ and k_0 lies within the range of amplified wave numbers (24) such that $A(\tau)$ is initially growing but it eventually decays when $k(\tau)$ has left this range. Substituting (25) into the quadratic term in equation (17) yields

$$(\nabla_\perp \psi)^2 = \frac{1}{2}A(\tau)^2 k(\tau)^2 \left(1 - \cos(2k(\tau)\hat{\eta})\right) . \tag{26}$$

so that a mode with the wave number $2k$ and another with zero wave number is generated. The amplitude of the generated $2k$ mode is growing as $A(\tau)$ is growing. Eventually $2k(\tau)$ enters the range of amplified wave numbers (24) so that it may grow without the support by nonlinear mode interaction. At the same time the initial k mode is decaying and becomes negligible so that there is a single non negligible mode and the spacial structure is identical to the initial structure. In figure 2a a numerical computation exhibiting this scenario is depicted where $\alpha = u^{(2)} = .013$. The horizontal direction in figure 2a is space η and the vertical direction is time τ. The gray scale is proportional to ψ

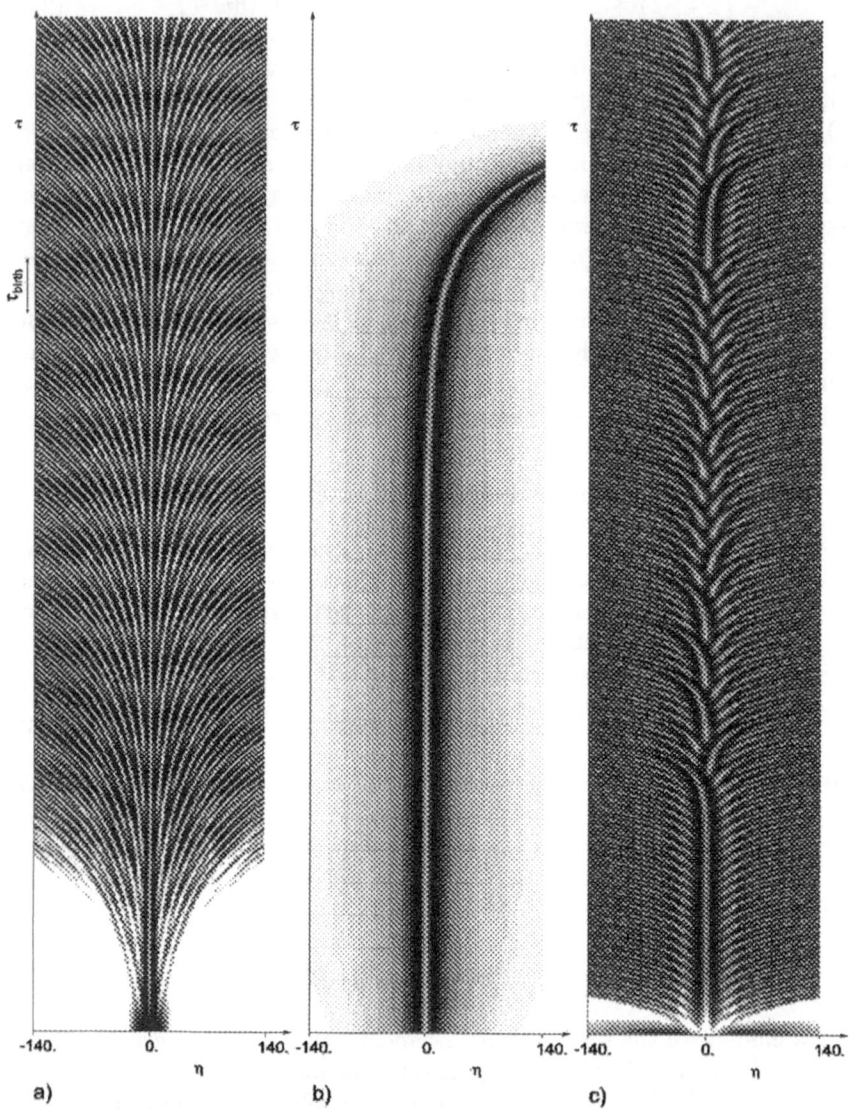

Fig. 2. Stretched patterns

where black stands for small $\psi = -10$ and white stands for $\psi = 0$. Following the black lines it is clear that the cells are transported with the flow in lateral direction. Only the crest at $\eta = 0$ stays in the center of the computational domain. As the cells are convected they expand in size, while the crests keep their size. When the cells have been stretched enough to accommodate a new cell then they split and a new cell is born. Since all cells give birth to new cells simultaneously the number of cells doubles at such an event. Since the stretching of the pattern continues the birth events repeat periodically in time where the time between successive birth events is the time required to stretch the pattern to twice its original size which is $\tau_{\text{birth}} = 2\ln 2/\alpha$. In the figure the evolution of an initial single cell perturbation is shown. The initial cell splits at the very beginning of the computation. Each cell is stretched by the flow field and later gives birth to a new cell doubling the total number of cells. This process continues doubling the number of cells repeatedly. This process resembles the growth process of an organism and we therefore call the mechanism the cell reproduction cascade (CRC).

3.3 Symmetry Breaking

Considering infinite transverse domains, (7) for a planar screen is invariant to translations in transversal direction. If $\psi(\eta, \zeta, \tau)$ satisfies the MKS-equation then a general shift also yields a solution $\psi(\eta - \eta_0, \zeta - \zeta_0, \tau)$.

This no longer holds for parabolic curved flames. Nevertheless, a modified translation invariance exists, if we allow for a time dependent shift $(\eta_0(\tau), \zeta_0(\tau))$.

Introducing the moving coordinates

$$
\begin{aligned}
\eta_I &= \eta - \eta_0(\tau) \ , \\
\zeta_I &= \zeta - \zeta_0(\tau) \ ,
\end{aligned}
\tag{27}
$$

we find that the MKS-equation for parabolic curved flames

$$
\frac{\partial \psi}{\partial \tau} + 4\nabla_\perp^4 \psi + \nabla_\perp^2 \psi + \frac{1}{2}(\nabla_\perp \psi)^2 + \mathbf{v}_{\perp,\text{eff}}.\nabla_\perp \psi - u^{(2)}\psi = 0
\tag{28}
$$

where

$$
\mathbf{v}_{\perp,\text{eff}} = \frac{1}{2}\alpha \cdot (\eta, \zeta)
\tag{29}
$$

transforms to the same equation with the modified flow field

$$
\mathbf{v}_{\perp,\text{eff}} = \left(\frac{1}{2}\alpha - \frac{\partial}{\partial \tau}\right) \cdot (\eta_I + \eta_0(\tau), \zeta_I + \zeta_0(\tau)) \ .
\tag{30}
$$

By requiring

$$
\frac{1}{2}\alpha\,\eta_0(\tau) - \frac{d\eta_0(\tau)}{d\tau} = 0 \quad \text{and} \quad \frac{1}{2}\alpha\,\zeta_0(\tau) - \frac{d\zeta_0(\tau)}{d\tau} = 0
\tag{31}
$$

the equations in the fixed coordinate system are identical to the equations in the moving coordinate system. The solution of equations (31) is

$$\eta_0\left(\tau\right) \propto \exp\left(\frac{1}{2}\alpha\tau\right) \quad \text{and} \quad \zeta_0\left(t\right) \propto \exp\left(\frac{1}{2}\alpha\tau\right). \tag{32}$$

If $\psi(\eta,\zeta,\tau)$ is a solution of the basic equations then there also exists a two parameter family of shifted solutions $\psi(\eta-\eta_0\exp\left(\frac{1}{2}\alpha\tau\right),\zeta-\zeta_0\exp\left(\frac{1}{2}\alpha\tau\right),\tau)$. The shifted solutions travel at exponentially growing or decaying speeds relative to the non shifted solution.

This result may be interpreted as an instability condition. Perturbing a given solution by a small displacement in the η_0 or ζ_0-direction, yields an instability if the displacement grows in time. Since the displacement grows exponentially we use the term exponential drift instability (EDI). On the other hand, we call the given solution stable with respect to translation if the displacement between the two solutions decays.

Therefore, solutions to the MKS-equation are stable with respect to the EDI if $\alpha < 0$, i.e. for effective inverse stagnation point flow. We find instability with respect to the EDI if $\alpha > 0$, i.e. for effective stagnation point flow.

For small $|\alpha|$ the displacement develops on a slow time scale $\mathcal{O}\left(1/\alpha\right)$. Such a solution may in fact seem to be stable for a long time but it will eventually start to travel.

Note, that the above arguments are valid for infinite domains only. Nevertheless, our computations for large but finite domains show the EDI in agreement with our arguments.

Figure 2b shows a density plot of a solution exhibiting the EDI. The meaning of the gray-scale and of the axis is similar as in figure 2a. The parameters in the computation are $\alpha = u^{(2)} = .014$ and the size of the domain is $\eta \in [-140, 140]$. Time is plotted in vertical direction. Initially the solution is stationary and forms a sharp ridge in the center which points towards the burnt gases. On either side of the ridge there is a symmetric cell (black) which points towards the fresh mixture. At the outer edge of these cells the solution shows a smooth $1/\eta$-profile such that non planar behavior may only be observed in a localized region around the origin $\eta = 0$. For later times the pattern starts to drift in lateral direction with increasing speed. When the region of localized behavior reaches the boundary the pattern is annihilated and the solutions becomes planar. Clearly the EDI breakes the initial reflection symmetry of the pattern.

The EDI was observed experimentally by Ishizuka and Law (1982). They investigate stagnation point flames in a circular geometry which may be modeled by (17) with $\alpha > 0$. In the experiment a pattern in form of a diametrically orientated groove is observed. This pattern is reported to be unstable to a drift instability, i.e. it travels in a direction normal to its length along the flame. When the pattern reaches the boundary it disappears. Later, a new groove forms with an arbitrary azimuthal orientation. We assume that the formation of new grooves is due to flow perturbations in form of turbulence.

Since the EDI is present for any pattern when $\alpha > 0$ we conclude that any localized pattern is eventually travelling towards the lateral boundary where it is destroyed. Therefore persistent patterns must be non localized.

The cell reproduction cascade represents such a non localized pattern. Figure 2c shows a computation where the initial condition was modified from the one shown in figure 2a. We use a initial perturbation that shifted it to the left by a very small amount. Following the center crest we see that the pattern starts to drift as a whole to the left with an exponentially growing speed. Again the EDI is breaking the reflection symmetry of the pattern. In contrast to the localized pattern shown in figure 2b, this pattern expands very rapidly in space so that the pattern fills the full domain persistently. As explained earlier the pattern is time periodic and after one period in time each cell splits forming a pair of new cells. Due to the EDI the positions where new cells are born pass at exponential speed the observer. At the same time the pattern expands at an exponential rate so that any cell that is born left of the origin will travel to the left while any cell that is born on the right of the origin will travel to the right. In order to characterize the pattern we check at each birth event whether the cell born closest to the origin travels to the right or to the left. In figure 2c this creates the pattern (26R)LLLRRRRLRRRLRRRRLRLLRL.... This pattern is apparently chaotic. Nevertheless we may calculate such a sequence by assuming that the pattern is perfectly time periodic and perfectly symmetric in a moving frame of reference where cells are born at fixed equally spaced locations $\eta_{\text{birth}} = (n + 1/2) \cdot \Delta\eta$ where $\Delta\eta$ is the spacing between neighboring birth locations and n is an integer. If the pattern drifts due to the EDI the location of birth events in the fixed coordinate system is

$$\eta_{\text{birth}} = \left(n + \frac{1}{2}\right) \cdot \Delta\eta - \eta_0 \exp\left(\frac{1}{2}\alpha\, m\, \tau_{\text{birth}}\right) \;, \qquad (33)$$

where $\tau_{\text{birth}} = 2\ln 2/\alpha$ i.e.

$$\eta_{\text{birth}} = \left(n + \frac{1}{2}\right) \cdot \Delta\eta - \eta_0 \cdot 2^m \;. \qquad (34)$$

The position η_{origin} of the cell which is born closest to the origin is in the fixed coordinate system

$$\eta_{\text{origin}}/\Delta\eta = \text{mod}\left(-\eta_0/\Delta\eta \cdot 2^m, 1\right) - 1/2 \;. \qquad (35)$$

The exponentially fast growing term 2^m in the modulo function causes the random like structure of the pattern. In addition small differences in the parameter η_0 will eventually be amplified exponentially, so that patterns with slightly different η_0 only resemble for a limited time.

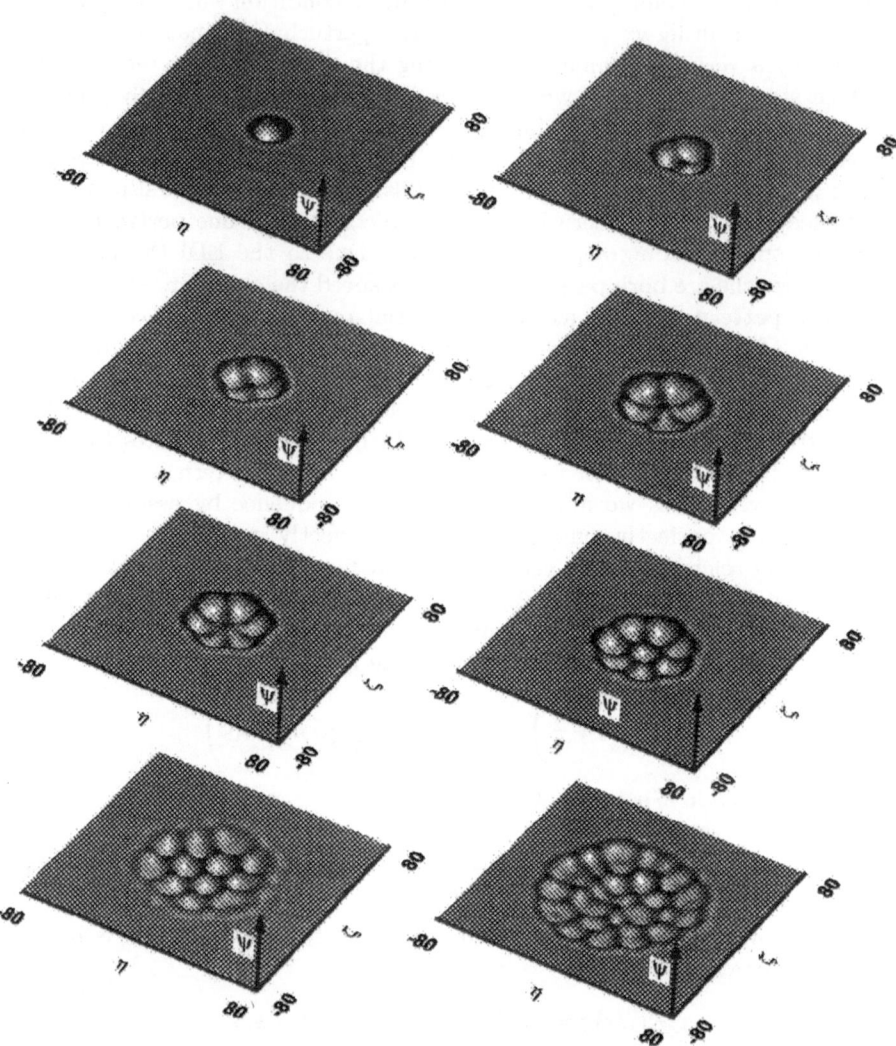

Fig. 3. Compressed patterns

3.4 Localization

We now consider a flame which is convex as viewed from the burnt gases i.e. a compressed flame where $\alpha < 0$. For this case the EDI is not present and the CRC-scenario is not working. Nevertheless, there are still disturbances that may grow for a finite time before decaying. Thus, we may conclude that the flame may well be unstable to finite amplitude disturbances. We present some results from numerical computations which exhibit solutions with a nonplanar behavior in a localized region of space. Note, that at least for stationary solutions, the dominant term in (17) for $|\eta|, |\zeta| \rightarrow \infty$ is the convection term, so that necessarily $\mathbf{v}_{\text{eff}}.\nabla_{\perp}\psi \rightarrow 0$ which is the case for localized solutions.

Fig. 3 shows a number of cellular patterns as viewed from the burnt gases. The gray scale corresponds to the local combustion temperature. All the patterns show an ordered array of cells. As $|\alpha|$ and $|u^{(2)}|$ is reduced the spatio temporal structure is becoming increasingly complex. For the patterns with 1 to 5 cells there is a range of parameters where the patterns are stationary. All other patterns showed dynamics where the patterns with 6 to 12 cells where time periodic in some range of parameters. The last pattern shows a "molten" spatial structure and is apparently chaotic.

All the patterns shown here, are similar to the ordered patterns experimentally observed by Gorman, el-Hamdi and Robbins (1994) on a circular porous plug burner.

4 Conclusions

Curved flames are subjected to flame stretch. For stretched cellular flames the cells expand until they split thus giving birth to new cells. Birth events occure repeatedly. Stretched patterns exhibit a spontanous symmetry breaking instability which gives rise to chaotic dynamics. Compressed cellular flames display pattern formation in a localized region about the origin.

References

Buckmaster J. D., Ludford G. S. S. (1982): *Theory of laminar flames.* Cambridge Uni. Press

Class A. G. (1995): Zellulare Strukturen laminarer Staupunktflammen. Forschungszentrum Karlsruhe, FZKA **5655**

Gorman M., el-Hamdi M., Robbins K. A.: Experimental observation of ordered states of cellular flames. Combust. Sci. and Tech., **98**, 37–45

Ishizuka S., Law C. K. (1982): An experimental study on extinction and stability of stretched premixed flames. 19th Symp. (Int.) Combust./Combust. Inst., 327–335

Matalon M., (1983): On flame stretch. Combust. Sci. and Tech., **31**:169–181

Matalon M., Erneux T. (1984): Expanding flames may delay the transition to cellular structures. SIAM J. Appl. Math., **44**(4),734—744

Margolis S. B., Sivashinsky G. I. (1984): Flame propagation in vertical channels: Bifurcation to bimodal cellular flames. SIAM J. Appl. Math., **44**(2), 344—368

Sivashinsky G. I. (1977): Diffusional-thermal theory of cellular flames. Combust. Sci. and Tech., **15,** 137

Sivashinsky G. I., Law C. K., Joulin G. (1982): On stability of premixed flames in stagnation point flow. Combust. Sci. and Tech., **28,** 155—159

Williams F. A. (1975): A review of some theoretical models of turbulent flame structure. Agar Conference Proceedings, 164

Nonlinear Flow Behavior and Shear-Induced Structure of Fluids

Siegfried Hess
Institut für Theoretische Physik,
Technische Universität Berlin
Hardenbergstr. 36, D – 10623 Berlin, Germany

Abstract

The viscous behavior of fluids becomes nonliear or non-newtonian, when the shear rate is larger than the reciprocal of a characteristic relaxation time. The method of nonequilibrium molecular dynamics (NEMD) is introduced with special emphasis on simulations of a plane Couette flow. Procedures to extract rheological properties, such as the (non-newtonian) viscosity, normal pressure differences as well as information on shear flow induced structural changes, are discussed. Firstly, fluids composed of spherical particles are considered. This comprises simple liquids and dense colloidal dispersions, where the states far away from equilibrium studied in NEMD are accessible in experiments. Secondly, as an axample for complex fluids, polymeric melts are investigated.

1 Introduction

The viscosity of "newtonian fluids" does not dependent on the shear rate over the experimentally accessible range of shear rates. Then the shear stress is a linear function of the shear rate. Gases and many liquids, e.g. water, belong to this kind of fluids. A nonlinear flow behavior, also referred to as "non-newtonian behavior", where the shear stress is a nonlinear function of the shear rate and where consequently the viscosity does depend on the shear rate is fairly common in complex fluids encountered in the kitchen and in substances of biological and of technical importance such as polymeric liquids, surfactant solutions and colloidal dispersions. Examples of experimental data obtained for dispersions containing spherical colloidal (latex) particles (diameter $165nm$) with different volume fractions (disolved in glycol) [1] are shown in Fig.(1). Both a decrease

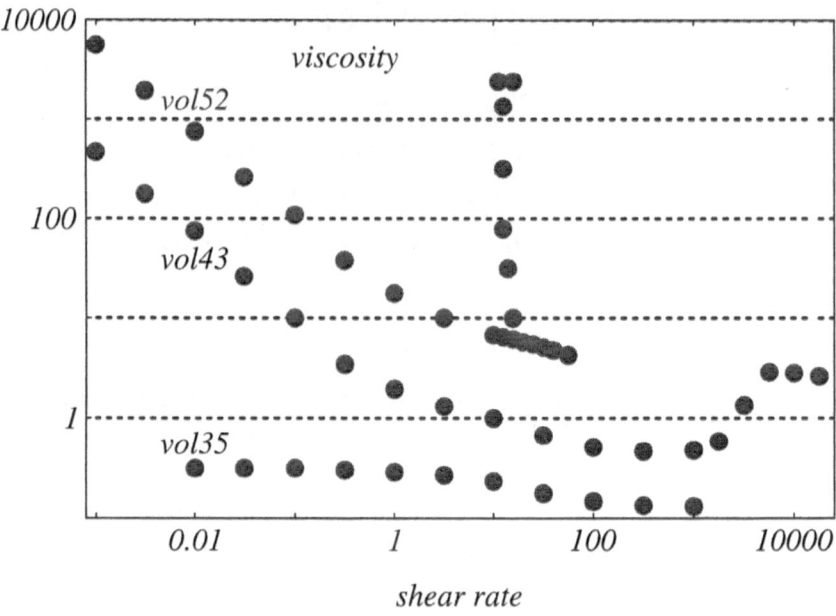

Figure 1: The viscosity of colloidal dispersions as function of the shear rate. The labels *vol*35, *vol*43 and *vol*52 refer to solutions where the volume fraction of the spherical colloidal particles is 0.355, 0.434 and 0.523 respectively.

and an increase of the viscosity can be observed with increasing shear rates. This behavior is termed "shear thinning" and "shear thickening", respectively. Also normal stress differences and volume changes occur in the non-newtonian flow regime. These are typical nonlinear phenomena since these effects are proportional to the second power of the shear rate in the small shear rate limit.

The nonlinear flow behavior is accompanied by shear-induced structural changes which can be detected by light scattering [2] or by neutron scattering experiments [1], [3]. A complementary method for the study of the dynamical processes involved in the nonlinear flow behavior is provided by "NonEquilibrium Molecular Dynamics" (NEMD) computer simulations. Here, basic informations on this method, some results obtained by it, as well as comparison with simple model calculations and experiments are presented.

The NEMD computer simulations devoted to the study of transport and relaxation processes in fluids have been developed over twenty years ago [4], [5]. By now the method is well established [6]-[17]. So during the last decade the emphasis was, on the one hand, on the investigation of physical phenomena in simple fluids far away from equilibrium, i.e. typical nonlinear phenomena, and, on the other hand, on the study of the material properties of complex fluids.

Here, a plane Couette flow, also referred to as "simple shear flow", is considered as an example of a stationary transport process. Results are presented for "simple" and for "complex" fluids. The method of NEMD simulations is firstly discussed for fluids composed of spherical particles. It turns out that the flow behavior of the so called simple fluids and the shear induced structural changes are not simple. A comparison with experimental results of (dense) colloidal dispersions of spherical particles is made. The complex fluids studied here are polymeric melts.

2 Fluids of Spherical Particles

2.1 Molecular dynamics

In a molecular dynamics computer simulation *Newtons equations of motion*

$$m \frac{d^2}{dt^2} \mathbf{r}^i = \mathbf{F}^i = \sum_j \mathbf{F}^{ij} \tag{1}$$

are integrated numerically for N particles with mass m, located at positions $\mathbf{r}^i$ in a volume V. The particle density is $n = N/V$. The particle i, ($i = 1, 2, ..., N$), feels the force $\mathbf{F}^i = \sum_j \mathbf{F}^{ij}$ which is the sum of the forces $\mathbf{F}^{ij}$ excerted by all other particles $j \neq i$ on particle i.

In order to avoid surface effects, periodic boundary conditions and the nearest image convention are used. This means, particle i either feels the force caused by particle j or by one of its images depending on which one is closest to it. The range of the force has to be shorter than half of the lenght of the basic (central) periodicity box.

The temperature T is linked with the kinetic energy:

$$\frac{3}{2} k_B T = N^{-1} \frac{1}{2} \sum_i m(\mathbf{c}^i)^2, \tag{2}$$

where k_B is the Boltzmann constant and $\mathbf{c}^i$ is the "peculiar velocity", i.e. the velocity of a particle relative to the flow velocity.

In order to simulate an isothermal system the temperature has to be kept constant. The simplest version of a "thermostat" consists in rescaling the the peculiar velocity after each time step by the factor $(T_{wanted}/T_{measured})^{1/2}$. Other thermostats, e.g. those referred to as "Gaussian" and "Nose-Hoover" are discussed in [10],[17].

A typical binary interaction potential Φ depending on the distance r is the *Lennard-Jones* (LJ) potential

$$\Phi = \Phi^{LJ} := 4\Phi_0 \left(\left(\frac{r_0}{r}\right)^{12} - \left(\frac{r_0}{r}\right)^6 \right) . \tag{3}$$

In the simulations dimensionless or "scaled" variables are used which are denoted by the same symbols as the physical variables when no danger of confusion exists. For a system of LJ-particles, lengths and energies are presented in units of the diameter r_0 and of the potential depth Φ_0. The units used for the particle density and for the temperature are r_0^{-3} and $k_B^{-1}\Phi_0$. The time is scaled with the reference time $t_0 = r_0 m^{1/2}\Phi_0^{-1/2}$, m is the mass of a particle. The pressure, the shear rate and the viscosity of the LJ-fluid are expressed in units of $r_0^{-3}\Phi_0$, t_0^{-1} and $r_0^{-3}\Phi_0 t_0 = r_0^{-2} m^{1/2}\Phi_0^{1/2}$. For many fluids composed of atoms or small molecules, the specific values of r_0 and Φ_0 are available which are needed to relate the theoretical results to the thermophysical properties of specific substances.

When only the repulsive r^{12}-part of the LJ interaction potential is taken into account one speaks of a "soft spheres" (SS) potential. The LJ-potential cut off at its minimum $r\, r_0^{-1} = 2^{1/6} \approx 1.1225$ which is also purely repulsive, is referred to as WCA-potential. A variety of other potentials have been used but the LJ-potential sets a standard for short range potentials.

The observables of interest, such as the internal energy and the components of the pressure or the stress tensor can be calculated from the known positions and velocities of the particles as time averages according to the rules of statistical physics. Typically, the required data are extracted after each 10th to each 100th time step. Similarly, more detailed information can be obtained from the simulation, such as the velocity distribution function, the pair correlation function or the static structure factor which can also be measured in scattering experiments.

In ordinary molecular dynamics (MD) simulations data are extracted when an equilibrium state with a specified density n (or pressure) and temperature T has been reached. In addition, dynamic quantities can be extracted from the temporal fluctuations, e.g. transport coefficients can be calculated from time correlation functions with the help of Green-Kubo relations. In non-equilibrium molecular dynamics (NEMD) simulations, on the other hand, transport and relaxation phenomena are investigated more directly and in close analogy to real experiments. Also states far away from equilibrium are studied. In the following, a stationary plane Couette flow is considered as a special case of a transport process.

2.2 Plane Couette Flow

For a simple shear flow in x-direction with the gradient in y-direction, the shear rate γ is given by

$$\gamma = \frac{\partial v_x}{\partial y}. \tag{4}$$

Such a flow can be either be generated by moving boundaries or by forces [4], [18], or as used here, by moving image particles undergoing an ideal Couette flow with the prescribed shear rate (homogeneous shear). Let the flow be switched on at $t = 0$. Then, at time t the image particles above (below) the basic (central) box have moved in x-direction to the right (left) by the distance $\gamma t L$ modulo(L) where L is the lenght of the periodicity box in y-direction. The periodic boundary conditions for the particles leaving and entering the basic box have to be modified (Lees-Edwards boundary conditions, [5]- [11]). For a system in a fluid state in equilibrium and for not too large shear rates, a linear velocity profile typical for a plane Couette flow is set up in the basic box (from which the data are extracted). At high shear rates where also plug-like flow occurs it is essential to use a velocity "profile unbiased thermostat" (PUT, [7], [20]). A shear flow can also be generated by modifying the equations of motion (SLLOD, [10], [11]).

2.3 Viscosity and other Rheological Properties

Rheological properties such as the (non-newtonian) viscosity and the normal pressure differences are obtained from the cartesian components of the stress tensor $\sigma_{\mu\nu} = -p_{\mu\nu}$ or of the pressure tensor $p_{\mu\nu}$ which is the sum of "kinetic" und "potential" contributions:

$$p_{\mu\nu} = p_{\mu\nu}^{kin} + p_{\mu\nu}^{pot}, \tag{5}$$

$$V p_{\mu\nu}^{kin} = \sum_i m_i c_\mu^i c_\nu^i, \tag{6}$$

$$V p_{\mu\nu}^{pot} = \frac{1}{2} \sum_{ij} r_\mu^{ij} F_\nu^{ij}. \tag{7}$$

Here c^i is the peculiar velocity of particle i, i.e. its velocity relative to the flow velocity $v(r^i)$, $r^{ij} = r^i - r^j$ is the relative position vector of particles i, j and F^{ij} is the force acting between them. The Greek subscript μ, ν which assume the values $1, 2, 3$ stand for cartesian components associated with the x, y, z-directions. In the simulations, the expression for the pressure tensor given is averaged over many (10^3 to 10^6) time steps.

For the present flow geometry, the (non-newtonian) viscosity η is obtained by dividing the $yx(21)$-component of the stress or pressure tensor by the shear rate γ:

$$\eta = \sigma_{yx}/\gamma = -p_{yx}/\gamma. \tag{8}$$

The kinetic and potential contributions to the pressure tensor and to the vis-

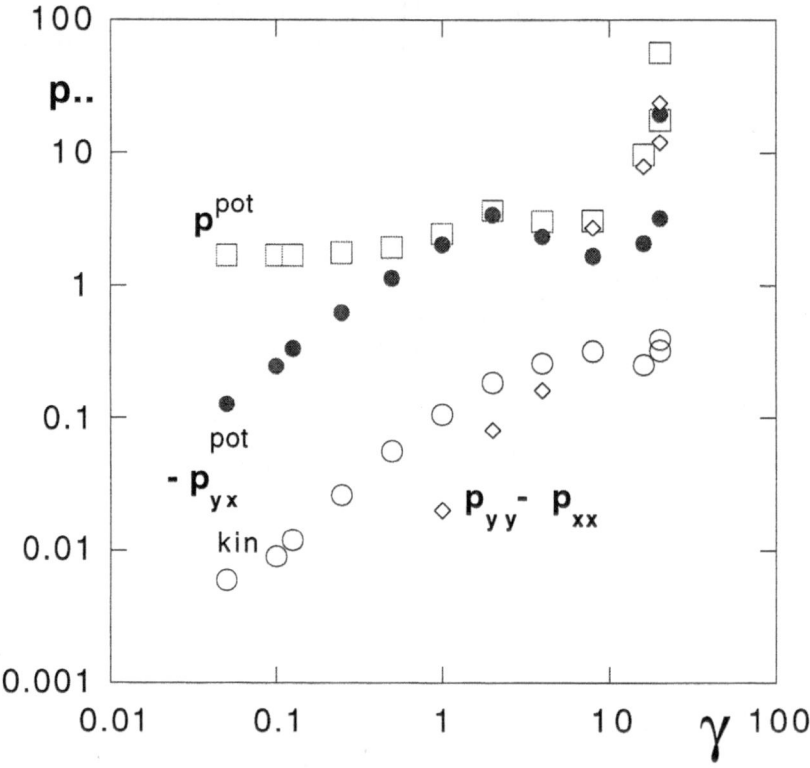

Figure 2: The kinetic (open circles) and the potential (closed circles) contributions to the shear stress $\sigma_{yx} = -p_{yx}$, the potential contributions to the scalar pressure p^{pot} (squares) and to the normal pressure difference $p_{yy} - p_{xx}$ (diamonds) for a LJ-fluid as functions of the shear rate γ.

cosity can be computed seperately from the simulation. Only the sum can be measured in a real experiment. The kinetic contribution to the viscosity is the dominating one in dilute gases [14]. In dense fluids (liquids) the potential contribution is the more important one, cf. Figs.(2,3). The data shown stem from simulations with $N = 512$ particles [15], the interaction has been cut off at $r = r_c = 2.5r_0$. The density $n = 0.84r_0^{-3}$ corresponds to the triple point density, the temperature $T = \Phi_0/k_B$ is somewhat higher than the tripel point temperature.

Normal stress or pressure differences, e.g. $\sigma_{xx} - \sigma_{yy} = p_{yy} - p_{xx}$ can be computed analagously, cf. Fig.(2). At small shear rates one has $-p_{yx} \sim \gamma$ and $p_{yy} - p_{xx} \sim \gamma^2$, as well as $p_{xx} + p_{yy} - 2p_{zz} \sim \gamma^2$.

In Fig.(3) also the total vicosity is shown as it follows from the entropy production which is proportional to $\eta\gamma^2$ and is determined by the heat removed from the system by the thermostat.

2.4 Structural Changes in the Various Flow Regimes

2.4.1 Qualitative Discussion

The shear rate dependence of the viscosity as displayed in the "flow curve" Fig.(3) shows four regimes: (*I*) The *newtonian flow* regime where the shear viscosity η is independent of the shear rate γ and where normal pressure differences practically vanish. In the present case the newtonian regime corresponds to $\gamma < 0.1$ (in LJ units). (*II*) A *weak shear thinning* for $0.2 < \gamma < 2$. (*III*) A *strong shear thinning* for $2 < \gamma < 20$. (*IV*) A *shear thickening* for $\gamma > 20$.

These qualitative differences of the flow behavior are linked with different flow induced structural changes in the fluid. In regimes (*I*) and (*II*) these can be noticed in the pair correlation function $g(\mathbf{r})$ or equivalently, in its spatial Fourier transform, the static structure factor $S(\mathbf{k})$ which determines the scattering intensity. Both quantities become anisotropic in the presence of a viscous flow. The structure factor shows distorted Debye-Scherrer rings. In regime (*III*) a long range partial positional ordering takes place which is apparent in real space and it is evident in snapshots [24, 25]. Of course, the long range ordering is also seen in $g(\mathbf{r})$ and it leads to Bragg-like peaks in $S(\mathbf{k})$, [1], [13]-[21].

Above, the various flow regimes have been distinguished by the shear rate expressed in LJ-units. The physically relevant variable, however, is the product $\gamma\tau$ of the shear rate and the Maxwell relaxation time τ which, in turn, is given by the small shear rate limit of the ratio η/G, i.e. of the viscosity and the (high frequency) shear modulus G. The latter quantity, which can also be extracted

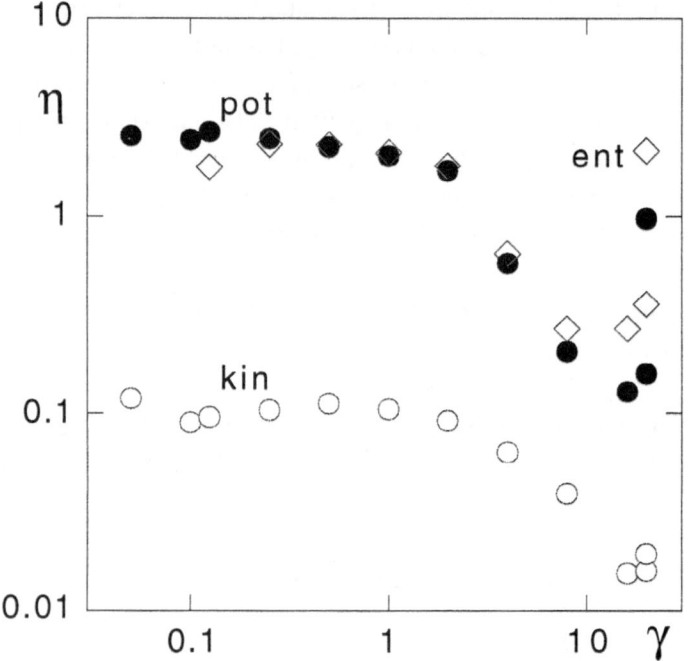

Figure 3: The kinetic (kin, open circles) and the potential (pot, closed circles) contributions to the viscosity $\eta = -p_{yx}/\gamma$, as well as the total viscosity inferred from the entropy production (ent, diamonds) for a LJ-fluid as functions of the shear rate γ for the same density and temperature as in the previous figure.

from the simulation, is approximately 25 for the present system and the relaxation time is $\tau \approx 0.1$ in LJ units. Thus non-newtonian flow phenomena can be observed for $\gamma > 0.1\tau^{-1}$. In simple fluids like liquid Argon this corresponds to a shear rate which is several orders of magnitude larger than $10^6 s^{-1}$ which can reasonably be reached in laboratory experiments. The situation is different in (dense) colloidal dispersions of spherical particles. There, considerably shorter relaxation times occur and non-newtonian effects can be noticed and are of importance for many applications.

The potential part of the pressure tensor is also determined by an integral over the pair correlation function g:

$$p_{\mu\nu}^{pot} = \frac{1}{2}n^2 \int r_\mu F_\nu \, g(\mathbf{r}) \, d^3r. \tag{9}$$

Here **F** is the force acting between two particles. The relation (9) is the basis
for the strong dependence of the vicous properties on the structure of a fluid.

2.4.2   Gereralized Stokes–Maxwell Model

The structural changes in the flow regimes I and II can be treated by starting
from a Kirkwood–Smoluchowki type of kinetic equation [19], [22], [23] for the
pair correlation function $g = g(\mathbf{r})$:

$$\partial g/\partial t + \gamma\, r_y\, \partial g/\partial r_x + \mathcal{D}(g) = 0\,. \tag{10}$$

Here **r** is the relative position vector between an arbitrary reference particle
and any other particle in the fluid. The "damping" term $\mathcal{D}(g)$ ensures that g
approches the equilibrium pair correlation function g_{eq} which is also referred to
as radial distribution function. With the relaxation time approximation

$$\mathcal{D}(g) = \tau^{-1}\,(g - g_{eq})\,, \tag{11}$$

where τ is the Maxwell relaxation time, and for a stationary situation, the kinetic
equation is equivalent to

$$g = g_{eq} - \gamma\tau\, r_y\, \partial g/\partial r_x\,. \tag{12}$$

Iteration of this equation leads to a power series in $\gamma\tau$, the first few terms are

$$g = g_{eq} - \gamma\tau\, r_x\, r_y\, r^{-1} g'_{eq} + (\gamma\tau)^2\, \left(r_y^2\, r^{-1} g'_{eq} + r_x^2\, r_y^2\, r^{-1}(r^{-1} g'_{eq})'\right)\ -..+... \tag{13}$$

Here the prime denotes differentiation with respect to $r = |\mathbf{r}|$. Note that g_{eq}
depends on r but not on the angles specifying the direction of the vector **r**.
The term linear in $\gamma\tau$ of (13) corresponds to a relation suggested by Stokes
and by Maxwell. The power series expansion (13) is referred to as generalized
Stokes-Maxwell model [19].

According to the linear Stokes–Maxwell relation, the function $g_+ = g_{45} -
g_{135}$, i.e. the difference between the pair correlation functions g_{45} and g_{135} along
the lines in the x-y plane which enclose the angles of 45° and 135° with the x-axis,
is given by $-\gamma\tau\, r g'_{eq}$. In Fig.(4), this relation is tested for a Lennard-Jones liquid
at the same temperature and density as used for the data presented in Fig.(3).
The shear rate $\gamma = 0.125$ is in the linear linear flow regime. The relaxation time
τ has been put equal to 0.14. The large gray dots mark the values of g_+ as
directly extracted from the simulation, the small black dots connected by lines
are the same quantity computed via the Stokes–Maxwell relation. Though the
agreement is far from perfect, the essential features are described rather well.

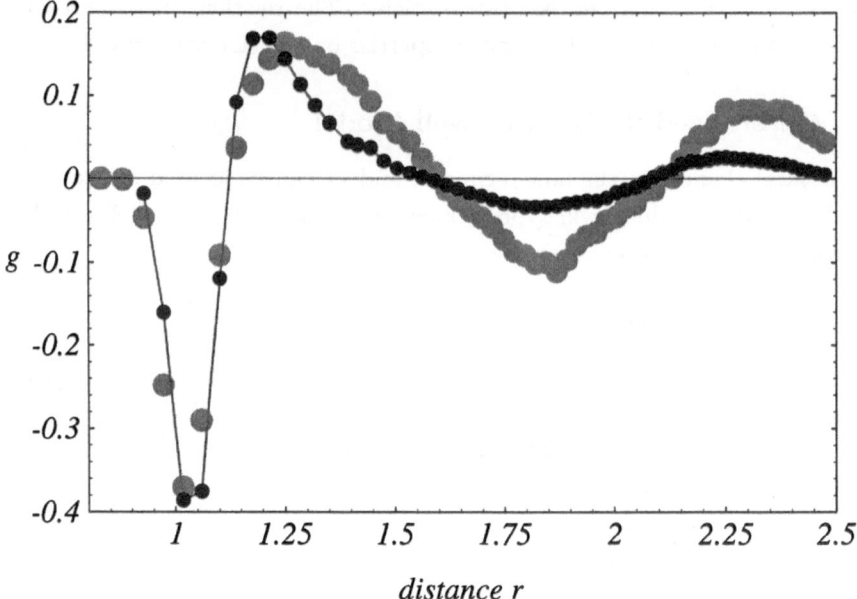

Figure 4: Test of the Stokes–Maxwell relation for a Lennard–Jones liquid in the linear flow regime. Data directly obtained in the NEMD simulation (large gray dots) are compared with those computed via the Stokes–Maxwell relation (small black dots connected by lines). The density, temperature and shear rate are $n = 0.84$, $T = 1.0$, $\gamma = 0.125$ (in LJ units), the relaxation time $\tau = 0.14$ was used.

The full solution of (12) can be written as

$$g = \int_0^\infty e^{-\alpha} g_{eq} \left(r_x - \gamma\tau\, r_y, r_y, r_z \right) d\alpha. \tag{14}$$

For the static structure factor $S = S(\mathbf{k})$ which is essentially the Fourier transform of $g - 1$, a power series expansion similar to (13) is obtained where, however, the components k_y and k_x of the scattering wave vector $\mathbf{k}$ play the role of r_x and r_y. The expression corresponding to (14) is

$$S = \int_0^\infty e^{-\alpha} S_{eq} \left(k_x, k_y + \gamma\tau k_x, k_z \right) d\alpha. \tag{15}$$

To show the basic features of the influence of the shear flow on the structure as described by (15), the simple model expression $1 - 0.9\exp(-k^2/10) +$

$20k^4 \exp(-2k^2)$ is used as an approximation for S_{eq} in the vicinity of its first maximum. The contour graphs Fig.(5) show $S(\mathbf{k})$ in the k_x–k_y plane in equilibrium (left) and for a plane Couette flow with $\gamma\tau = 0.1$ (right). This shear rate is in the linear or newtonian flow regime. Under shear, the Debye–Scherrer ring is elliptically distored and its long axis encloses the angle $-45°$ (or $135°$ with the flow direction. In Fig.(6), $S(\mathbf{k})$ is displayed in the k_x–k_y plane (left) and k_x–k_z plane (right) for $\gamma\tau = 0.5$. The shear rate is in the weak shear thinning regime. Nonlinear effects are the modulation of the intensity around the Debye–Scherrer rings and the rotation of the long axis of the ellipse in the k_x–k_y plane from the diagonal towards the y-axis.

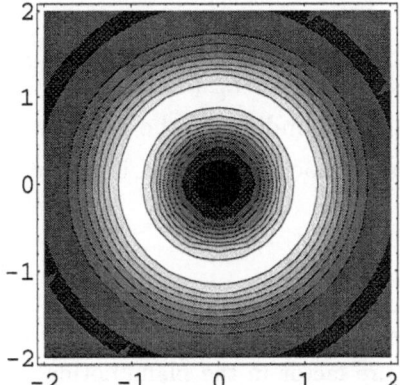 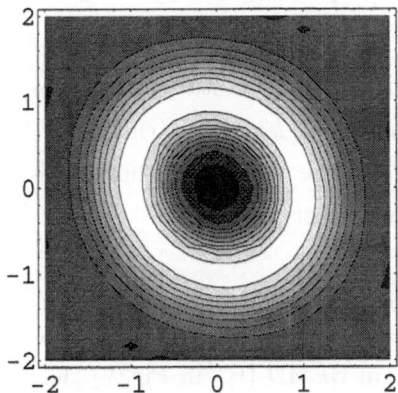

Figure 5: The static structure factor $S(\mathbf{k})$ in the k_x–k_y plane, in equilibrium (left) and under shear with $\gamma\tau = 0.1$ (right), corresponding to the linear or newtonian flow regime. Bright (dark) regions indicate high (low) intensities. The flow velocity is in the horizontal (x) direction.

2.4.3 Long Range Partial Positional Ordering

A long range partial positional ordering, in paricular the formation of layers and strings of particles has been found in the simulations [24], [25] for shear rates where a strong shear thinning occurs. Although actual snap shots of the positions of the particles depend on the details of the simulation and are affected by the boundary conditions [20], the phenomenon as such seems to be generic.

Dense colloidal dispersions of spherical particles exhibit flow curves, cf. Fig.(1), which are qualitatively similar to those presented above. In the extreme shear thinning regime III where partial positional order is observed in

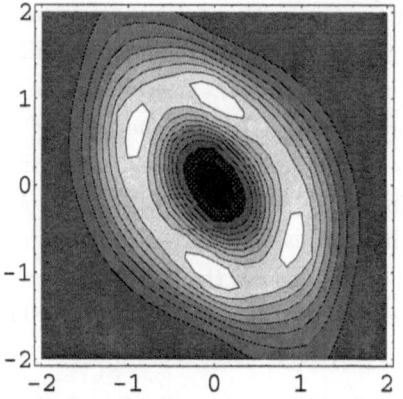 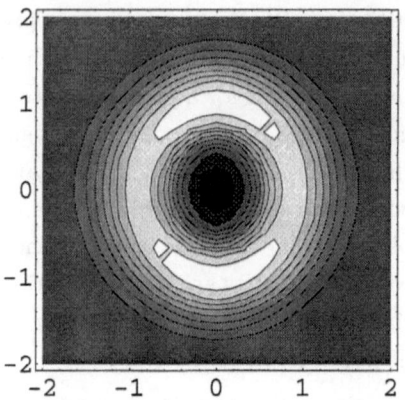

Figure 6: The static structure factor $S(\mathbf{k})$ in the k_x-k_y plane (left) and in the k_x-k_z plane (right) for $\gamma\tau = 0.5$, corresponding to the nonlinear, weak shear thinning flow regime. The flow velocity is in the horizontal (x) direction.

the NEMD simulations, the static structure factor as measured in small angle neutron scattering (SANS) experiments agrees very well with that computed from NEMD [1]. In Fig.(7) the static structure factor in the plane normal to

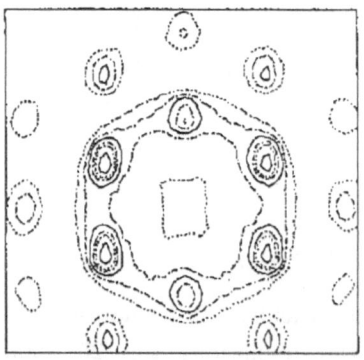 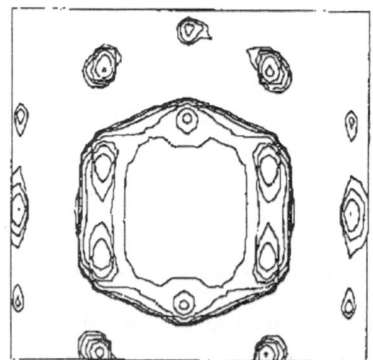

Figure 7: The static structure factor in the plane normal to the direction of the velocity gradient for the SANS experiment (left) and the NEMD simulation (right).

the direction of the velocity gradient is shown for the SANS experiment (left) and the NEMD simulation [20] for a soft shere fluid (right). In the latter case, the form factor of the spherical particles with the known radius (0.55 in reduced units) has been taken into account. The dispersion corresponds to the substance labelled by *vol43* in Fig.(1), it was subjected to a Couette flow with a shear rate $1 s^1$.

The generalized Stokes–Maxwell model dicussed above does not apply to the flow regimes *III* and *IV*. Further theoretical work is needed for a better understanding of the dynamic processes in these regimes. The transition from the weak to the strong shear thinning, however, follows from a stability analysis within a generalized hydrodynamic theory [20].

2.5 Colloidal Dispersions

In addition to the direct interaction between the dispersed particles and the thermostating influence of the solvent already considered in the MD- and NEMD-simulations, particles in a dispersion feel a friction and the pertaining (Brownian) fluctuating forces. These additional forces are taken into account in the Brownian dynamics (BD) simulations. Results from such a BD-simulation are presented in [16]. Furthermore, the particles experience hydrodynamic interactions and possess rotational degrees of freedom.

The friction is characterized by the friction coefficient

$$\beta = k_B T (m D_0)^{-1} = 6\pi a \eta_F m^{-1}. \tag{16}$$

This coefficent also determines the magnitude of the fluctuating forces. In (16), m and $a = r_0/2$ are the mass and the radius of a particle, D_0 is the diffusion coefficient in the dilute dispersion, η_F is is the viscosity of the solvent. In order to have the connection with the MD-results, firstly the dimensionless MD-variables, now marked with an asterix ..*, are also used in the BD-simulation. Simulations have been performed [16] of "soft-spheres" (SS) with the interaction potential $\Phi^{SS} = \Phi_0^{SS}(r_0/r)^{12}$ with a cut off at $2.5 r_0$ at a number density $n^* = n r_0^3 = 0.84$ (corresponding to a packing fraction of $y = \pi n^*/6 \approx 0.44$) and at a temperature $T^* = (1/4)(\Phi_0^{SS}/k_B)$ with the values $\beta^* = 50, 1600$ (BD) and $\beta^* = 0$ (MD) for the dimensionless friction coefficient

$$\beta^* = \beta t_{ref} = 3\pi \eta_F^*, \quad \eta_F^* = \eta_F/\eta_{ref}, \tag{17}$$

with $t_{ref} = r_0 (\Phi_0^{SS}/m)^{-1/2}$ and $\eta_{ref} = r_0^3 \Phi_0^{SS} t_{ref}$. The reference value for the shear rate is $\gamma_{ref} = t_{ref}^{-1}$, thus $\gamma^* = \gamma/\gamma_{ref}$.

For particles with the diameter $r_0 = 165 nm$ (as used in [1]) and a mass density $1 g cm^{-3}$ now $\beta^* = 1600$ corresponds to a fluid viscosity $\eta_F = 1.2 \cdot$

$10^{-3}\,Pa\,s$, which is that of water at room temperature. The value $\beta^* = 50$ corresponds to a fictious solvent with a viscosity smaller by a factor $1/300$, finally $\beta^* = 0$ is the "frictionless" MD-simulation.

The dimensionless shear rates γ^* relevant for the BD-simulation differ from those of the MD-simulation by orders of magnitude. For this reason, the BD-data are expressed as function of the dimensionless shear rate, also referred to as Peclet number,

$$\Gamma = \gamma\,\tau_B = \gamma^*\,\tau_B^* = \gamma^*\,\beta^*/(24\pi\,T^*) \tag{18}$$

with the characteristic time coefficient

$$\tau_B = a^3\,\eta_F/(k_B\,T)\,. \tag{19}$$

For particles with $2a = r_0 = 165nm$ this time is $0.15\cdot 10^{-3}\,s$ for a solvent with the viscosity $10^{-3}\,Pa\,s$ such as water. For the MD-simulation the dimensionless shear rate

$$\Gamma = \gamma\,a/c_0 = \gamma^*\,(4\,T^*)^{-1/2} \tag{20}$$

is used where $c_0 = (k_B\,T/m)^{1/2}$ is a thermal velocity, then one has $\Gamma = \gamma^*$. The viscosity increases strongly with increasing friction. At the state point studied, the viscosity does not reach a newtonian limit for small shear rates. Such a behavior is typical for dense dispersions, cf. Fig.(1). The corresponding relative viscosity is defined by

$$H = \eta/\eta_F = \eta^*/\eta_F^* = 3\pi\,\eta^*/\beta^*\,. \tag{21}$$

For the MD-data where no solvent viscosity γ exists, the expression (21) was used with $\beta^* = 1$. For small shear rates (analogous to the flow regime II of the LJ-system) all relative viscosities coincide. This scaling behavior is not perfect at higher shear rates (flow regime III) where the BD-simulations show partial positional ordering very similar to that one observed in the MD-simulations. As expected, the scaled BD50-results are much closer to the MD-data than the BD1600-values. In the BD-simulations, the transition from flow regimes II to III (where a long range order sets in) is smoother and less sensitive to the size and shape of the periodicity box than in the MD-simulations, see Fig.(8), which will be discussed below.

The contributions to the (relative) viscosity in a dispersion considered so far do not comprise hydrodynamic effects nor the Einstein contribution $1 + 2.5\,y$, $y = \pi\,r_0^3\,n/6$ already present in dilute solutions. It is conjectured to relate the dimensionless variables Γ and H used here with the shear rate and viscosity of a dense dispersion by:

$$\gamma = \tau_B^{-1}(D_s/D_0)\,\Gamma\,, \tag{22}$$

$$\eta = \eta_F(H+1+2.5\,y)(D_0/D_s)\,, \tag{23}$$

where $(D_s/D_0) \approx 1 - 1.83\,y$ is the ratio of the short time diffusion coefficient D_s in a dense dispersion and of corresponding coefficient in the dilute solution. For the example discussed above, this means $r_0 = 165nm$, $y = 0.44$, $\eta_F = 21 \cdot 10^{-3}\,Pa\,s$, furthermore $1+2.5\,y = 2.1$, $D_s/D_0 = 0.2$, $\tau_B^{-1}(D_s/D_0) = 65\,s^{-1}$ and $\eta_F(D_0/D_s) = 0.1\,Pa\,s$. The viscosities resulting from the MD and BD simulations with this scaling are displayed in Fig.(8) as functions of the shear rate. These values are of the same order of magnitude as those of the data points labelled by "vol43", in the shear thinning regime, in Fig.(1).

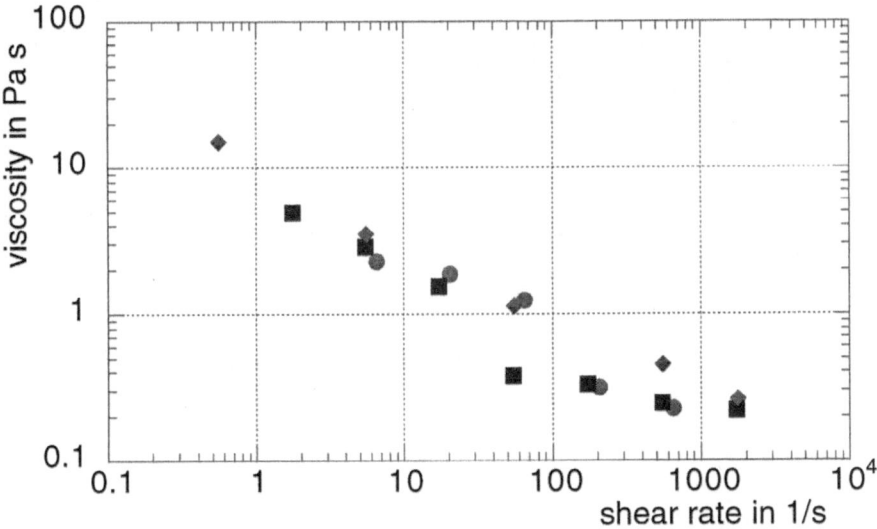

Figure 8: The viscosity, in $Pa\,s$, obtained from the MD (circles) and BD (squares: $\beta^* = 50$, diamonds: $\beta^* = 1600$) simulations as function of the shear rate, in s^{-1}.

3   Polymer Melts

To model a polymer melt, one starts from a simple fluid of spherical particles and introduces extra binding forces or constraints [27] in order to form molecular chains with a prescribed chain length of N_{ch} beads. Rheological studies for LJ-fluids where the binding was achieved by increasing the energy parameter Φ_0 for neighbors in a chain by a factor showed many features of the nonlinear flow behavior typical for polymeric melts [26], [28]. Also thermal degradation and shear induced breaking of chains could be observed. The results to be presented

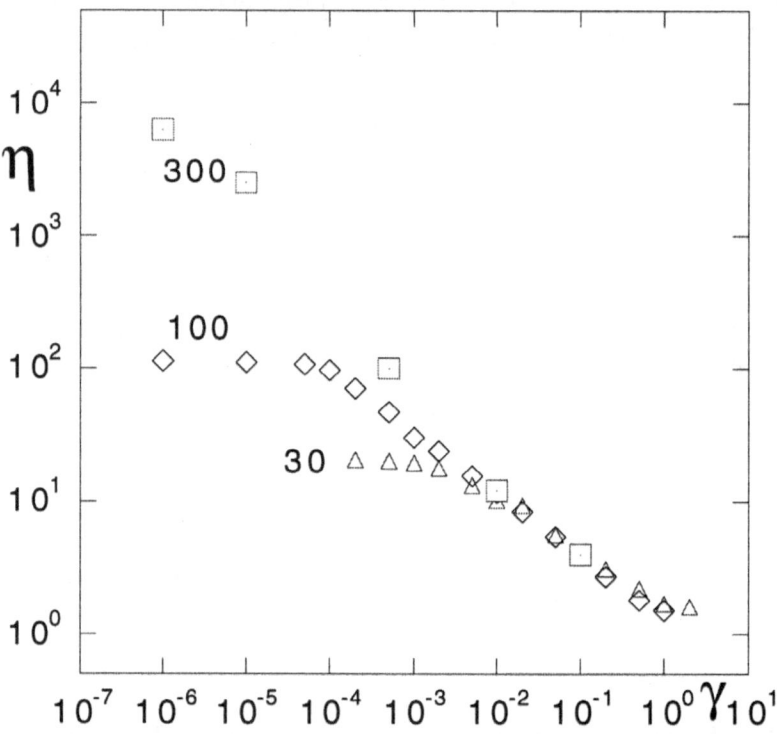

Figure 9: The viscosity of melts consisting of polymer molecules with the chain lengths 30 (triangles), 100 (diamonds), and 300 (squares) as functions of the shear rate γ. The density and the temperature are $n = 0.84$ and $T = 1$ in LJ-units (M. Kröger).

here [30] follow from an extension of the previous simulations [29] for a system where all particles interact with the repulsive part of the LJ-potential (WCA) and an attractive FENE potential with the maximal bond length R_0 is used for the binding within the chains. More specifically,

$$\Phi = \Phi^{WCA} := 4\Phi_0 \left[\left(\frac{r_0}{r} \right)^{12} - \left(\frac{r_0}{r} \right)^6 + \frac{1}{4} \right], \quad r \leq 2^{1/6} r_0, \tag{24}$$

and $\Phi^{WCA} = 0$ for $r > 2^{1/6} r_0$;

$$\Phi = \Phi^{FENE} := -0.5 \, k^* \, \Phi_0 \frac{R_0^2}{r_0^2} \ln \left[1 - \frac{r^2}{R_0^2} \right], \quad r \leq R_0, \tag{25}$$

and $\Phi^{FENE} = \infty$ for $r > R_0$. For this potential with $R_0 = 1.5$, $k^* = 30$, $T = \Phi_0/k_B$, and $n r_0^3 = 0.85$ extensive equilibrium MD-studies have been made by Kremer and Grest [31]. In [29, 30] and for the data to be presented here, the same potential parameters and the same state point is used except for a slightly smaller density of $n r_0^3 = 0.84$. Molecules with chain lengths $N_{ch} = 10, 30, 60, 100, 150, 200, 300$ and 400 were studied. The systems contained $N = 6000, 8400$ and 30000 monomers. In Fig.(9), the viscosity η is displayed as function of the shear rate γ for $N_{ch} = 30, 100, 300$. Notice that the values of η and γ, both expressed in LJ-units, span a much wider range than the data shown in Fig.(3) for a simple LJ-liquid. The newtonian limit η_0 of the viscosity, for long molecules only reached at extremely small shear rates, is presented in Fig.(10) as function of the chain length N_{ch}. Two regimes, referred to as Rouse regime, where $\eta_0 \sim N_{ch}$ and as reptation regime, where $\eta_0 \sim N_{ch}^{3.5}$ can be distinguished. The transition between these regimes occures at $N_{ch} \approx 100$. This value is, as expected, about three times the entanglement lenght of ≈ 35 inferred from equilibrium studies [31]. A procedure to analyze and measure entanglements in MD-simulations has recently been invented [32]. Other rheological properties, such as the first and second normal stress differnces and the viscometric functions can and have been computed [29, 30]. The shear induced bond orientation as it can be measured in flow birefringence, as well as the static structure factor of the whole melt or of marked chains and the shape of single polymer chains were analyzed and found to be in good agreement with experiments [33].

Other geometries, such as extensional flow of polymer melts (M. Kröger) and the deformation of polymers in a glassy state, as well as systems with prescribed components of the pressure tensor (H. Voigt) have been simulated. Also dilute polymer solutions have been studied under nonequilibrium conditions (C. Aust) [34].

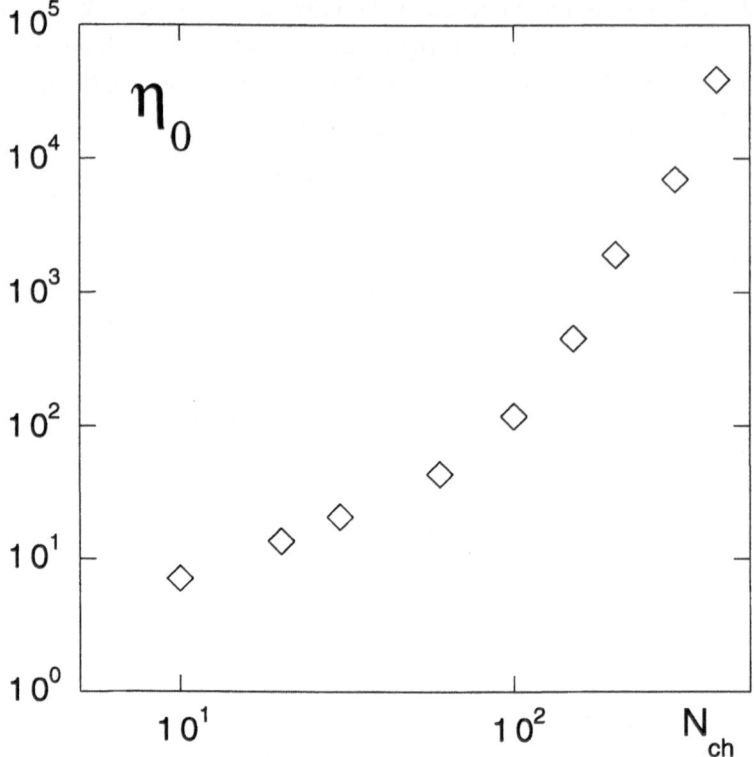

Figure 10: The newtonian viscosity η_0 of polymer melts as function of the chain length N_{ch} (M. Kröger).

4 Anisotropic Fluids

Anisotropic fluids, in particular nematic, nematic discotic and smectic liquid crystals oriented by the application of external electric or magnetic fields, as well as ferro–fluids and magneto–rheological fluids possess a rather complex rheological behavior which can also be treated by NEMD simulations [16], [34]-[41]. Strong shear induced structural changes occur in the vicinity of the nematic–smectic phase transition in liquid crystals and in magneto–rheological fluids where layers and chains of particles are broken and reformed during the flow. Further theoretical work and extensive simulation are needed for a better understanding of these highly nonlinear phenomena.

5 Acknowledgements

Financial support by the Deutsche Forschungsgemeinschaft (DFG) via the Son-derforschungsbereich (SFB) 335 "Anisotrope Fluide" and the Graduiertenkolleg "Polymerwerkstoffe", by the BASF AG (Ludwigshafen), the substantial help of M. Kröger and H. Voigt and many enlightning discussions with H.M. Laun (BASF), as well as the generous donation of computer time by the Konrad-Zuse-Zentrum für Informationstechnik (Berlin) and by the Höchstleistungsrechen-zentrum der Kfa Jülich GmbH are gratefully acknowledged.

References

[1] H.M. Laun, R. Bung, S. Hess, W. Loose, O. Hess, K. Hahn, E. Hädicke, R. Hingmann, F. Schmidt, and P. Lindner J. Rheol. **36** (1992) 743

[2] N.A. Clark and B.J. Ackerson, Phys.Rev.Lett. **44** (1980) 1005; B.J. Ackerson, Physica A **174** (1991) 15

[3] P. Lindner, Physica A **174** (1991) 74

[4] W.T. Ashurst and W.G. Hoover, Am.Phys.Soc. **17** (1972) 1196; Phys.Rev.Lett. **31** (1972) 206

[5] A.W. Lees and S.F. Edwards, J.Phys.C **5** (1972) 1921

[6] G. Ciccotti, G. Jacucci, and I.R. McDonald, Phys. Rev. A **13** (1975) 426; J. Stat. Phys. **21** (1979) 1; C. Trozzi and G. Ciccotti, Phys. Rev. A **29** (1984) 916

[7] W.G. Hoover, Annu. Rev. Phys. Chem. **34**, 103 (1983) D.J. Evans and G.P. Morriss, Comp. Phys. Rep. **1** (1984) 287 ; D.J. Evans and W.G. Hoover, Ann. Rev. Fluid Mech. **18** (1986) 243

[8] B.D. Holian and D.J. Evans, J. Chem. Phys. **78** (1983) 5157

[9] D.M. Heyes, J.Chem.Soc.Faraday. II, **82** (1986) 1365

[10] W.G. Hoover, *Molecular Dynamics*, Springer, Berlin 1986

[11] M.P. Allen and D.J. Tildesley, *Computer Simulation of Liquids*, Claren-don, Oxford (1987)

[12] R. Haberlandt, S. Fritzsche, G. Peinel, und K. Heinzinger, *Molekular-Dynamik*, Vieweg, Braunschweig 1995

94

[13] S. Hess and W. Loose, *Molecular dynamics: Test of microscopic models for the material properties of matter,*in D. Axelrad and W. Muschik (eds.) *Constitutive laws and microstructure,* Springer, Berlin 1988, p. 92

[14] W. Loose and S. Hess, Phys. Rev. Lett. **58** (1988) 2443; Phys. Rev. A **37** (1988) 2099

[15] S. Hess, Rheology and shear induced structure of fluids in J. Casas-Vázques and D. Jou (eds.) *Rheological Modelling: Thermodynamical and statistical approaches,* Lecture Notes in Physics **381**, Springer, Berlin 1991, p. 51

[16] S. Hess, M. Kröger, W. Loose, C. Pereira Borgmeyer, R. Schramek, H. Voigt, and T. Weider, *Simple and Complex Fluids under Shear,* in K. Binder and G. Ciccotti (eds.) *Monte Carlo and Molecular Dynamics of Condensed Matter Systems,* IPS Conf. Proc. **49**, Bologna 1996, p. 825 - 841

[17] S. Hess, *Constraints in Molecular Dynamics, Nonequilibrium Processes in Fluids via Computer Simulations,* in K.H. Hoffmann and M. Schreiber (eds.) *Computational Physics,* Springer, Berlin 1996, p. 268 - 293

[18] S. Hess and W. Loose, Physica A **162** (1989) 138

[19] S. Hess and H.J.M. Hanley, Phys.Lett. **A 98** (1983) 35; H.J.M. Hanley, J.C. Rainwater, and S. Hess, Phys.Rev. **A 36** (1987) 1795; J.C. Rainwater, H.J.M. Hanley, and S. Hess, Phys.Lett. **A 36** (1988) 450;

[20] W. Loose and S. Hess, Rheol. Acta **28** (1989) 91; W. Loose and S. Hess, in *Microscopic simulation of complex flows,* M. Mareschal (ed.), ASI series, Plenum Press (1990); S. Hess and W. Loose, Ber. Bunsenges., Phys. Chem. **94** (1990) 216

[21] O. Hess, W. Loose, T. Weider, and S. Hess, Physica B **156/157** (1989) 505; T. Weider, U. Stottut, W. Loose, and S. Hess, Physica A **174** (1991) 1; S. Hess, D. Baalss, O. Hess, W. Loose, J. F. Schwarzl, U. Stottut and T. Weider in G. A. Maugin (ed) *Continuum models and discrete systems,* Longman, Essex, 1990, vol. 1, p. 18 - 30

[22] S. Hess, Physica A **118** (1983) 79

[23] J.F. Schwarzl and S. Hess, Phys.Rev. **A 33** (1986) 4277

[24] J.J. Erpenbeck, Phys. Rev. Lett. **52** (1984) 1333

[25] S. Hess in D. Quemada (ed.) *Nonlinear behavior of dispersive media*, J. Mécanique Théor. Appl., Numéro spécial 1985, p. 1 - 19; Int.J. Thermophys. **6** (1985) 657; J.de Physique **46**, Colloque 3 (C3), Supplement 3 (1985) 191

[26] S. Hess, J. Non-Newtonian Fluid Mech. **23** (1987) 187

[27] J.P. Ryckaert, Mol.Phys. **55** (1985) 549

[28] M. Kröger and S. Hess, Physica **A 195** (1993) 336

[29] M. Kröger, W. Loose and S. Hess, J. Rheol. **37** (1993) 1057;
M. Kröger, *Rheologie und Struktur von Polymerschmelzen*, W&T, Berlin 1995

[30] M. Kröger, Rheology 95 (1995) 66

[31] K. Kremer and G.S. Grest, J.Chem.Phys. **92** (1990) 5057

[32] M. Kröger and H. Voigt, Macromol.Theory Simul. **3** (1994) 639

[33] R. Muller, J.J. Pesce, and C. Picot, Macromol. **26** (1993) 4356

[34] S. Hess, C. Aust, L. Bennett, M. Kröger, C. Pereira Borgmeyer, T. Weider, Physica **A 240** (1997) 126

[35] S. Hess, J. Non-Equilibr. Thermodyn. **11** (1986) 176

[36] D. Baalss and S. Hess, Phys. Rev. Lett. *57* (1986) 86; Z. Naturforsch. **43 a** (1988) 662 ;
H. Sollich, D. Baalss, and S. Hess, Mol.Cryst.Liq.Cryst. **168** (1989) 189

[37] S. Hess, J. Schwarzl, and D. Baalss, J.Phys. Condens.Matter *2* (1990) SA 279

[38] S. Hess, D. Frenkel, M. Allen, Mol.Phys. **74** (1991) 765

[39] H. Ehrentraut and S. Hess, Phys. Rev. E **51** (1995) 2203

[40] S. Sarman and D.J. Evans, J.Chem.Phys. **99** (1993) 9021; S. Sarman, Physica **A 240** (1997) 160

[41] C. Cozzini, L.F. Rull, G. Ciccotti, G.V. Paolini, Physica **A 240** (1997) 173

Experimental Study of Horizontally Shaken Granular Matter – The Swelling Effect

Thorsten Pöschel and Dirk E. Rosenkranz

Humboldt-Universität zu Berlin, Institut für Physik,
Invalidenstraße 110, D-10115 Berlin, Germany

Abstract. We report on a new effect observed in horizontally shaken granular material in a rectangular box. Within a certain region of parameters of the oscillation the granular material swells recurrently with a characteristic frequency (see http://summa.physik.hu-berlin.de/GranMat/). This frequency is significantly smaller than the frequency of forcing. We claim that this new effect is caused by the the interplay between Reynolds dilatancy due to convective motion and mechanical stability of the diluted granular material in the box. Measurements of the dissipated energy per time supports this explanation.

1 Introduction

Shaken granular materials reveal different types of interesting macroscopic effects including surface fluidization (Warr et al. (1995)), structure formation (Melo et al. (1995), Metcalfe et al. (1997)), localized excitations (Umbanhowar et al. (1996)), compaction, (e.g. Ayer and Soppet (1965/1966), Esipov et al. (1996)), segregation, (e.g. Williams (1976), Knight et al. (1993), Gallas et al. (1994)), convection and others.

When granular material in a rectangular container is shaken in either direction, vertically or horizontally, one observes convective motion. Convection due to vertical oscillations has been subjected to *many* experimental (e.g. Rátkai (1976), Laroche et al. (1989), Fauve et al. (1989), Jaeger and Nagel (1992), Ehrichs et al. (1995), Jaeger et al. (1996)) and analytical (e.g. Bourzutschky and Miller (1995), Hayakawa et al. (1995), Salueña et al. (1997), Esipov et al. (1996)) investigations. The typical convection cells could be reproduced by molecular dynamics simulations in two (Gallas et al. (1992), Taguchi (1992), Pöschel and Herrmann (1995)) as well as in three dimensions (Gallas et al. (1994)).

In contrast there are relatively few results on horizontally shaken granular material. Numerical molecular dynamics simulations and analytical investigations predicted that convection occurs in horizontally shaken containers too (Esipov et al. (1996), Salueña et al. (1997), Liffman and Metcalfe (1997)). Recently this effect has been investigated quantitatively (Tennakoon and Behringer (1997), Rosenkranz and Pöschel (1997)). Despite of convective motion we know that horizontally shaken granular material reveals a rich variety of complex patterns (Iwashita et al. (1988), Strassburger et al.

(1997), Ristow et al. (1997)). Up to now we are far from a real understanding of these complex phenomena.

2   The Swelling Effect

In the present paper we want to report on a new effect in horizontally shaken granular matter, which we call the "swelling effect": Suppose a rectangular container filled with granular material up to a height z_f^0 is shaken sinusoidal in horizontal direction $x(t) = A\cos(2\pi ft)$ with amplitude A and frequency f. In a certain interval of parameters A, f and z_f^0 we observe time dependent variation of the volume of the material and hence of its overall density. This effect can be measured by recording the filling height $z_f(t) > z_f^0$. While there are several experimental investigations of exciting time dependent phenomena in vertically vibrated containers (Clement et al. (1996), Umbanhowar et al. (1996), Melo et al. (1995)) there is a main difference between those effects and the swelling effect: All these effects act on time scales which are comparable with the period of forcing $1/f$. The swelling effect, however, has a period which is up to several ten to a hundred times larger than the period of oscillation $1/f$. Furthermore we do *not* observe a period doubling scenario as seen e.g. by Melo et al. (1995). Video sequences of the new effect are presented in the internet at URL: http://summa.physik.hu-berlin.de/GranMat/.

3   Experimental Setup

The experimental setup we have used for our investigations, is shown in Fig. 1. The probe carrier with the rectangular container was mounted to a precisely balanced horizontal linear bearing. A second linear bearing was driven by a stepping motor via a crankshaft. In between both bearings there was a piezo force sensor (Kistler, type 9203 with Kistler charge amplifier, type 5001) which measured the force acting from the motor driven bearing on the bearing which holds the probe carrier. The force was registered by a computer with a sampling rate of 2 kHz. The stepping motor was computer controlled, i.e. per precisely 2,000 impulses the motor axis revolves once. This high angular resolution provides a quasi steady motion, the finite step size per computer signal does not influence the convective behavior. The amplitude of oscillation was adjustable by changing the eccentricity of the crankshaft with a precision of ±0.05 mm. The entire mechanical device was fixed on a oscillation damping table. For a detailed description of the equipment containing all technical details see (Rosenkranz (1997)).

The container of size 6 cm × 10 cm consists of transparent plastic material. It was illuminated from the top and a camera was located in a distance of 4 m in a direction perpendicular to the axis of oscillation. The camera was levelled exactly with the upper surface of the granular material at rest. The convex shape of the upper surface of the shaken granular matter forms a bright

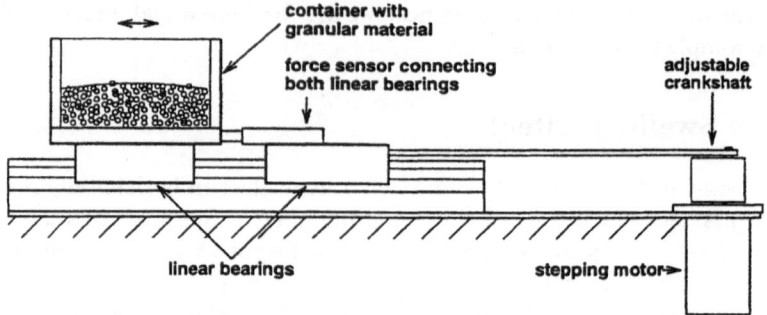

Fig. 1. Experimental setup

region in the camera plane, because of the top lighting. Hence the position and width of this region give us measures of the height and the curvature of the upper material surface. A second camera was mounted about 1 m above the equipment to monitor the motion of the material at the upper surface. The output of both cameras has been digitized and recorded by standard digital equipment and has been processed afterwards.

4 Qualitative Description

Figure 2 displays a time series of snapshots of the middle part of the surface region taken with time delay $\Delta t = 0.2$ sec. The container was horizontally shaken with amplitude $A = 0.15$ cm and frequency $f = 25$ sec^{-1}. The dashed line is at fixed height. The displayed series covers slightly more than one swelling oscillation period which is about 2.2 sec, i.e. about 55 times the period of the driving oscillation. One sees clearly the height oscillation the surface, which is in its dependence on the forcing parameters A and F and the filling height z_f^0 the subject of our interest.

The swelling oscillation shown in Fig. 2 could be observed for a wide range of frequency, amplitude and filling height and for a variety of materials and grain sizes ($r = 20\,\mu$m $\cdots 100\,\mu$m). The swelling was accompanied by a rapid convective motion of the material at the surface. This flow was not uniform but its intensity varied with the same characteristic frequency as the swelling itself.

5 Quantitative Results

To analyze the effect quantitatively we recorded time series of several observables, i.e.

– the height of the material surface in the middle of the container

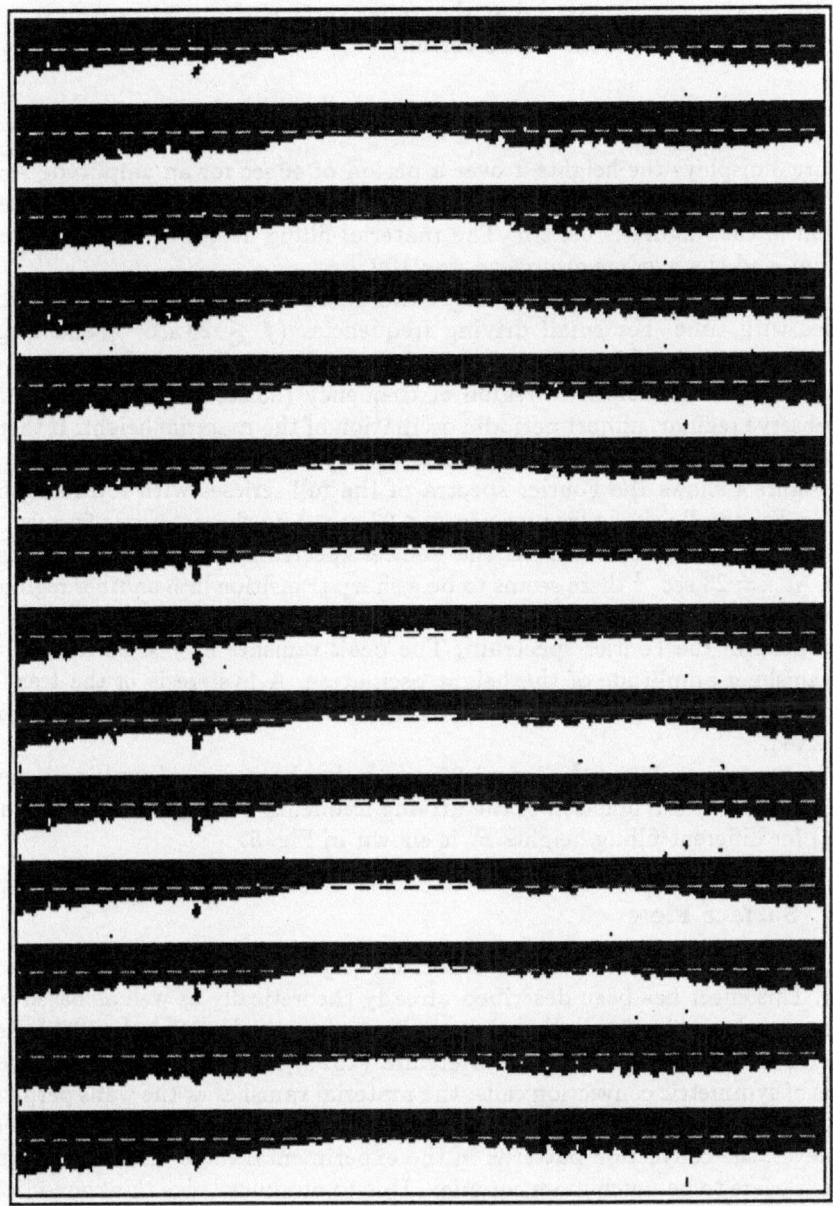

Fig. 2. Time series of snapshots of the surface region (time grows from top to bottom). The dashed line leads the eye to a fixed height above the container bottom.

- the motion of the material at the upper surface
- the force acting on the probe carrier.

5.1 Height Oscillations

Figure 3 displays the heights z over a period of 40 sec for an amplitude $A = 0.15$ cm and different driving frequencies f. The total height of the displayed region in each figure is 0.4 cm. The material filling height at rest was $z_f^0 = 2.5$ cm, and the average grain size was $100\,\mu m$.

For all frequencies shown in Fig. 3 we observe, that the material height varies with time. For small driving frequencies ($f \lesssim 23\ \mathrm{sec}^{-1}$) and large driving frequencies ($f \gtrsim 30\ \mathrm{sec}^{-1}$) the height changes irregularly with time, whereas for a characteristic region of frequency ($23\ \mathrm{sec}^{-1} \lesssim f \lesssim 29\ \mathrm{sec}^{-1}$) we observe regular, almost periodic oscillation of the material height. If there exists a characteristic frequency, it should be found in the Fourier spectra

Figure 4 shows the Fourier spectra of the full serieses with total lengths 60 sec. For small driving frequencies $f < 23\ \mathrm{sec}^{-1}$ no characteristic frequency of swelling can be observed, i.e. the Fourier spectrum reveals many frequencies. At $f = 23\ \mathrm{sec}^{-1}$ there seems to be a sharp transition into another regime, where we find a characteristic frequency of the swelling oscillation indicated by a peak in the Fourier spectrum. The peak vanishes at $f \approx 30\ \mathrm{sec}^{-1}$ due to vanishing amplitude of the height oscillation. A hysteresis of the transition point when increasing and decreasing the driving frequency has not be observed.

In the region $23\ \mathrm{sec}^{-1} \lesssim f \lesssim 29\ \mathrm{sec}^{-1}$ the frequency of swelling is not constant, but it is a function of the driving frequency f (Fig. 4). The function $F(f)$ for different filling heights z_f^0 is shown in Fig. 5.

5.2 Surface Flow

When granular material is shaken horizontally one observes convective motion. This effect has been described already theoretically as well as based on molecular dynamics simulations in two dimensions (Salueña et al. (1997), Esipov et al. (1996), Liffman and Metcalfe (1997)). Essentially one observes pairs of symmetric convection cells: the material vanishes at the walls perpendicular to the shaking direction and reappears in the center of the container. However, the convection patterns in the experimental three dimensional system appears to be much more complex. Due to our knowledge this convection has never been described in detail and this work is in progress (Rosenkranz and Pöschel (1997)).

In the region of driving parameters A, f and z_f^0, where one observes the swelling effect, there is an intensive convective motion, however, the intensity of the convective motion oscillates as well. Viewing the container from the top (second camera) one finds that the oscillation of material height (Fig. 3) corresponds to a varying particle velocity at the upper surface. When the

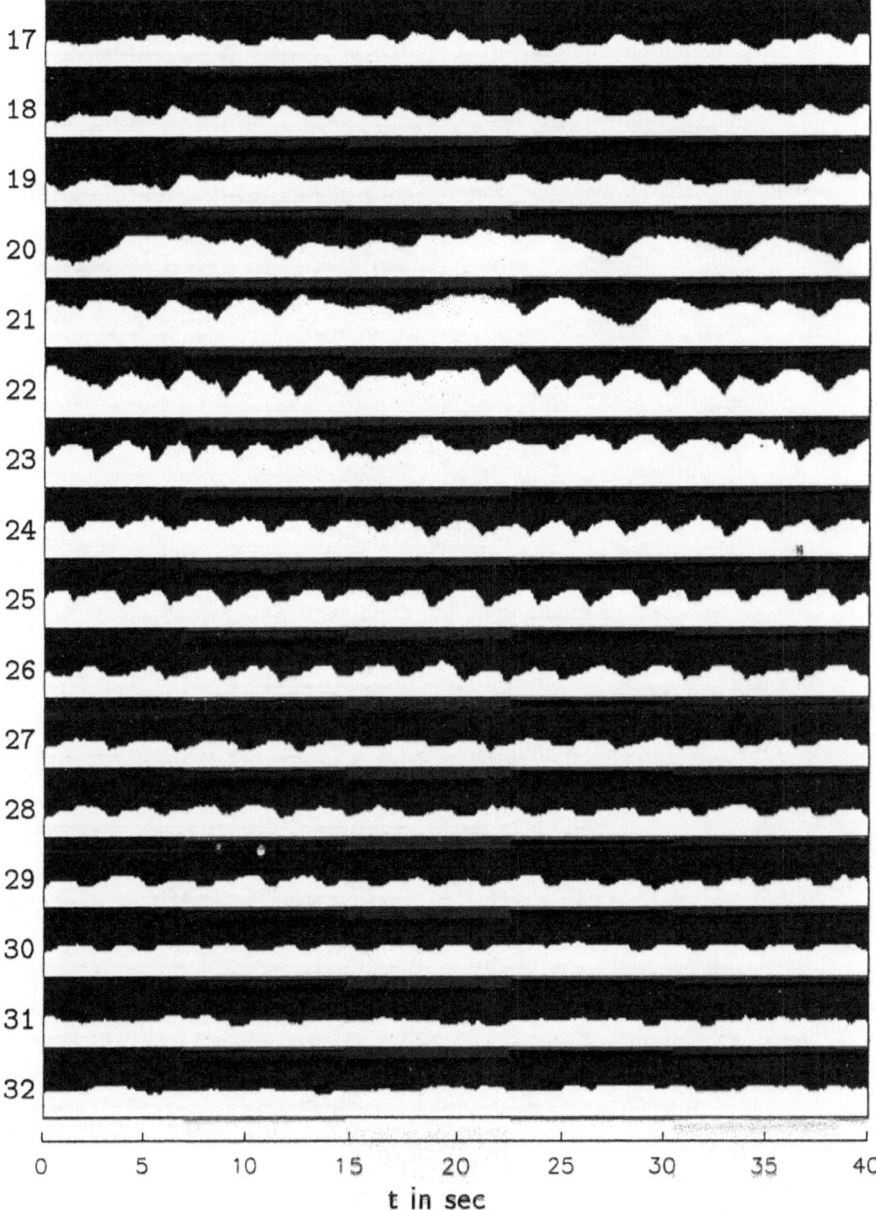

Fig. 3. Height of the granular material in the middle of the container over time for driving frequencies $f = 17 \sec^{-1} \ldots 32 \sec^{-1}$.

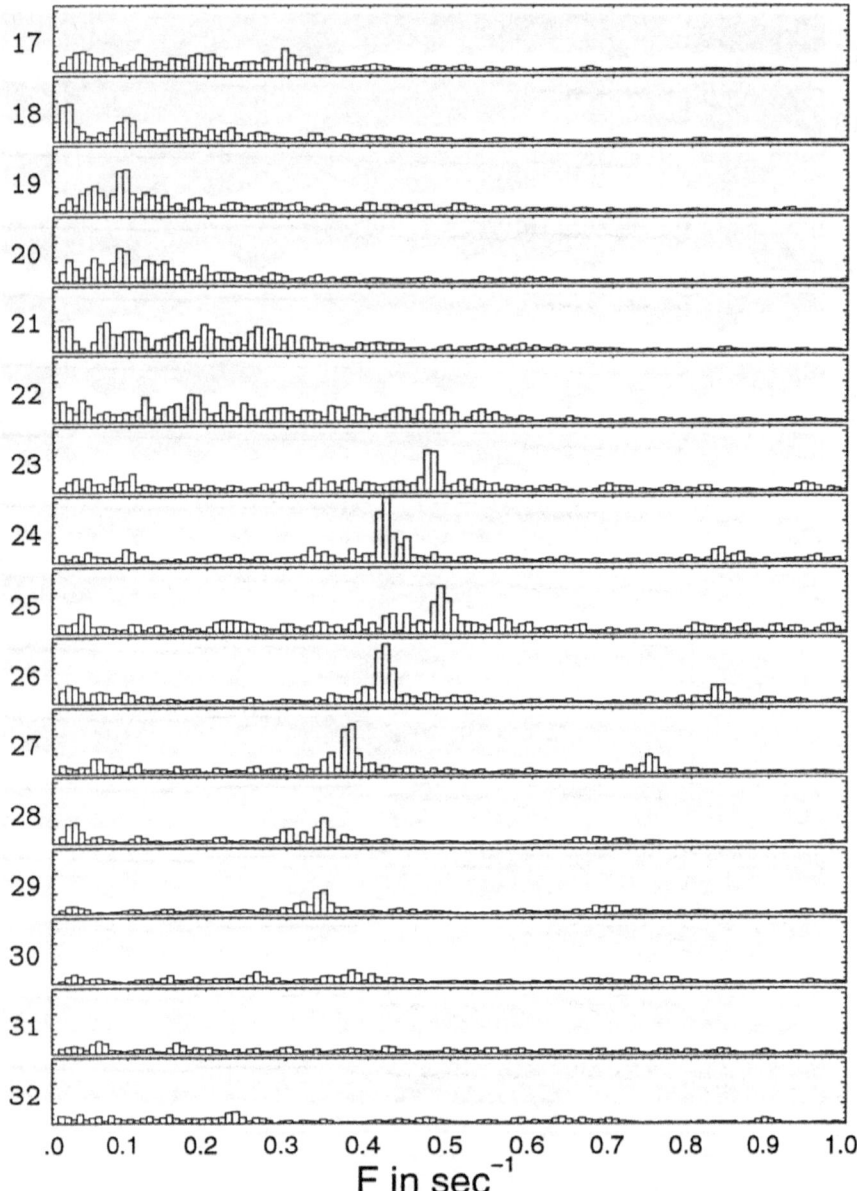

Fig. 4. Fourier spectra of the time serieses shown in Fig. 3.

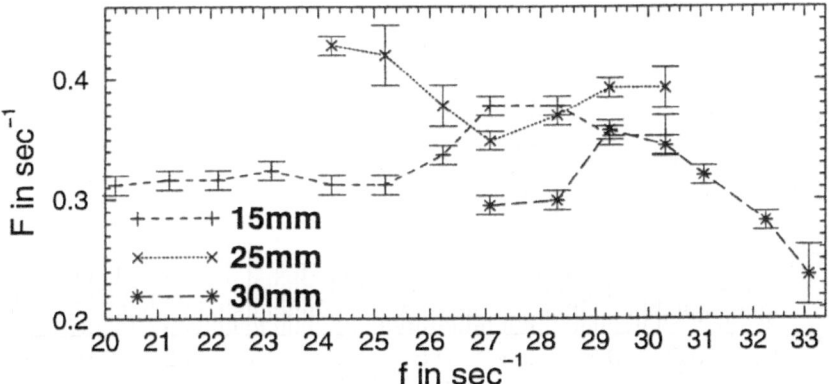

Fig. 5. Swelling frequency F as a function of the frequency of forcing f for various filling heights z_f^0.

material is swelled there is an intensive flow, whereas the flow comes almost to rest instantly when the material is collapsed.

We want to show now that the frequency of the character of surface flow coincides with the frequency of swelling. To this end we put tracer particles into the granular system and take a series of snapshots of the container viewing from top. Then we calculate a corresponding sequence of pictures by subtracting the gray scale values of the pixels of consecutive snapshots. The average value ΔG of the gray scale of the pixels of the difference pictures provides a measure of the material flow at the surface. If there would be no flow, the difference pictures appear to be equal coloured, i.e $\Delta G = 0$. If there would be a steady flow, we would get $\Delta G \approx const.$ For the case of a time varying flow we find a characteristic time series $\Delta G(t)$ of which we expect to have the same characteristic frequencies as the swelling itself.

Figure 6 shows the Fourier transform of ΔG for a driving frequency of $f = 25 \sec^{-1}$. The Fourier spectrum of the average pixel differences ΔG of the gray scale of difference pictures reveals a significant peak. One notes that the variation of the surface flow has the same characteristic frequency as the swelling (c.f. Fig. 4). Hence, we claim that convective motion in the horizontally shaken container and swelling are closely related effects.

5.3 Dissipation of Mechanical Energy

Using the force sensor, which measures the force $F(t)$ acting on the probe carrier, we calculated the energy applied by the motor to the probe carrier

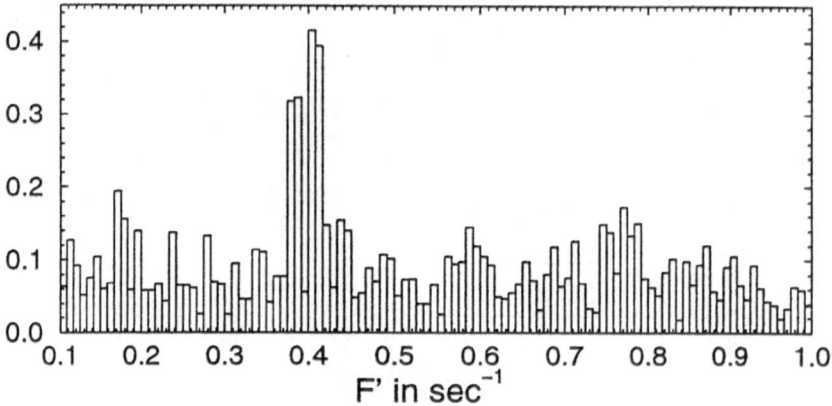

Fig. 6. Fourier spectrum of the oscillation of the material flow at the surface over time for $f = 25 \text{ sec}^{-1}$.

during the nth period

$$E_n = 2\pi f A \int\limits_{n/f}^{(n+1)/f} F(t) \cos(2\pi ft) \, dt .\tag{1}$$

E_n is the energy dissipated during the nth period by the entire system, i.e. by the linear bearing and by the motion of the granular material in the container. This time series E_n can be analyzed by Fourier analysis similar to the surface flow intensity and the height of the material. Figure 7 shows the Fourier transform of the dissipated energy, again for amplitude $A = 0.15$ cm and a range of driving frequencies f.

Comparing Figs. 4 and 7 we notice that both figures agree well, i.e. the characteristic frequencies of swelling oscillation and of energy dissipation coincide.

5.4 The Influence of Air on The Swelling Effect

For the cases of other dynamical effects there has been a controversial discussion about the influence of air. Whereas for the effect of irregular motion and clogging flow of granular material through a pipe there is experimental evidence that surrounding air has major influence on the motion of the sand (Raafat et al. (1996), Horikawa et al. (1995)), the influence of air on the spontaneous heap formation of vertically vibrated granular material seems to be unclear yet (Laroche et al. (1989), Evesque and Rajchenbach (1989), Evesque (1990), Savage (1988)).

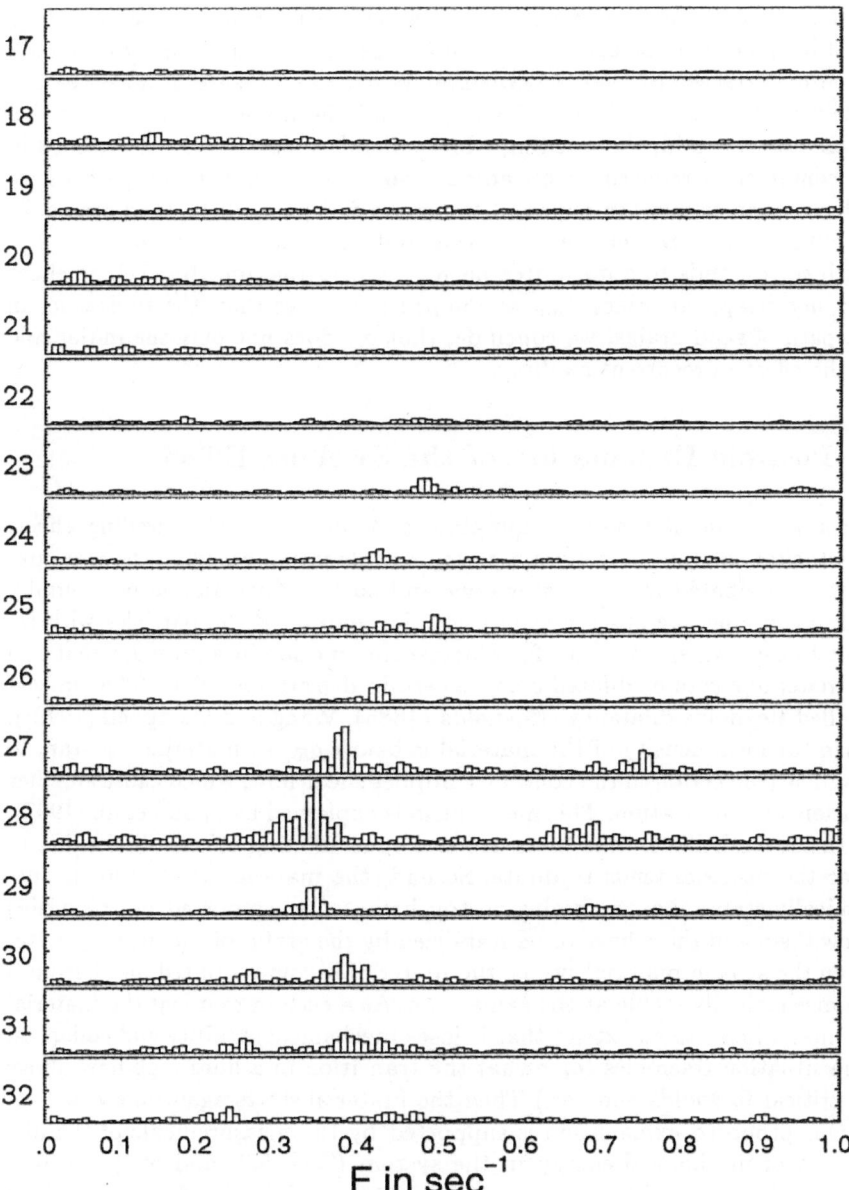

Fig. 7. Fourier spectra of the dissipated energy E for a range of frequencies.

Hence, the influence of air on the swelling effect has to be discussed. One could imagine that the convective motion of sand "pumps" air into the bulk of the material which leads to swelling. When the pressure of air exceeds a certain critical value a "bubble" escapes and the material collapses. If this explanation is valid, the swelling effect should vanish if the air pressure in the container is reduced to an amount so that the mean free path of air molecules comes into the region of the mean free path of the sand grains. In the experiment, however, we found that reducing the pressure to $p = 50$ Pa, which corresponds to a mean free path of $\lambda_{air} \approx 130$ μm, the swelling effect does *not* disappear. Since $\lambda_{air} \approx 130$ μm is smaller than the typical mean free path of sand grains, we conclude, that air does not play the major role for the effect of recurrent swelling.

6 Possible Explanation of the Swelling Effect

Presently we do not have a complete explanation of the swelling effect. Nonetheless we want to suggest a possible explanation based on the measurement of dissipated mechanical energy and surface flow: In the horizontally shaken container we observe convection, i.e. motion of the particles with respect to each other. To allow for macroscopic motion in a granular material the material has to be diluted below a certain density ρ_R before. This process is called Reynolds dilatancy (Reynolds (1885), Wang and Campbell (1992)). When the local density of the material is below ρ_R the material can start to flow. The convection in the container implies shear flow, which causes further dilution and fluidization. This mechanism is explained by Spahn et al. (1997).

Hence we believe, that there are two contradicting effects. First, due to shear the material tends to dilute. Second, the material has to remain mechanically stable, i.e. the grains on top have to be supported by the grains below them and these have to be stabilized by the grains of the next layer etc. When the shaken material swells, the material becomes diluted and becomes less mechanically stable at the same time. At a certain moment the material becomes diluted to an extend that it loses mechanical stability and collapses. (The situation resembles somewhat the transition in a fluid pipe flow above the critical Reynolds number.) Then the material starts again to swell.

Our proposed explanation is supported by the measurements of the dissipation of mechanical energy in the system (Sect. 5.3) and of the surface material flow (Sect. 5.2). If the material is collapsed, i.e. if a large part of the granular material has a density above ρ_R, the grains in these regions cannot move with respect to each other due to the Reynolds dilatancy effect. Hence, in these regions there is no shear motion and therefore no or low dissipation of mechanical energy. From this consideration we conclude, that the collapsed material should dissipate less energy per time than the swelled material. This behavior coincides with what we found in the measurement of the dissipated energy in Sect. 5.3. Moreover the results in Sect. 5.2 show that the intensity of surface flow oscillates with the same frequency F. Therefore the experimental results on energy dissipation and surface flow support our hypothesis on the origin of the swelling effect.

If this explanation is valid it explains the swelling effect itself and the saw teeth shape of the height over time curves (Fig. 3). However, it does not explain the transition from periodic to irregular behavior of swelling and the characteristic shift of swelling frequency with driving frequency in the periodic regime.

7   Conclusion

We have described a new effect observed in a container filled with granular material which is subjected to sinusoidal horizontal oscillations. Within a certain range of parameters of driving the material swells recurrently. Due to our knowledge this effect has not been reported in the literature so far.

For fixed amplitude of driving A we find for low as well as for high frequencies f, that the material swells irregularly, whereas in an intermediate frequency interval swelling occurs almost periodically with a certain frequency F. In this interval the swelling frequency F is about 50 to 100 times smaller than the driving frequency f, hence we claim that the effect reported in this paper is not comparable with periodic effects reported by other authors (e.g. Clement et al. (1996), Metcalfe et al. (1997), Umbanhowar et al. (1996), Melo et al. (1995)), where the frequency of the effect is the same as the frequency of driving or small multiples of it. In the periodic interval the swelling frequency F is a characteristic function of the driving frequency f.

Similar as the height of the material, the intensity of the convective flow of the granular material in the container as well as the energy dissipated by the granular system vary in time too. Comparing the characteristic frequencies by Fourier analysis we noticed, that all three oscillating observables are closely related.

Finally we proposed a possible explanation of the effect which is based on the effect of Reynolds dilatancy in flowing granular matter and which is supported by the coincidence of the measured time serieses of material height, surface flow intensity and energy dissipation.

We did experiments with several types of containers (aspect ratio), with different grain material and with various amplitudes. While the detailed properties of swelling depend strongly on the details of the experiment we observed the effect in almost all cases.

Acknowledgement

The authors thank R. P. Behringer, L. Schimansky-Geier and J. A. Freund for helpful discussion, V. Buchholtz, K. Reinhardt and M. Stock for help with mechanical equipment and the Deutsche Forschungsgemeinschaft (grants DFG Po 472/2 and DFG Ro 548/5) for financial support.

References

Ayer, J. E. and Soppet, F. E. (1965/1966): Vibratory Compaction, Part I: Compaction of Spherical Shapes and Part II: Compaction of Angular Shapes. *J. Am. Ceramic Soc.* **48**, 180–183 and **49**, 207–210.

Behringer, R. P. and Jenkins, J. T. (eds.) (1997): Powders and Grains'97, Balkema, Rotterdam.

Bourzutschky, M. and Miller, J. (1995): "Granular" Convection in a Vibrated Fluid. *Phys. Rev. Lett.* **74**, 2216–2219.

Clément, E., Vanel, L., Rajchenbach, J. and Duran, J. (1996): Pattern formation in a vibrated 2d granular layer. *Phys. Rev. E*, **53**, 2972–2975.

Ehrichs, E. E., Jaeger, H. M., Karczmar, G. S., Knight, J. B., Kuperman, V. Yu. and Nagel, S. R. (1995): Granular Convection Observed by Magnetic Resonance Imaging, *Science* **267**, 1632–1634.

Esipov, S. E., Salueña, C. and Pöschel, T. (1996): Glassy Behavior of Granular Media and Fluctuational Hydrodynamics. *subm. Phys. Rev. E* (1996).

Evesque, P. (1990): Comment on (Laroche et al. (1989)), *J. Phys. France* **51**, 697–699 and reply by Laroche, C., Douady, S. and Fauve, S. *J. Phys. France* **51**, 700.

Evesque, P. and Rajchenbach, J. (1989): Instability in a Sand Heap. *Phys. Rev. Lett.* **62**, 44-46.

Fauve, S., Douady, S. and Laroche, C. (1989): Collective behaviours of granular masses under vertical vibrations. *J. Physique* **C3**, 187–191.

Gallas, J. A. C., Herrmann, H. J. and Sokołowski, S. (1992): Convection Cells in Vibrating Granular Media. *Phys. Rev. Lett.* **69**, 1371–1374.

Gallas, J. A. C., Herrmann, H. J., Pöschel, T. and Sokołowski, S. (1994): Molecular dynamics simulation of size segregation in three dimensions. *J. Stat. Phys.* **82**, 443–450.

Hayakawa, H., Yue, S. and Hong, D. C. (1995): Hydrodynamic description of granular convection. *Phys. Rev. Lett.* **75**, 2328–2331.

Horikawa, S., Nakahara, A., Nakayama, T., and Matsushita, M. (1995): Self-Organized Critical Density Waves of Granular Material Flowing through a Pipe. *J. Phys. Soc. Jpn.* **64**, 1870–1873.

Iwashita, K., Tarumi, Y., Casaverde, L., Uemura, D., Meguro, K. and Hakuno, M. (1988): Granular assembly simulation for ground collapse. in: Satake, M. and Jenkins, J. T. (eds) *Micromechanics of Granular Material*, Elsevier, Amsterdam, 125–132.

Jaeger, H. M. and Nagel, S.R. (1992): Physics of the Granular State. *Science* **255**, 1523–1531.

Jaeger, H. M., Nagel, S. R. and Behringer, R. P. (1996): Granular Solids, Liquids and Gases. *Rev. Mod. Phys.* **68**, 1259–1273.

Knight, J. B., Jaeger H. M., and Nagel S. R. (1993): Vibration-Induced Size Separation in Granular Media: The Convection Connection. *Phys. Rev. Lett.* **70**, 3728–3731.

Laroche, C., Douady, S. and Fauve, S. (1989): Convective Flow of Granular Masses under Vertical Vibration. *J. Physique* **50**, 699–706.

Liffman, K. and Metcalfe, G. (1997): Convection due to horizontal shaking. in: (Behringer and Jenkins (1997)), 405–408.

Melo, F., Umbanhowar, P. and Swinney, H. L. (1995): Hexagons, kinks, and disorder in oscillated granular layers. *Phys. Rev. Lett.* **75** 3838–3841.

Metcalfe, T., Knight, J. B., and Jaeger, H. M. (1997): Surface Patterns in Shallow Beds of Vibrated Granular Material. *Physica A* **236**, 202–210.

Pöschel, T. and Herrmann, H. J. (1995): Size Segregation and Convection. *Europhys. Lett.* **29**, 123–128.

Raafat, T., Hulin, J. P. and Herrmann, H. J. (1996): Density Waves in Dry Granular Media Falling through a Vertical Pipe. *Phys. Rev. E* **53**, 4345–4350.

Rátkai, G. (1976): Particle Flow and Mixing in Vertically Vibrated Beds. *Powder Technol.* **15**, 187–192.

Reynolds, O. (1885): On the dilatancy of media composed of rigid particles in contact. *Phil. Mag. Soc.* **20** 469.

Ristow, G., Strassburger, G. and Rehberg, I. (1997): Phase Diagram and Scaling of Granular Material under Horizontal Vibrations. *preprint.*

Rosenkranz, D. (1997): Dynamic Phenomena in Granular Materials, PhD Thesis, Humboldt-Universität zu Berlin (in preparation).

Rosenkranz, D. and Pöschel, T. (1997): Recurrent swelling of horizontally shaken granular material. *subm. Phys. Rev. Lett.* (1997).

Rosenkranz, D. and Pöschel, T. (1997): Convective motion in three dimensional horizontally shaken granular material. *(in preparation).*

Salueña, C., Esipov, S. E. and Pöschel, T. (1997): Hydrodynamic fluctuations and averaging problems in dense granular flows. in: (Behringer and Jenkins (1997)), 341–344.

Savage, S. B. (1988): Streaming motions in a bed of vibrationally fluidized dry granular material. *J. Fluid Mech.* **194**, 457–478.

Spahn, F., Schwarz, U. and Kurths, J. (1997): Clustering of granular Assemblies with temperature dependent restitution under Keplerian differential rotation. *Phys. Rev. Lett.* **78**, 1596–1599.

Strassburger, G., Betat, A., Scherer, M. A. and Rehberg, I. (1996) in: Wolf, D. E. and Schreckenberg, M. and Bachem, A. (eds.) *Traffic and Granular Flow*, World Scientific (Singapore), 329–333.

Taguchi, Y-h. (1992): New Origin of a Convective Motion: Elastically Induced Convection in Granular Materials. *Phys.Rev.Lett.* **69**, 1367–1370.

Tennakoon, S. G. K. and Behringer, R. P. (1997): Liquefaction of a horizontally vibrated granular bed. *preprint.*

Umbanhowar, P. B., Melo, F. and Swinney, H. L. (1996): Localized excitations in a vertically vibrated granular layer. *Nature* **382**, 793–796. *Nature* **382** 793 (1996).

Wang, D. G. and Campbell, C. S. (1992): Reynolds' Analogy for a Shearing Granular Material. *J. Fluid Mech.* **224**, 527–546.

Warr, S., Huntley, J. M., and Jacques, G. T. H. (1995): Fluidization of a two-dimensional granular system: Experimental study and scaling behavior. *Phys. Rev. E*, **52**, 5583–5595.

Williams, J. C. (1976): The Segregation of Particulate Materials. A Review. *Powder Technology* **15**, 245–251.

On Relaxational Granular Compaction

Stefan J. Linz

Theoretische Physik I, Institut für Physik,
Universität Augsburg, D-86135 Augsburg, Germany

Abstract. Granular materials differ from ordinary solids and fluids in many respects. One particularly fascinating aspect of granular systems is their ability to
reduce their packing fraction under tapping. In this contribution, we review and
partly extend recent progress in this area with special emphasis on the nonlinear
dynamical modeling of the compaction process. In particular, we also present and
discuss a novel minimal model for relaxational granular compaction.

1 Introduction

The static and dynamical properties of granular systems such as sand, powders, and beads are a largely unexplored part of classical physics. Only recently, physicists have recognized the lack of understanding of these macroscopic dissipative many–particle systems and started considerably intense
research activities in this field (for an overview, cf. Jaeger and Nagel (1992),
Jaeger et al. (1996)). Granular systems consist of dry cohesionless grains. The
grains are macroscopically extended particles that can vary in shape and size.
They are packed together because of gravity. Micromechanically, a granular
system consists of two separate phases: the solid phase or grain space which
is build up by the heterogeneous network of particles and the porous phase
or void space between these particles. The total volume of the solid phase is
basically constant. The volume of the void space, however, can be changed by
external forcing such as tapping or pressure, and with it the packing fraction
of the granular system. The packing fraction ρ is defined by the ratio of the
volume of the solid phase and the total volume of the granular system. *Compaction* means nothing else but the reduction of the void volume between the
grains of a granular system. Due to their macroscopic extension and packing,
the grains can not perform any Brownian motion. In order to reduce the void
volume and to relax the granular system to a closer packed state, external
perturbations such as successive tapping or shaking are necessary.

Compaction of granular systems seems to be a sort of folklore. Young
children at the beach figure out after a while that they can fill more dry sand
into a pail if they tap it on the ground from time to time. Since granular
materials are part of our every-day lives, compaction also possesses wide
practical applications in powder technology. A simple problem is how the
volume of a granular material can be minimized to reduce the costs of packing,
shipping and handling.

Systematic experimental and theoretical explorations of the compaction dynamics of shaken or tapped granular systems have only recently started. In particular, the question of how fast the compaction happens and which final packing state can be achieved due to tapping seems to be highly non–trivial. Surprisingly, Knight et al. (1995) experimentally found a comparably simple, monotonic relaxation of the packing fraction to a fully packed state that does not belong to the standard exponential, stretched–exponential, or algebraic relaxation laws known from condensed matter physics. The experiment of Knight et al. (1995) has triggered several theoretical works by Ben-Naim et al. (1996), Linz (1996), Nicodemi et al. (1997a-c), Peng and Ohta (1997) that seek a deeper theoretical understanding of the compaction dynamics of granular systems under tapping. In this contribution, we mainly discuss and extend the phenomenological approach started by Linz (1996).

2 Experimental Facts

In their seminal study, Knight et al. (1995) explored the dynamics of the compaction process using well-determined experimental conditions. They studied the relaxation dynamics of monodisperse spherical glass particles with a diameter of 2mm in a long thin vertical tube of diameter 1.88cm under the influence of a large number of vertical periodic well separated taps (typically of the order 10^4) with a well controlled tapping intensity Γ. The latter is determined by the ratio of the peak intensity of a tap and the gravitational acceleration.

The experimental results obtained by Knight et al. (1995) can be summarized as follows. Above a critical value $\Gamma_c \simeq 1$, a peculiar dynamics for the successive compaction process from tap to tap happens. For $1.8 < \Gamma < 5.4$, the authors found that the most satisfactory fit of their data for the time dependence of packing fraction ρ_n at rest after successive taps, $n = 1, 2, 3, ...$, has the functional form

$$\rho_n = \rho_\infty - \frac{\rho_\infty - \rho_0}{1 + B\ln(1 + n/\tau)}. \tag{1}$$

The parameters B and τ, as well as the initial and the final packing fraction ρ_0 (slightly larger than the loose packed limit) for fixed Γ, and ρ_∞ (denoted ρ_f by Knight et al. (1995)), can be extrapolated from the experiment by Knight et al. (1995). Note that ρ_∞ is the fully compacted limit that can be reached from a loose packed state for a given constant tapping intensity, and not the absolute closest packed limit that can be realized in a granular material. Knight et al. (1995) also reported measurements of the strong dependence of B, τ and ρ_∞ on the applied tapping intensity Γ. The functional form, Eq.(1), given by Knight et al. (1995) without theoretical motivation, however, seemed to be incompatible with previous theoretical approaches on the compaction problem (Barker and Mehta (1993), Hong et al. (1994)). Recently, Nowak et al. (1997a,b) have shown experimentally that an even more compacted state for a given $\Gamma < 3.5$ can be reached. First, relax the system to its fully compacted limit, then increase the tap intensity to values above $\Gamma = 3.5$ and decrease it again. By that procedure, an even higher, but still Γ-dependent limit of compaction can be achieved.

3 A Simple Stroboscopic Model of Compaction

Our focus is a macromechanical theory for relaxational granular compaction. By that, we mean an approach that does not start from basic micromechanical modeling of the individual grain dynamics, but a rheologically inspired continuum description that is based on comparably simple phenomenological modeling. To understand the formula by Knight et al. (1995), Eq.(1), it seems to be necessary to understand the response of a granular material to *one* tap first. Unfortunately, this has not been experimentally investigated. Nevertheless, it is plausible that the response to tapping basically consists of (a) a decompaction (or elongation of the granular material along the height of the tube) up to a maximum decompacted state, and followed (b) by a recompaction (a collapse of the granular material) due to gravity until a new rest state has been reached. This new rest state is more compacted than the one before the tap.

Next, we propose a simple, qualitative and exactly solvable model that is based on viscoplastic arguments. It mimics the decompaction and recompaction processes that occur from tap to tap. We assume that

- the granular material behaves viscoplastically,
- the changes of the packing fraction during decompaction and recompaction are basically homogeneous along the height of the system,
- the friction between the granular material and the side walls is negligible,
- there is no lift-off of the grains at the bottom of the container while decompacting.

Although all three assumptions are only approximately valid, they seem to be an appropriate starting point for our modeling (for a detailed discussion cf. Linz (1996)).

We also introduce at this point the *compaction ratio*

$$\alpha(t) = \frac{\rho(t) - \rho_\infty}{\rho_0 - \rho_\infty} \tag{2}$$

as a convenient dynamical variable for the dynamic packing problem. This is the difference of the packing fraction $\rho(t)$ at time t and its fully compacted limit $\rho_\infty = \rho(t \to \infty)$ reduced by its initial difference. Therefore, it is experimentally measurable. Using $\alpha(t)$ instead of $\rho(t)$, allows us to eliminate the fitting parameters, $\rho_0 = \rho(t = 0)$ and ρ_∞, from the subsequent discussion. By definition, the initial compaction ratio $\alpha_0 = \alpha(t = 0)$ equals unity, and $\alpha(t)$ approaches zero for the limit $t \to \infty$. Therefore, the initial packing fraction and the fully compacted limit of a granular system correspond to compaction ratios unity or zero, respectively. During the decompaction and recompaction process, $\alpha(t)$ can also reach values larger than unity. In terms of the compaction ratio, the response of the granular material to successive tapping, as discussed above, should basically look as shown in Fig. 1.

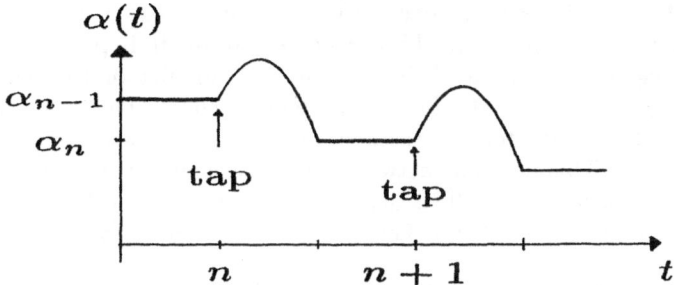

Fig. 1. Sketch of the time evolution of the compaction rate $\alpha(t)$ as a function of time t for two subsequent taps. After a tap at time $t = n$, a decompaction and a subsequent recompaction happens until the compaction rate stops at $\alpha_n < \alpha_{n-1}$.

Using the assumptions stated above, $\alpha(t)$ is basically proportional to the rescaled and non–dimensionalized height of the granular system. So, we can model its dynamics using Newton's equation.

Suppose the granular system is at rest after the $(n - 1)$-th tap and possesses a compaction ratio α_{n-1}. At time $t = n$, the n–th tap occurs. Viscoplastic yield requires that decompaction from rest only occurs if the tap intensity overcomes a certain value. In the experiment by Knight et al. (1995), the excitation of decompaction (the tap) has a complicated pulse profile. This can be mimicked by a decompaction rate $v_d = \dot{\alpha}(n)$ being the initial condition which starts the n-th decompaction process and overcomes a yield decompaction rate $v_c = \dot{\alpha}_c > 0$. It is the same for each tap since the tapping is periodic with the same tap intensity. Since the grains are not elastically coupled, there is no restoring force except the downwards acting gravity. Therefore, the decompaction process $\dot{\alpha}(t) > 0$ can be modeled by

$$\ddot{\alpha} = -kg\,\Theta(v_d - v_c), \quad \text{if } \dot{\alpha} > 0 \tag{3}$$

with $\Theta(x)$ denoting the Heaviside function, $g > 0$ being the non-dimensionalized gravity constant, and k as a positive multiplicative constant. If $v_d > v_c$, the system decompacts according to

$$\alpha(t) = -\frac{1}{2}kg(t-n)^2 + v_d(t-n) + \alpha_{n-1} \tag{4}$$

until the maximum decompaction $\alpha_{max} = v_n^2/2kg + \alpha_{n-1} > \alpha_{n-1}$ has been reached at time $t = n + v_d/kg$. Then the recompaction process $\dot{\alpha}(t) < 0$ driven by gravity begins. To obtain a saturation of the recompaction at a finite positive value of α, an additional counteracting "force" that can be derived from a potential $V_n(\alpha)$ is necessary. The physics of the problem suggests that the potential $V_n(\alpha)$ acts as a compaction barrier with $V_n(\alpha \to 0) \to \infty$ as the ultimate limit for compaction. This corresponds to the fully packed limit. The potential $V_n(\alpha)$ depends explicitly on the tap number or the history of the compaction process in the form of a memory effect. This is because $V_n(\alpha)$ originates from the internal resistance of the granular network against further compaction. This granular network, however, changes and compacts more and more from tap to tap. This also implies that potential V_n becomes steeper and steeper from tap to tap. Therefore, the recompaction process can be modeled by

$$\ddot{\alpha} = -kg - \partial_\alpha V_n(\alpha), \quad \text{if } \dot{\alpha} < 0 \tag{5}$$

and

$$\ddot{\alpha} = 0, \quad \text{if } \dot{\alpha} = 0. \tag{6}$$

Given an appropriate potential V_n as discussed below, Eq.(5) can be solved with the initial conditions $\alpha(t = n + v_d/kg) = \alpha_{max}$ and $\dot{\alpha}(t = n + v_d/kg) = 0$. The result is a decrease of $\alpha(t)$ that eventually reaches $\dot{\alpha} = 0$ in finite time. Then the recompaction stops. This new rest state defines the compaction ratio α_n. Due to the viscoplastic yield condition, the granular system stays at rest until the next sufficiently strong tap occurs.

As in the case of decompaction, one can assume the recompaction is basically driven by the gravational term $-kg$ until the compaction ratio of the previous compacted rest state, α_{n-1}, has been reached again, so that $V_n(\alpha > \alpha_{n-1}) = 0$. At $\alpha(t) = \alpha_{n-1}$, the potential V_n is suddenly present and slows down further recompaction.

Next we have to specify the structure of the potential barrier $V_n(\alpha)$. As discussed above, $V_n(\alpha)$ must diverge as α goes to zero. Therefore, it should contain a contribution that decays like α^{-p} with p being positive. As it turns out later, a decay of V_n proportional $1/\alpha$ with a potential steepness f_n being explicitly dependent on the tap number n is appropriate. The dependence on the tap number reflects the intrinsic resistance of the tapped granular material against further compaction. Moreover, V_n must also contain a linear term in α that compensates for the gravitational recompaction. Therefore, a phenomenological model for the compaction barrier is

$$V_n(\alpha) = \begin{cases} 0 & \text{if } \alpha > \alpha_{n-1} \\ -kg\alpha + f_n\dfrac{1}{\alpha} & \text{if } \alpha \le \alpha_{n-1} \end{cases} \tag{7}$$

with $k > 0$ and $f_n > 0$ being a tap dependent coefficient. Since Eq.(5) has a first integral,

$$\frac{1}{2}\dot{\alpha}^2 + kg\alpha + V_n(\alpha) = \text{constant}, \tag{8}$$

one can take advantage of Eq.(8) by using that at time $t_r = n + 2v_d/kg$, $\alpha(t_r) = \alpha_{n-1}$ and $\dot{\alpha}(t_r) = -v_d$ holds. The next rest state determines α_n. Inserting this in Eq.(8) yields

$$\frac{1}{2}v_d^2 + \frac{f_n}{\alpha_{n-1}} = \frac{f_n}{\alpha_n} \tag{9}$$

or equivalently,

$$\frac{1}{\alpha_n} - \frac{1}{\alpha_{n-1}} = h_n \quad \text{with} \quad h_n = \frac{v_d^2}{2f_n}. \tag{10}$$

Eq.(10) constitutes a difference equation or a stroboscopic law for the inverse of the compaction ratio α_n that can be solved recursively if the h_n is specified. The right hand side of Eq.(10), h_n, is determined by the ratio of the square of the initial condition of the n-th tap, v_d, and therefore, of the square of the tap intensity, and the inverse of the tap–dependent steepness f_n of the potential $V_n(\alpha)$. Within our phenomenologial approach, the latter has to be specified appropriately.

From the view point of nonlinear dynamics, the stroboscopic law, Eq.(10), corresponds to the nonlinear non-autonomous iterated map

$$\alpha_n = f(\alpha_{n-1}, n) = \frac{\alpha_{n-1}}{1 + h_n\alpha_{n-1}}. \tag{11}$$

Supposing that h_n equals a constant C *independent* of n leads to an algebraic decay, $\alpha_n = 1/(1 + Cn)$. To obtain a slower decay as found by Knight et al. (1995), h_n in Eq.(11) must be *time–dependent* and, moreover, *decay–weakening*. Therefore, h_n must decrease in time. As we show below, an ansatz of h_n that allows for different decays for short and long times,

$$h_n = \frac{C}{1 + n/\nu} \tag{12}$$

with C and ν being positive, is an appropriate form for h_n that can reproduce the empirical result, Eq.(1). Here, C and the characteristic decay time ν do not depend on the time n, but can depend, at least in general, on the vibration intensity Γ of the shaking process and on micromechanical properties of the granular system such as grain size, grain material, grain shape. Eq.(12) also implies that the steepness of the compaction barrier $V_n(\alpha)$ increases linearly from tap to tap.

Combining Eq.(10) and Eq.(12) we arrive at the *stroboscopic decay law*,

$$\frac{1}{\alpha_n} - \frac{1}{\alpha_{n-1}} = \frac{C}{1 + n/\nu}, \quad n = 1, 2, 3, \ldots \tag{13}$$

which is our *central model* for the compaction process studied by Knight et al. (1995). A simple phenomenological interpretation of the dynamics of the packing process generated by *periodic* tapping follows. The difference of successive reciprocal compaction ratios decays (i) linearly in time n for times that are short in comparison to a characteristic decay time ν, and (ii) proportional to the inverse of time, $1/n$, in the long time limit.

An exact solution of Eq.(13) can be obtained by using the properties of harmonic series. As shown by Linz (1996), the compaction ratio α_n after the n–th tap is determined by

$$\alpha_n = \frac{1}{1 + C\nu[\Psi(n+1+\nu) - \Psi(1+\nu)]}, \tag{14}$$

for $n = 0, 1, 2, .., \infty$. Here, $\Psi(x) = -\sum_{k=0}^{\infty}[1/(k+x)-1/(k+1)]-\gamma$ denotes the Digamma function and $\gamma = 0.57721...$ the Euler–Mascheroni constant. Note that (i) the solution (14) decreases monotonically to zero, (ii) the compaction ratio α_n only depends on ν, the product of C and ν, and the shake number n, and (iii) both C and ν must be non–zero for a "true" compaction process.

To show that Eq.(14) reproduces the experimental result of Knight et al. (1995), one has to perform asymptotic considerations of the Digamma function for large n and match these with the initial value (for details cf. the paper by Linz (1996)). Then, one obtains for the second term in the denominator of Eq.(14)

$$\Psi(n+1+\nu) - \Psi(1+\nu) \simeq \ln(1 + n/n_c) \tag{15}$$

with a very high accuracy for all n and the values of ν that are relevant in the experiment of Knight et al. (1995). Inserting Eq.(15) in Eq.(14) and rewriting this in terms of the packing fraction ρ_n, $\rho_n = \rho_\infty + (\rho_0 - \rho_\infty)\alpha_n$, yields the experimental result, Eq.(1), with a characteristic relaxation time $n_c = \exp\{\Psi(1+\nu)\}$. One can directly relate the coefficients C and ν in Eq.(13) to the coefficients B and τ measured by Knight et al. (1995), yielding

$$\tau = n_c = \exp\{\Psi(1+\nu)\} \quad \text{and} \quad B = C\nu. \tag{16}$$

All dependences on the tapping intensity Γ are hidden in the parameters C and ν. Knight et al. (1995) have also reported the dependence of $\log B$ and $\log \tau$ as a function of the vibration intensity Γ in the range $1 < \Gamma < 5$. They found that both quantities decay rapidly from very large values $B \sim O(10^3)$ and $\tau \sim O(10^5)$ at $\Gamma \simeq 1$ to comparably small values $B \simeq 10^{-1}$ and $\tau \simeq 1.8$ at $\Gamma \simeq 3$. For larger Γ (at least up to $\Gamma \simeq 5$), B and τ are almost constant. Using their data for $\Gamma \geq 1.8$, Linz (1996) conjectured a relationship $\tau = KB$ with $K \simeq 18$ being basically independent of the vibration intensity Γ. As explained in detail in this paper, this might indicate that for large enough decay rates $\nu \geq 2$, $C \simeq 1/K$. Therefore, C is practically independent of the vibration intensity. Below the threshold shaking intensity $\Gamma_c \simeq 1$, compaction does not take place. The granular system behaves like a solid body. This is reflected in Eq.(13) by a jump of ν to zero for $\Gamma \leq \Gamma_c$ and then implies $\alpha_n = \alpha_0$ for all n.

Our model offers a *macroscopic* interpretation of the compaction dynamics under periodic tapping found by Knight et al. (1995). Between the initial packing fraction ρ_0 (or $\alpha_0 = 1$) and the close packed limit ρ_∞ (or $\alpha_\infty = 0$), a periodically tapped granular system goes through an infinite sequence of metastable packing states or compaction ratios α_n until the final compacted state $\alpha = 0$ has been reached. Starting with $\alpha_0 = 1$, the decay of the compaction ratio due to periodic tapping happens on two distinct time regimes with different decay behavior which reflect the two contributions in Eq.(13). For short times $n/\nu \ll 1$, the decay law, Eq.(13), can be approximated by $1/\alpha_n - 1/\alpha_{n-1} \simeq C(1 - n/\nu)$. This leads to a short–time compaction dynamics $\alpha_n \sim 1 - Cn$ for $n/\nu \ll 1$ being independent of ν at this order. For longer times $n/\nu \gg 1$, the dynamics is governed by $\alpha_n \sim 1/(C\nu \ln n)$. This implies a fast relaxation of the compaction ratio decaying linearly with n for short times and a slow $1/\ln n$ relaxation for longer times. The crossover between these two time ranges typically occurs at about $n/\nu \sim O(0.1)$. Consequently, for values of ν of unit order (or vibration intensities Γ larger than 3), the crossover happens immediately at the beginning of the compaction process. The relaxation dynamics basically shows the $1/\ln n$-behavior. For large ν (vibration intensities close to $\Gamma \simeq 2$), the crossover occurs at times of the order 10^2 to 10^3. Then, both relaxation dynamics can be observed. In Fig. 2, we demonstrate this effect by showing the time evolution of the compaction ratio α_n, Eq.(14), for a large and a small vibration intensity Γ.

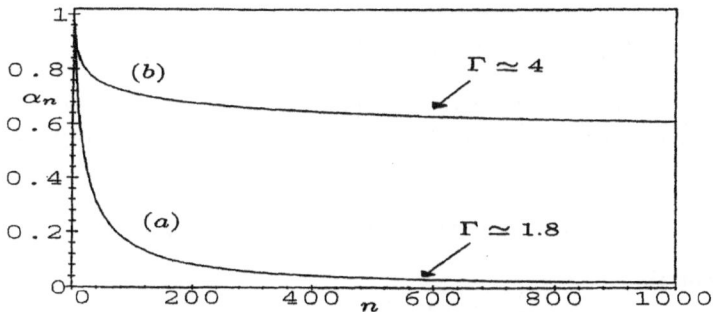

Fig. 2. Dependence of α_n on the time n, Eq.(14). The curve (a) with $\nu = 10000$ and $C = 0.056$ corresponds to a vibration intensity $\Gamma = 1.8$, and the curve (b) with $\nu = 1.3$ and $C = 0.077$ to $\Gamma = 4$. These values for C and ν are extrapolations from the experimental results for B and τ by Knight et al. (1995). This figure is adapted from Linz (1996).

4   A Time-Continuous Compaction Model

As we have seen in the previous section, the dynamics of the compaction ratio and, therefore, of the packing fraction after each tap (as measured in the experiment by Knight et al. (1995)) must be represented in terms of a stroboscopic (iterated) map. Nevertheless, to make contact with alternative theoretical proposals (cf. Hong et al. (1994), Ben-Naim et al. (1996)) that are formulated in terms of monotonic time-continuous compaction laws, one might be tempted to obtain a time-continuous model equation for the compaction ratio. Such a coarse-grained description neglects the actual decompaction and recompaction process from tap to tap and just interpolates continuously between the discrete successive compaction ratios α_n. A time-continuous compaction model can be found from Eq.(11) by replacing α_{n-1} by the coarse grained compaction ratio $A(t)$, the difference between successive compaction ratios, $\alpha_n - \alpha_{n-1}$, by its derivative $\dot{A}(t)$, and h_n by an appropriate time-continuous function $\chi(t)$. Then, the non-autonomous nonlinear relaxator equation

$$\dot{A}(t) = \frac{A(t)}{1 + \chi(t)A(t)} - A(t) \tag{17}$$

or, equivalently

$$\dot{A}(t) = \sum_{n=1}^{\infty}(-1)^n \chi^n(t)A^{n+1}(t) \tag{18}$$

results. Taking into account that one has to expect that the non-autonomous term $\chi(t)$ should have a modulus that is small in comparison to unity (cf. also the discussion in the last sections) and that $A(t)$ is also unity or even much less for all times t, we can approximate Eq.(18) by

$$\dot{A}(t) = -\chi(t)A^2(t), \tag{19}$$

being the *minimal time-continuous counterpart* of the stroboscopic law, Eq. (11). Nonlinear dynamically speaking, Eq.(19) determines the relaxatory evolution of variable $A(t)$ according to

$$\dot{A} = -\partial_A \Phi(A) \tag{20}$$

in a cubic potential

$$\Phi(A) = \frac{1}{3}\chi(t)A^3 \tag{21}$$

with a time-dependent steepness $\chi(t)$. Therefore, a minimal mechanical analogue for the relaxational granular compaction process consists of the approach of $A(t)$ from $A(0) = 1$ towards the fixed point $A = 0$ in the potential $\Phi(A)$. The time dependence of this approach and the fact that the fully compacted limit $A(t \to \infty) = 0$ can be reached at all strongly depends of the actual form of $\chi(t)$. As stated in the last section, h_n and therefore, also

$\chi(t)$ are directly related to the inverse of the resistance of the granular system with respect to further compaction. We consider this *inverse compaction resistance* $\chi(t)$ as the major relevant intrinsic macroscopic property of the granular system as far as compaction is concerned. If this resistance is infinite corresponding to $\chi(t)$ equals zero, one obtains $\dot{A}(t) = 0$ and, therefore, $A(t) = 1$ for all times t. This is the limit of a granular system that cannot be compacted by tapping at all. For a system that can be successively compacted to the fully compacted limit $A(t \to \infty) = 0$, either a nonzero positive constant χ or a decay of the inverse compaction resistance $\chi(t)$ from a finite value C at the beginning, $t = 0$, to a smaller positive value or even zero at the end of the compaction process $t \to \infty$ must be expected.

Since Eq.(19) can be written as $(d/dt)A^{-1}(t) = \chi(t)$, its general solution for the relevant initial condition $A(t = 0) = 1$ and arbitrary inverse compaction resistance $\chi(t)$ is given by

$$A(t) = \left[1 + \int_0^t \chi(u)du\right]^{-1}. \tag{22}$$

It can be evaluated if $\chi(t)$ is specified. As a important consequence of Eq.(22), the integral of the inverse compaction resistance χ over the whole compaction process must diverge, $\lim_{t\to\infty} \int_0^t \chi(u)du \to \infty$, to guarantee that the compaction ratio can reach the fully compacted limit $A(t \to \infty) = 0$. Next, we discuss several special cases.

Constant inverse compaction resistance: Assuming that $\chi(t) = C > 0$ is constant in time, the solution of Eq.(19) reads

$$A(t) = \frac{1}{1 + Ct}. \tag{23}$$

Therefore, a constant $\chi(t) = C$ leads for times large in comparison to $1/C$ to a purely algebraic decay, $A(t) \propto t^{-1}$. It is important to note that Hong et al. (1994) found in a numerical simulation a decay of the filling height of the container proportional to t^{-1} under tapping. Since the filling height is proportional to the compaction ratio, the result of Hong et al. (1994) can be recovered in our model (19) as the special case $\chi(t)$ being independent of time.

Algebraically decaying inverse compaction resistance: Assuming that $\chi(t)$ is given by the continuum analogue of h_n, Eq.(12), $\chi(t) = C(1 + t/\mu)^{-1}$ with C and μ being positive and independent of time, the solution of Eq.(19) reads

$$A(t) = \frac{1}{1 + C\mu \ln(1 + t/\mu)}. \tag{24}$$

By setting $B = C\mu$ and $\tau = \mu$, one does not only recover the $1/\ln t$ long-time dynamics found in the experiment by Knight et al. (1995) in the coarse-grained description of the compaction process, but also the exact analytical form of Eq.(1). Moreover, one also recovers the result of Hong et al. (1994) as the limit of *very large* μ since then $C\mu \ln(1 + t/\mu) = Ct[1 - t/2\mu + O(t^2/\mu^2)]$ holds for a wide range of the times, $0 \leq t << 2\mu$. This is a direct consequence of the fact that the prefactor of the logarithm in (24) is proportional to the characteristic relaxation time μ, and is not independent of μ. Therefore, the result of Hong et al. (1994) is not really incompatible with the experimental result as stated by Knight et al. (1995).

Other inverse compaction resistances: One might think of other simple analytical forms of $\chi(t)$. Assuming that $\chi(t) = Ce^{-t/\mu}$ decays exponentially with time, one also obtains an exponential decay of $A(t)$ which is given by $A(t) = 1/(1 + C\mu - C\mu e^{-t/\mu})$. The resulting decay, however, saturates in the long-time limit at $A(t \to \infty) = 1/(1 + C\mu)$ being nonzero for finite C and/or μ. Therefore, the compaction process does not reach the fully compacted limit $A = 0$ and an exponentially decaying inverse compaction resistance $\chi(t)$ does not seem to be compatible with the basic assumption that $A(t \to \infty) = 0$. Similarly, algebraic decays of $\chi(t)$ of the form $C/[1 + (t/\mu)^\beta]$ with integer $\beta \geq 2$ also lead to decays of $A(t)$ that stop before $A = 0$ has been reached. In all three cases, this is caused by the fact that $\chi(t)$ flattens the potential Φ in Eq.(21) so fast that $A(t)$ cannot reach $A = 0$. In order to obtain an exponential decay of $A(t)$ to $A(t \to \infty) = 0$, one has to assume that $\chi(t) = Ce^{t/\mu}$. This is counterintuitive, since it corresponds to an increasing inverse compaction resistance. It might be the reason why such decay laws have not been observed. The constant and the algebraically decaying inverse compaction resistances discussed above are not the only choices for $\chi(t)$ that can lead to "true" compaction. For example, any $\chi(t)$ that decays from a finite positive value C at $t = 0$ to a nonzero positive value $C_\infty < C$ for $t \to \infty$ leads to a true compaction.

What do we learn from the above discussion? Provided that the compaction model, Eq.(19) is an appropriate description of the compaction process, the experimental result of Knight et al. (1995) and the numerical result of Hong et al. (1994) can be explained from *one root* and basically only *one macromechanical parameter*, the inverse compaction resistance. Therefore, it is apparently the *minimal* model for the relaxational granular compaction. Knowing the analytical form of $A(t)$ from an experiment or a numerical simulation, one can easily relate it to the inverse compaction resistance by using $\chi(t) = (d/dt)A^{-1}(t)$.

5 Summary and Open Problems

Introducing the compaction ratio α as an appropriate macromechanical variable, we have argued why the non–autonomous map, Eq.(11), might be the appropriate phenomenological description of the compaction dynamics found by Knight et al. (1995). We have shown that this map with the special non-autonomous term h_n given in Eq.(12) leads to the empirical result of Knight et al. (1995). Based on this map, we have also given a simple novel coarse-grained time-continuous model for the compaction dynamics that contains the numerical result of Hong et al. (1994) and the experimental result of Knight et al. (1995) as special cases.

The structural simplicity of the map, Eq.(11), and of the time-continuous model, Eq.(19), raises the question whether there is some underlying universality of granular compaction dynamics. For example, can it be that Eq.(19) holds for any type of granular system (ultrafine and fine powders, granules, spheres etc.) and that it is only in the inverse compaction resistance $\chi(t)$ where the micromechanical details of granular material enter? This speculation is obviously very challenging and shows the need for a micromechanical theory of the inverse compaction resistance $\chi(t)$.

There are two interesting alternative approaches to explain the experimental result of Knight et al. (1995). Ben-Naim et al. (1996) explained the compaction dynamics by a stochastic adsorption - desorption process. They attribute the inverse logarithmic increase of the packing fraction in time to the fact that an increasing number of particles must be rearranged in order to increase the density in the long time limit. Nicodemi et al. (1997a-c) have drawn analogies of the slow granular relaxation process, Eq.(1), with the dynamics of a frustated Ising lattice gas subject to gravity and vibrations. Their Monte–Carlo simulations of the latter also reproduce Eq.(1). At first sight, these two approaches look rather distinct from ours. Whether they are related to each other or not remains open.

Acknowledgements. It is my pleasure to thank Karen Swalin for helpful comments on the manuscript.

References

Barker G.C., Mehta A. (1993): Phys. Rev. **E 47**, 184

Ben-Naim E., Knight J.B., Nowak E.R. (1996): preprint (cond-mat/9603150)

Hong D.C., Yue S., Rudra J.K., Choi M.Y., Kim Y.W. (1994): Phys. Rev. **E 50**, 4123

Jaeger H.M., Nagel S.R. (1992): Science **255**, 1523

Jaeger H.M., Nagel S.R., Behringer R.H. (1996): Rev. Mod. Phys. **68**, 1259

Knight J.B., Fandrich C.G., Lau C.N., Jaeger H.M., Nagel S. (1995): Phys. Rev. **E 51**, 3957

Linz S.J. (1996): Phys. Rev. **E 54**, 2925

Nicodemi M., Coniglio A., Herrmann H. (1997a-c): Phys. Rev. **E 55**, 3962; J. Phys. **A 30**, L379; Physica **A 240**, 405

Nowak E.R., Knight J.B., Povinelli M., Jaeger H.M., Nagel S.R. (1997a): to appear in Powder Technology

Nowak E.R., Povinelli M., Jaeger H.M., Nagel S., Knight J.B., Ben-Naim E. (1997b): Powders and Grains '97, 377

Peng G., Ohta T. (1997): preprint (condmat/9707237)

From Microscopic to Macroscopic Traffic Models

Dirk Helbing[1,2]

[1] II. Institute of Theoretical Physics, University of Stuttgart, Pfaffenwaldring 57/III, 70550 Stuttgart, Germany
[2] Department of Fluid Mechanics and Heat Transfer, Tel Aviv University, Tel Aviv, 69978, Israel

Abstract. The paper presents a systematic derivation of macroscopic equations for freeway traffic flow from an Enskog-like kinetic approach. The resulting fluid-dynamic traffic equations for the spatial density, average velocity, and velocity variance of vehicles are compared to equations, which can be obtained from a microscopic force model of individual vehicle motion. Simulation results of the models are confronted with empirical traffic data.

1 Introduction

During the last five years, modelling and simulating traffic dynamics has found a large and rapidly growing interest in physics. This is due to

1. similarities of traffic dynamics with flows of gases, fluids, and granular media,
2. instability phenomena and critical behavior of traffic (cf. Fig. 1),
3. interesting applications of cellular automata and molecular dynamics simulation methods,
4. the need of efficient traffic optimization methods in order to keep or increase the level of mobility.

Usually, one distinguishes three levels of modelling: The microscopic level of description delineates the dynamics of the single driver-vehicle units [1], [2], [3], [4], [5]. This allows to consider different vehicle characteristics and driving styles, so that many of these models aim at a high-fidelity description of traffic flow, e.g. [3], [5]. They are mostly used for detail studies (e.g. of on-ramp traffic, bottlenecks, effects of traffic optimization measures), but they consume an enourmous amount of CPU time due the the large number of variables involved. An alternative approach are cellular automata, which allow to simulate a minimal model of traffic dynamics faster than real-time [6], [7], [8], [9].

Computational efficiency can also be reached by macroscopic traffic models, but at a higher degree of accuracy [10], [11], [12], [13], [5]. Macroscopic traffic models consist of equations for a few aggregate quantities like the spatial density ρ, the average velocity V, and (in some cases) additional velocity moments. These equations are similar to fluid-dynamic equations, but

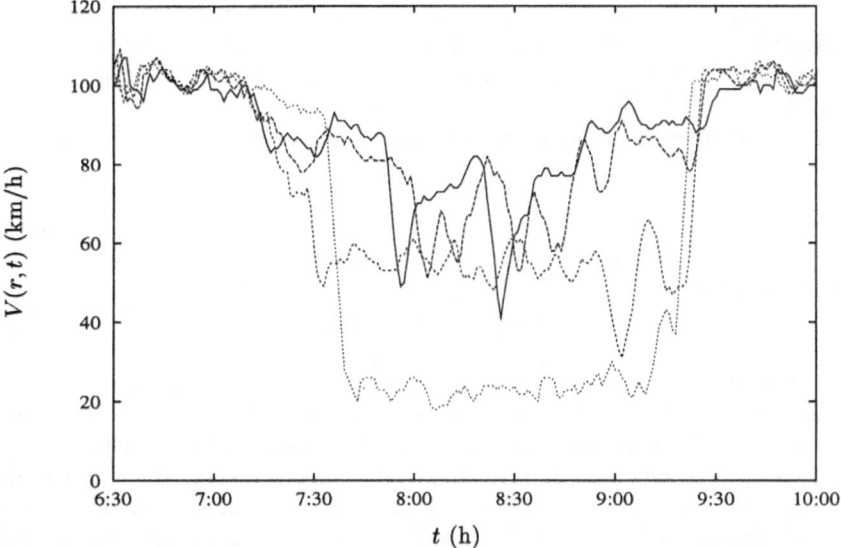

Fig. 1. Temporal evolution of the average velocity $V(r,t)$ at subsequent cross-sections of the Dutch highway A9 from Haarlem to Amsterdam at October 14, 1994 (five minute averages of single vehicle data) [14], [5]. The prescribed speed limit is 120 km/h. We observe a breakdown of velocity during the rush hours between 7:30 am and 9:30 am due to the overloading of the highway at $r = r_0 := 41.8\,\text{km}$ ($\cdots$). At the subsequent cross-sections the traffic situation recovers (- - -: $r = r_0 + 1\,\text{km}$; – –: $r = r_0 + 2.2\,\text{km}$; —: $r = r_0 + 4.2\,\text{km}$). Nevertheless, the amplitudes of the small velocity fluctuations at r_0 become larger and larger, leading to so-called stop-and-go waves.

some fundamental differences with respect to the dynamics of ordinary fluids have recently been recognized [5], [13]. For congested conditions, their detailled form is not at all obvious. Therefore, it has been suggested to derive the macroscopic traffic equations from a kinetic, i.e. mesoscopic level of description, which delineates the spatio-temporal evolution of the velocity distribution [5], [13], [15], [16], [17], [18].

2 Enskog-like Kinetic Traffic Model

In the following, we define the coarse-grained phase-space density $\tilde{\rho}(r, v, t)$ of vehicles per lane with velocity v at place r and time t by

$$\tilde{\rho}(r,v,t) := \frac{1}{(2\,\Delta r)(2\,\Delta v)} \sum_{\alpha} \int\limits_{r-\Delta r}^{r+\Delta r} dr' \int\limits_{v-\Delta v}^{v+\Delta v} dv'\, \delta(r' - r_\alpha(t))\, \delta(v' - v_\alpha(t)). \quad (1)$$

$r_\alpha(t)$ is the location and $v_\alpha(t)$ the velocity of vehicle α at time t. We do not distinguish different lanes, here, although this is possible [5], [19]. Instead, we treat the overall cross section of an n-lane freeway in an effective way [5], [20].

Let us assume an acceleration equation of the form

$$\frac{dv_\alpha}{dt} = f_0(v_\alpha) + \sum_{\beta(\neq\alpha)} f_{\alpha\beta}(r_\alpha, v_\alpha, r_\beta, v_\beta) + \xi_\alpha(t)\,, \tag{2}$$

where the function

$$f_0(v_\alpha) := \frac{V_0 - v_\alpha(t)}{\tau} \tag{3}$$

describes an adaptation of the actual velocity $v_\alpha(t)$ to the desired velocity V_0 within a (possibly density-dependent) relaxation time τ. The second term in (2) delineates the effect of interactions with vehicles β, and $\xi_\alpha(t)$ reflects velocity fluctuations due to imperfect driving. We will assume $\langle \xi_\alpha(t)\xi_\beta(t')\rangle = 2D\,\delta_{\alpha\beta}\,\delta(t-t')$, where the diffusion function D is density- and velocity-dependent [5], [13], [26]. For reasons of simplicity, the desired velocity V_0 and the relaxation time τ were taken identical for all vehicles, but it is also possible to generalize this model [16], [17], [18].

From (1) and (2), the following dynamical equation for the phase-space density can be derived:

$$\frac{\partial\tilde\rho}{\partial t} + \frac{\partial(\tilde\rho v)}{\partial r} + \frac{\partial}{\partial v}[\tilde\rho\, f_0(v)] = \left(\frac{\partial\tilde\rho}{\partial t}\right)_{\text{fl}} + \left(\frac{\partial\tilde\rho}{\partial t}\right)_{\text{int}}. \tag{4}$$

The fluctuation term gives a contribution

$$\left(\frac{\partial\tilde\rho}{\partial t}\right)_{\text{fl}} = \frac{\partial^2(\tilde\rho D)}{\partial v^2}\,. \tag{5}$$

In addition, we have used the abbreviation

$$\left(\frac{\partial\tilde\rho}{\partial t}\right)_{\text{int}} := -\frac{\partial}{\partial v}(\tilde\rho f_{\text{int}}) \tag{6}$$

with the average interaction force

$$f_{\text{int}}(r, v, t) := \frac{1}{4\tilde\rho\,\Delta r\,\Delta v}\sum_{\alpha\neq\beta}\int_{r-\Delta r}^{r+\Delta r} dr' \int_{v-\Delta v}^{v+\Delta v} dv'\, f_{\alpha\beta}\,\delta(r'-r_\alpha(t))\,\delta(v'-v_\alpha(t))\,. \tag{7}$$

The interaction term (6) reflects deceleration processes. In analogy to the Enskog theory of dense gases [21] and granular media [22], [23], [24], but with an interaction law typical for vehicles [5], [13], it is approximated by

$$\left(\frac{\partial\tilde\rho}{\partial t}\right)_{\text{int}} = (1-p)\chi(r+l, t)B(v) \tag{8}$$

with the Boltzmann-like interaction function

$$B(v) = \int\limits_{w>v} dw\,(w-v)\,\tilde{p}(r,w,t)\tilde{\rho}(r+s,v,t)$$

$$- \int\limits_{v>w} dw\,(v-w)\tilde{p}(r,v,t)\tilde{\rho}(r+s,w,t)\,. \tag{9}$$

According to this, the phase-space density $\tilde{\rho}(r,v,t)$ increases due to deceleration of vehicles with velocities $w > v$, which cannot overtake vehicles with velocity v. The density-dependent probability of immediate overtaking is represented by p. A decrease of the phase-space density $\tilde{\rho}(r,v,t)$ is caused by interactions of vehicles with velocity v with slower vehicles driving with velocities $w < v$. The corresponding interaction rates are proportional to the relative velocity $|v-w|$ and to the phase space densities of both interacting vehicles. By $s(V) = l_0 + l(V)$ ($\approx$ vehicle length + safe distance) it is taken into account that the distance of interacting vehicles is given by their velocity-dependent space requirements. These cause an increase of the interaction rate, which is described by the pair correlation function $\chi(r) = [1 - \rho(r,t)s]^{-1}$ at the 'interaction point' $r+l$. A more detailed discussion of the above kinetic traffic model is presented elsewhere [5], [13].

Now, we will focus on the the macroscopic equations for the spatial density

$$\rho(r,t) = \int dv\,\tilde{\rho}(r,v,t)\,, \tag{10}$$

the average velocity

$$V(r,t) = \int dv\,v\frac{\tilde{\rho}(r,v,t)}{\rho(r,t)}\,, \tag{11}$$

and the velocity variance

$$\theta(r,t) = \int dv\,[v - V(r,t)]^2\frac{\tilde{\rho}(r,v,t)}{\rho(r,t)}\,. \tag{12}$$

These are obtained by multiplying the kinetic equation with v^k and integrating with respect to v. In order to obtain a closed system of equations, we assume that the velocity distribution $P(v;r,t)$ has a Gaussian form, i.e.

$$P(v;r,t) := \frac{\tilde{\rho}(r,v,t)}{\rho(r,t)} = \frac{e^{-[v-V(r,t)]^2/[2\theta(r,t)]}}{\sqrt{2\pi\theta(r,t)}}\,. \tag{13}$$

According to empirical data, this approximation is well justified (cf. Figs. 2 and 3).

After some straightforward calculations, the following equations are obtained:

$$\frac{\partial\rho}{\partial t} + \frac{\partial(\rho V)}{\partial r} = 0\,, \tag{14}$$

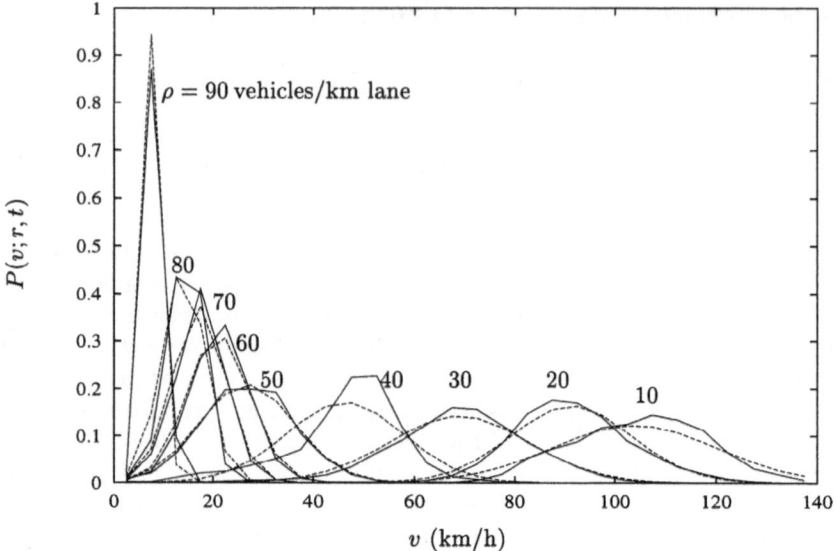

Fig. 2. Comparison of empirical velocity distributions at different densities (—) with frequency polygons of grouped Gaussian velocity distributions with the same mean value and variance (– –) [25]. A significant deviation of the empirical relations from the respective discrete Gaussian approximations is only found at a density of $\rho = 40$ vehicles/km lane, where the temporal averaging period of $T = 2\,\mathrm{min}$ may have been too long due to rapid stop-and-go waves.

$$\frac{\partial(\rho V)}{\partial t} + \frac{\partial}{\partial r}[\rho(V^2 + \theta)] = \frac{\rho}{\tau}(V_0 - V) + (1 - p)\chi(r + l, t)\int dv\, v\mathcal{B}(v)\,, \quad (15)$$

$$\frac{\partial}{\partial t}[\rho(V^2 + \theta)] + \frac{\partial}{\partial r}[\rho(V^3 + 3V\theta)] = \frac{2\rho}{\tau}(V_0 V + \tau D - V^2 - \theta)$$

$$+ (1 - p)\chi(r + l, t)\int dv\, v^2\mathcal{B}(v)\,. \quad (16)$$

Equations (14) to (16) are similar to the Euler equations of ordinary fluids. In particular, the density equation (14) agrees with the continuity equation, reflecting that the number of vehicles is conserved (on a circular road). However, equations (15) and (16) contain some additional terms compared to the hydrodyamic equations for momentum density and energy density, which are essential for the instability of traffic flow. The respective first terms on the right-hand sides of (15) and (16) originate from the acceleration towards the desired velocity V_0 and from velocity fluctuations. The respective last terms reflect interaction (deceleration) effects. In contrast to ordinary fluids, they do not vanish, since vehicular interactions do not conserve momentum and energy.

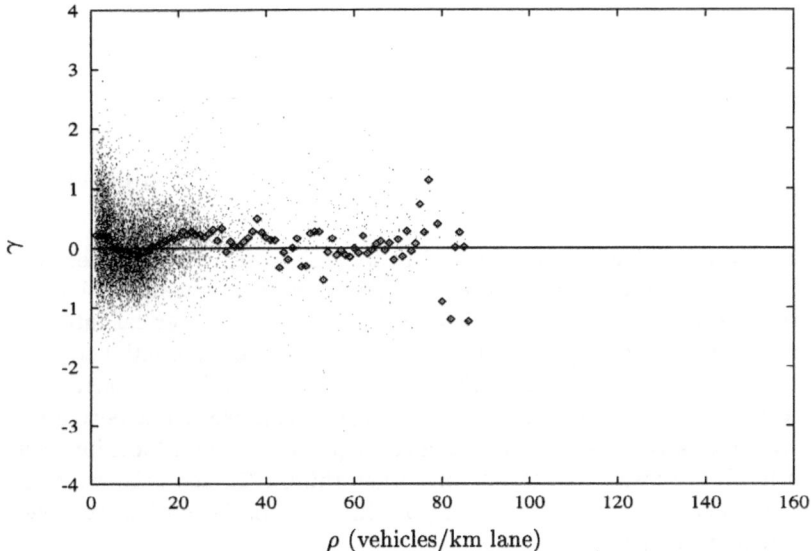

Fig. 3. Density-dependence of the skewness γ of the velocity distribution ($\cdot$: 1-minute data; $\diamond$: respective mean values) [25]. The large variation of the 1-minute data at low densities is due to the small number of vehicles which pass a cross section during the time interval $T = 1 \, \mathrm{min}$, whereas the large variation of their mean values at high densities comes from the few 1-minute data, over which could be averaged. The 1-minute data of the skewness scatter around the zero line (—) and mostly lie between -1 and 1, so that it is negligible most of the time.

To obtain the explicit form of the interaction terms, one has to carry out a number of lengthy calculations. Using the abbreviations

$$\rho_+(r,t) := \rho(r+s,t) \,, \quad V_+(r,t) := V(r+s,t) \,, \quad \theta_+(r,t) := \theta(r+s,t) \,, \quad (17)$$

and introducing the Gaussian error function

$$\Phi(x) = \int\limits_{-\infty}^{x} dy \, \frac{e^{-y^2/2}}{\sqrt{2\pi}} \,, \tag{18}$$

one finally finds

$$\int dv \, vB(v) = -\rho\rho_+ \left\{ \left[(\theta + \theta_+) + (V - V_+)^2 \right] \Phi\left(\frac{V - V_+}{\sqrt{\theta + \theta_+}} \right) \right.$$
$$\left. + (V - V_+)(\theta + \theta_+) \frac{e^{-(V-V_+)^2/[2(\theta+\theta_+)]}}{\sqrt{2\pi(\theta + \theta_+)}} \right\} \tag{19}$$

and

$$\int dv\, v^2 \mathcal{B}(v) = -2\rho\rho_+(\theta - \theta_+) \left[(\theta + \theta_+) \frac{e^{-(V-V_+)^2/[2(\theta+\theta_+)]}}{\sqrt{2\pi(\theta + \theta_+)}} \right.$$

$$\left. + (V - V_+)\, \Phi\left(\frac{V - V_+}{\sqrt{\theta + \theta_+}}\right) \right] + (V + V_+) \int dv\, v\mathcal{B}(v)\,. \quad (20)$$

The macroscopic traffic equations (14) to (16) were written as equations for fluxes with sink/source terms (the terms on the right-hand side). The flux representation is very advantageous, since many numerical integration algorithms have been developed for this type of partial differential equations. However, due to (19) and (20), the flux representation is non-local. This is caused by the finite space requirements of cars, i.e. a driver reacts to another car already at a certain distance. As a consequence, the non-local interaction terms imply viscosity effects, among other things. To see this, we expand them up to second order. Neglecting products of spatial derivatives, we get the continuity equation

$$\frac{\partial \rho}{\partial t} + V\frac{\partial \rho}{\partial r} = -\rho \frac{\partial V}{\partial r}\,, \quad (21)$$

the velocity equation

$$\frac{\partial V}{\partial t} + V\frac{\partial V}{\partial r} = a_1 \frac{\partial \rho}{\partial r} + a_2 \frac{\partial V}{\partial r} + a_3 \frac{\partial \theta}{\partial r}$$

$$+ b_1 \frac{\partial^2 \rho}{\partial r^2} + b_2 \frac{\partial^2 V}{\partial r^2} + b_3 \frac{\partial^2 \theta}{\partial r^2} + \frac{V_e - V}{\tau}\,, \quad (22)$$

and the variance equation

$$\frac{\partial \theta}{\partial t} + V\frac{\partial \theta}{\partial r} = c_1 \frac{\partial \rho}{\partial r} + c_2 \frac{\partial V}{\partial r} + c_3 \frac{\partial \theta}{\partial r}$$

$$+ d_1 \frac{\partial^2 \rho}{\partial r^2} + d_2 \frac{\partial^2 V}{\partial r^2} + d_3 \frac{\partial^2 \theta}{\partial r^2} + \frac{2(\theta_e - \theta)}{\tau} \quad (23)$$

(which corresponds to the equation of heat conduction in ordinary fluids). Here, we have used the abbreviations

$$a_1 = -[\tfrac{1}{\rho} + (1 - p)\chi s(1 + \rho\chi l)]\theta\,, \qquad a_2 = (1 - p)\chi\rho\left(2s\sqrt{\tfrac{\theta}{\pi}} - \rho\theta\chi\tfrac{l^2}{V}\right),$$

$$a_3 = -[1 + (1 - p)\rho\chi\tfrac{s}{2}]\,, \qquad b_1 = -(1 - p)\chi s\left(\tfrac{s}{2} + \rho\chi\tfrac{l^2}{2}\right)\theta\,,$$

$$b_2 = (1 - p)\chi\rho\left(s^2\sqrt{\tfrac{\theta}{\pi}} - \rho\theta\chi\tfrac{l^3}{2V}\right), \qquad b_3 = -(1 - p)\chi\rho\tfrac{s^2}{4}\,,$$

$$c_2 = -[2 + (1 - p)\chi\rho s]\theta\,, \qquad c_3 = 2(1 - p)\chi\rho s\sqrt{\tfrac{\theta}{\pi}}\,,$$

$$d_2 = -(1 - p)\chi\rho s^2 \tfrac{\theta}{2}\,, \qquad d_3 = (1 - p)\chi\rho s^2\sqrt{\tfrac{\theta}{\pi}}\,,$$

$$(24)$$

and
$$V_e = V_0 - \tau(1-p)\rho\chi\theta, \qquad \theta_e = \tau D. \tag{25}$$

Note that $c_1 = 0$ and $d_1 = 0$, which is a consequence of the assumed interaction law of vehicles. It is one of the advantages of a kinetic derivation of macroscopic traffic equations, that the above functions can be analytically calculated. For example, we have obtained an expression for the equilibrium velocity V_e. According to (25), it is given by the desired velocity V_0, diminished by a term due to decelerating interactions. The latter is proportional to the vehicle density and to the velocity variance, which is very plausible. The function $\partial P/\partial\rho := -\rho\, a_1$ can be interpreted as the partial derivative of the "traffic pressure" P with respect to density. The quantity $\eta := \rho\, b_2$ has the meaning of a viscosity, which smoothes out sudden spatial changes of the velocity profile $V(r,t)$. Both quantities are non-negative and diverge at maximum density $\rho_{\max} := 1/l_0$, as it should be for reasons of consistency [5], [13]. Previous macroscopic traffic models did not describe these important facts correctly, since they have neglected the terms in (24) which explicitly contain l or s. Therefore, they are not valid for high vehicle densities, i.e. for congested conditions. Finally, note that it is possible to calculate Navier-Stokes corrections of the coefficients a_i, b_i, c_i, and d_i [26].

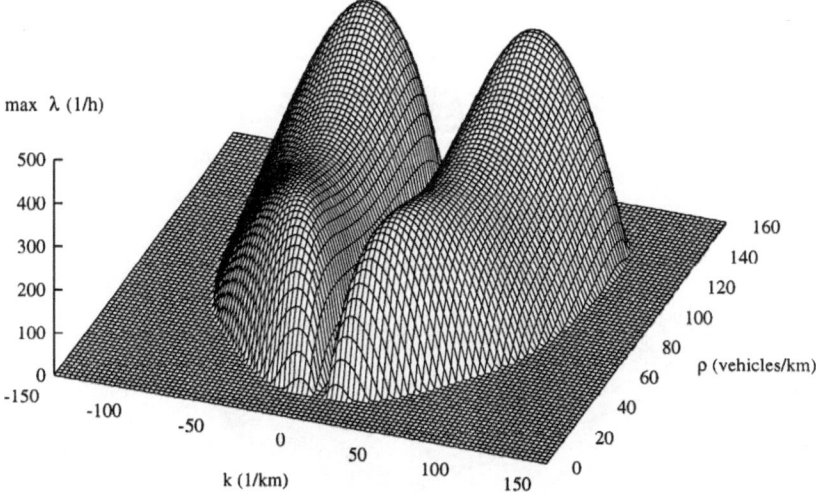

Fig. 4. Instability diagram for the Euler-like macroscopic traffic equations, including the dynamic variance equation [26].

According to our approximations, equations (21) to (23) are valid for small gradients of ρ, V, and θ. Therefore, they allow to investigate the evolution of small disturbances of the (stationary and spatially homogeneous) equilibrium solution. Figure 4 depicts the result of a linear instability analysis, showing that traffic flow is only stable at small and extreme densities as well as large wave numbers $|k|$ (i.e. small wave lengths $\ell = 2\pi/|k|$). This is in agreement with empirical findings.

The instability diagram is obtained by

1. assuming a small periodic perturbation $\delta g(r,t) = g_0 \exp[ikr + (\lambda + i\omega)t]$ of the macroscopic traffic quantities $g \in \{\rho, V, \theta\}$ relative to the stationary and spatially homogeneous equilibrium solution $g_e(\rho)$ (g_0 being the amplitude, k the wave number, λ the growth rate, and ω the frequency of the perturbation),
2. inserting $g(r,t) = g_e + \delta g(r,t)$ into the macroscopic traffic equations,
3. neglecting quadratic terms in the small perturbations $\delta g(r,t) \ll g_e$,
4. determining the three complex eigenvalues $\tilde{\lambda} = \lambda + i\omega$ of the linearized equations in dependence of ρ and k.

An explicit example for this procedure is discussed in [5], [13]. Equilibrium traffic flow is unstable if at least one of the growth rates is positive, i.e. $\max \lambda > 0$. Therefore, the instability diagram shows $\max \lambda(k, \rho)$ if this is greater than zero, otherwise 0. Figure 5 depicts a simulation result which demonstrates emerging stop-and-go traffic.

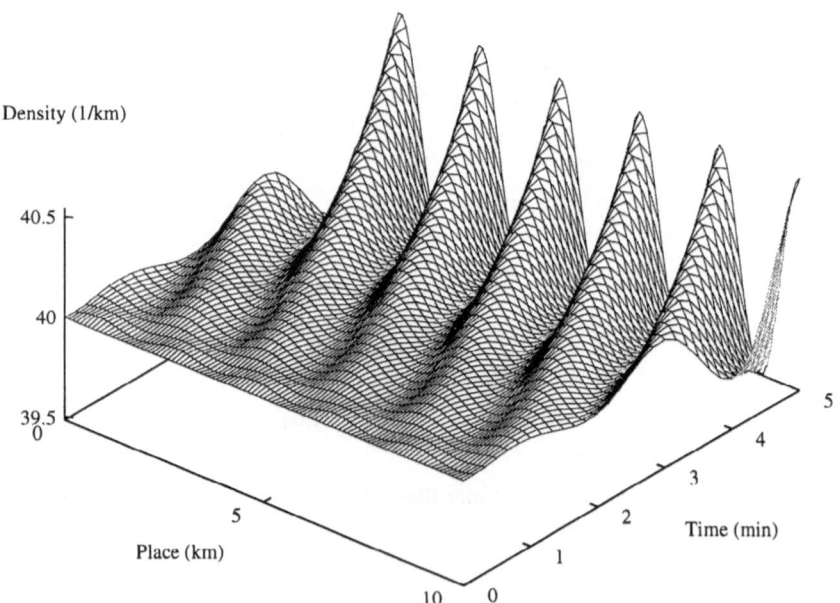

Fig. 5. Above a certain density, traffic flow is unstable. This gives rise to the development of stop-and-go waves. In the represented simulation, we applied periodic boundary conditions (which corresponds to a circular road).

2.1 Non-linear Phenomena

As a consequence of the inherent non-linearity of the macroscopic traffic equations, they display the self-organization of a number of collective patterns of motion. This includes the formation of density clusters ('traffic jams'), anti-clusters, dipole layers, cascades of density clusters ('stop-and-go traffic'), and merging of density clusters. Moreover, one finds subcritical instabilities and non-linear wave selection phenomena [11], [12].

3 An Alternative Approach

The Boltzmann-like formula (9) for vehicle interactions implicitly assumes, that deceleration maneuvers happen instantaneously. This approximation is only valid, if the average duration of deceleration maneuvers is considerably smaller than the time scale of the macroscopic traffic dynamics. However, one can also derive fluid-dynamic traffic equations without this assumption. We will illustrate this for a one-lane microscopic traffic model without possibilities of overtaking.

3.1 A Concrete Microscopic Model

Let us start with the *social force model* of vehicle dynamics, given by $dr_\alpha/dt = v_\alpha(t)$ and

$$\frac{dv_\alpha}{dt} = \underbrace{\frac{V_0 - v_\alpha(t)}{\tau}}_{\text{Acceleration}} + \underbrace{f_{\alpha(\alpha+1)}(r_\alpha, v_\alpha, r_{\alpha+1}, v_{\alpha+1})}_{\text{Deceleration}} + \xi_\alpha(t) . \qquad (26)$$

It is known that models of this kind are able to describe the emergence of stop-and-go traffic [5], [4] (cf. Figure 6).

The advantage of the social force concept is, that it allows a very intuitive modelling of decision processes which are related to continuous changes in some (possibly abstract) space of behavioral alternatives [27]. According to this, the different motivations which influence an individual at the same time, are described by additive, force-like quantities. These generalized forces are, of course, no Newtonian forces. For example, they do not obey the law *actio = reactio*. The social force concept is well compatible with theoretical concepts from the social sciences and has been elaborated in detail [27], [5]. It has already been successful in describing various self-organization phenomena in pedestrian crowds [5], [28], but it was also applied to opinion formation processes [27], [29].

In the case of driver behavior, we have two conflicting motivations: On the one hand, the driver likes to accelerate towards his desired velocity V_0. On the other hand, he wants to keep a safe distance to the car in front. The latter is described by a repulsive deceleration force $f_{\alpha(\alpha+1)}$. Effects $f_{\alpha\beta}$

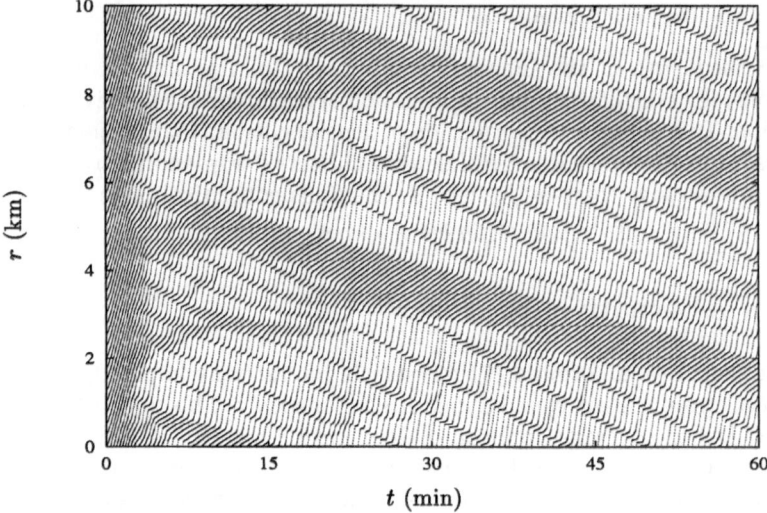

Fig. 6. Representation of the vehicle trajectories of each 10th car on a circular road [5]. The slopes of the trajectories indicate the respective vehicle velocities, whereas their density reflects the spatial vehicle density. The simulation starts with a homogeneous traffic situation (i.e. all vehicles have initially the same distance to the car in front). In the course of time, density clusters emerge. These are often called 'phantom traffic jams', since they do not originate from any localized bottleneck.

of interactions with other vehicles $\beta \neq (\alpha + 1)$ have been assumed to be negligible, here. However, they could easily be included in accordance with Eq. (2).

As Fig. 7 shows, a good agreement with empirical data of driver-vehicle behavior can be achieved with the following form of the repulsive interaction force:

$$f_{\alpha(\alpha+1)} := \frac{V_e'(r_{\alpha+1} - r_\alpha) - V_0}{\tau} + f_{\alpha(\alpha+1)}' \qquad (27)$$

with

$$f_{\alpha(\alpha+1)}' := -\exp\left(-\frac{r_{\alpha+1} - r_\alpha - s(v_\alpha)}{R}\right) \frac{v_\alpha - v_{\alpha+1}}{\tau'} \Theta(v_\alpha - v_{\alpha+1}). \qquad (28)$$

Here, $\Theta(\Delta v)$ is the Heaviside step function.

If we would restrict the model to the first term of (27) (i.e. in the case $\tau' \to \infty$), we would arrive at the microsimulation model by Bando et al. [4]. $V_e'(\Delta r)$ is the equilibrium velocity, which is a function of the distance $\Delta r := r_{\alpha+1} - r_\alpha$ to the next car ahead. The additional term (28) takes into account that

1. drivers brake stronger, when the relative velocity $\Delta v := v_\alpha - v_{\alpha+1}$ is large or when the distance Δr to the car in front is small,

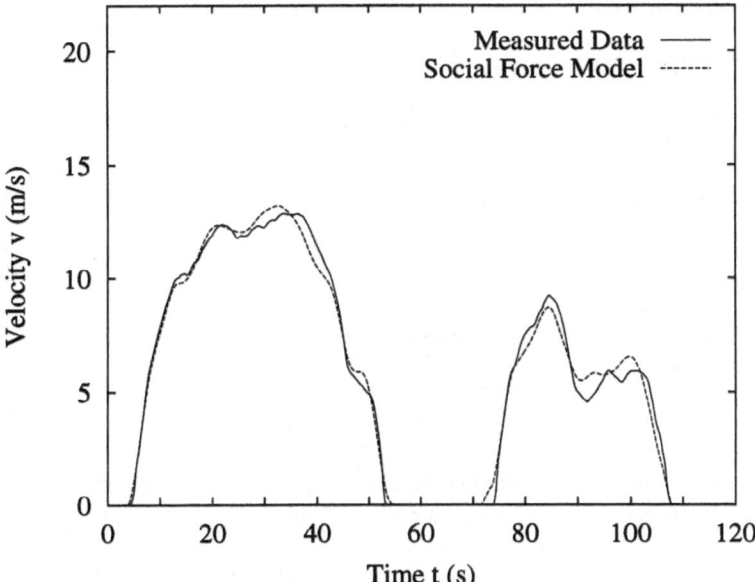

Fig. 7. Time-dependent velocity of a car which follows another car in city traffic. The simulation results for the social force model treat the velocity of the vehicle ahead and its initial distance as given. We find a good agreement with the empirical follow-the-leader data.

2. the deceleration time τ' is shorter than the acceleration time τ,
3. drivers begin to brake at a larger distance, if they drive fast. This is described by the velocity-dependence of the safe distance $s(v)$. R is the range of the repulsive effect of a car.

In most microsimulations, the relations $V_e'(\Delta r)$ and parameters τ, τ', R are specified individually (i.e. in an α-dependent way).

It can be shown that the above force model is consistent in the limiting cases. For large distances or $v_\alpha \approx v_{\alpha+1}$, vehicle α approaches the distance-dependent equilibrium velocity V_e':

$$\frac{dv_\alpha}{dt} \approx \frac{V_e'(r_{\alpha+1} - r_\alpha) - v_\alpha(t)}{\tau} + \xi_\alpha(t). \tag{29}$$

For small distances and $v_\alpha > v_{\alpha+1}$, it decelerates to the velocity $v_{\alpha+1}$ of the car in front:

$$\frac{dv_\alpha}{dt} \approx \frac{v_{\alpha+1}(t) - v_\alpha(t)}{\tau' e^{[r_{\alpha+1} - r_\alpha - s(v_\alpha)]/R}} + \xi_\alpha(t). \tag{30}$$

With decreasing distance it brakes stronger.

3.2 Derivation of Macroscopic Traffic Equations

In the following, we write the acceleration relation in the form

$$\frac{dv}{dt} = f(\Delta r, v, w) + \xi(t) := \frac{V_e'(\Delta r) - v}{\tau} + f'(\Delta r, v, w) + \xi(t) \qquad (31)$$

with the abbreviations $v := v_\alpha$, $w := v_{\alpha+1}$, $\Delta r := r_{\alpha+1} - r_\alpha$, $\xi := \xi_\alpha$, and

$$f' := f'_{\alpha(\alpha+1)} = f_{\alpha(\alpha+1)} + \frac{V_0 - V_e'(\Delta r)}{\tau}. \qquad (32)$$

Relation (31) is now inserted into the kinetic equation

$$\frac{\partial \tilde{\rho}}{\partial t} + \frac{\partial (\tilde{\rho} v)}{\partial r} + \frac{\partial (\tilde{\rho} f)}{\partial v} = \left(\frac{\partial \tilde{\rho}}{\partial t} \right)_{\text{fl}}, \qquad (33)$$

which is again a direct consequence of definition (1). Note that the interaction effects were absorbed into the function f, here. Next, we multiply this equation with v^k and $P'(w, \Delta r | v, r, t)$, which denotes the probability that, given a car with velocity v is located at place r, the car in front drives with velocity w at a distance Δr. Finally, the resulting equation is integrated with respect to w and Δr. This gives the macroscopic equations

$$\frac{\partial \rho}{\partial t} + \frac{\partial (\rho V)}{\partial r} = 0, \qquad (34)$$

$$\frac{\partial (\rho V)}{\partial t} + \frac{\partial}{\partial r} [\rho(V^2 + \theta)] = \frac{\rho}{\tau}(V_e^* - V) + \rho \mathcal{F}_1, \qquad (35)$$

$$\frac{\partial}{\partial t} [\rho(V^2 + \theta)] + \frac{\partial}{\partial r} [\rho(V^3 + 3V\theta)] = \frac{2\rho}{\tau}(V_e^* V + \tau D - V^2 - \theta) + \rho \mathcal{F}_2 \qquad (36)$$

with

$$\mathcal{F}_k(r, t) := k \int d\Delta r \int dv \int dw \, v^{k-1} f'(\Delta r, v, w) P'(\Delta r, w | r, v, t) \frac{\tilde{\rho}(r, v, t)}{\rho(r, t)}$$

$$= -k \int d\Delta r \int dv \int_{w<v} dw \, v^{k-1} \exp\left(-\frac{\Delta r - s(v)}{R} \right) \frac{v - w}{\tau'}$$

$$\times P'(\Delta r, w | r, v, t) P(v; r, t) \qquad (37)$$

and

$$V_e^* := \int d\Delta r \int dv \int dw \, V_e'(\Delta r) P'(\Delta r, w | r, v, t) P(v; r, t). \qquad (38)$$

To get (37), we made use of partial integration.

In the following, we will apply the factorization approximation

$$P'(\Delta r, w | r, v, t) \approx P_*(\Delta r; r, t) P(w; r + \Delta r, t), \qquad (39)$$

which is even exact, if the distributions of the velocities w and the headways Δr are statistically independent, and independent of v. Furthermore, we assume that the headway distribution $P_*(\Delta r; r, t)$ is a function of the density ρ and average velocity V at a certain place $r + \delta r$ between r and $r + \Delta r$:

$$P_*(\Delta r; r, t) \equiv P_*\Big(\Delta r; \rho(r + \delta r, t), V(r + \delta r, t)\Big). \qquad (40)$$

Then, we can expand (37) and (40) in the small quantities $\delta v := v - V$ and δr, respectively. In this way, we obtain

$$\mathcal{F}_k(r, t) \approx -k \int d\Delta r \int dv \int\limits_{w < v} dw\, v^{k-1} \frac{v - w}{\tau'} \exp\left(-\frac{\Delta r - s(V)}{R}\right) \left\{ \frac{ds}{dV} \frac{\delta v}{R} \right.$$
$$\left. + \frac{1}{2}\left[\frac{d^2 s}{dV^2} + \left(\frac{ds}{dV}\right)^2\right] \frac{\delta v^2}{R^2} \right\} P_*(\Delta r; r, t) P(w; r + \Delta r, t) P(v; r, t)$$

$$(41)$$

and

$$P_*(\Delta r; r, t) \approx P_*\Big(\Delta r; \rho(r, t), V(r, t)\Big) + \frac{\partial P_*}{\partial \rho}\left(\frac{\partial \rho}{\partial r}\delta r + \frac{\partial^2 \rho}{\partial r^2}\frac{\delta r^2}{2}\right)$$
$$+ \frac{\partial P_*}{\partial V}\left(\frac{\partial V}{\partial r}\delta r + \frac{\partial^2 V}{\partial r^2}\frac{\delta r^2}{2}\right), \qquad (42)$$

where we again neglected products of partial derivatives $\partial g/\partial r$. After carrying out the integrations over v, w, and Δr, the resulting macroscopic traffic equations can again be written in the form of Eqs. (21) to (23). However, the coefficients a_i, b_i, c_i, and d_i are different, since we did not apply the approximation of sudden deceleration maneuvers. The problem of this method is, that it does not provide the functional form of the headway distribution $P_*(\Delta r; \rho, V)$, which is needed for the explicit evaluation of the coefficients. Nevertheless, the use of the above results will be presented by a simple example.

3.3 Relation Between Bando's Microscopic and Payne's Macroscopic Traffic Model

The microsimulation model by Bando et al. [4] is obtained by neglecting the fluctuation term and the second term in (27), i.e. by setting $D := 0$ and $f' := 0$. In order to calculate the corresponding macroscopic traffic equations, we make a very simple assumption, here, namely that the headways Δr are given by the inverse of the density:

$$P_*(\Delta r; r, t) := \delta\left(\Delta r - \frac{1}{\rho(r + \delta r, t)}\right). \qquad (43)$$

Inserting this into the above equations, we finally arrive at the continuity equation

$$\frac{\partial \rho}{\partial t} + V\frac{\partial \rho}{\partial r} = -\rho\frac{\partial V}{\partial r}, \tag{44}$$

the velocity equation

$$\frac{\partial V}{\partial t} + V\frac{\partial V}{\partial r} \approx -\frac{1}{\rho}\frac{\partial(\rho\theta)}{\partial r} + \frac{1}{\tau}\left[V_e^*\left(\frac{1}{\rho}\right) - V\right] - \frac{1}{\tau\rho^2}\frac{\partial V_e^*}{\partial\Delta r}\left(\frac{\partial\rho}{\partial r}\delta r + \frac{\partial^2\rho}{\partial r^2}\frac{\delta r^2}{2}\right)$$

$$\approx -\frac{1}{\rho}\frac{\partial(\rho\theta)}{\partial r} + \frac{1}{\tau}[V_e(\rho) - V] + \frac{1}{\tau}\frac{\partial V_e}{\partial\rho}\left(\frac{\partial\rho}{\partial r}\delta r + \frac{\partial^2\rho}{\partial r^2}\frac{\delta r^2}{2}\right), \tag{45}$$

and the variance equation

$$\frac{\partial\theta}{\partial t} + V\frac{\partial\theta}{\partial r} = -2\theta\frac{\partial V}{\partial r} - \frac{2}{\tau}\theta, \tag{46}$$

where

$$V_e(\rho) := V_e^*\left(\frac{1}{\rho}\right). \tag{47}$$

Close to the equilibrium solution, the variance equation can be neglected due to $\theta \approx 0$. The instability condition of the remaining equations (44) and (45) reads

$$\rho\left|\frac{dV_e}{d\rho}\right| \overset{!}{>} \frac{\delta r}{\tau} \tag{48}$$

(cf. [5], [13]). This is only compatible with the instability condition

$$\frac{dV_e^*}{d\Delta r} \overset{!}{>} \frac{1}{2\tau} \qquad \text{or} \qquad \rho^2\left|\frac{dV_e}{d\rho}\right| \overset{!}{>} \frac{1}{2\tau} \tag{49}$$

of the Bando model [4], if we choose

$$\delta r \overset{!}{=} \frac{1}{2\rho} \approx \frac{\Delta r}{2}, \tag{50}$$

which is very plausible. In this case, the macroscopic equations (44) and (45) agree with the traffic model by Payne [10], but they contain the additional term $[\delta r^2/(2\tau)](\partial V_e/\partial\rho)\partial^2\rho/\partial r^2$, which describes a smoothing of sudden spatial changes in density and velocity. However, as soon as the approximation $\theta \approx 0$ becomes invalid, Payne's model does not anymore reflect the traffic dynamics according to Bando's model.

4   Summary and Conclusions

We have presented microscopic and macroscopic traffic flow models for freeways, which were successfully confronted with empirical data (cf. also [5], [13], [14]). Moreover, it has been shown, how macroscopic traffic models can be systematically derived from the equations of motion for single vehicles. This is essential for increasing the speed of traffic simulations. Apart from the kinetic approach to this problem, which based on the assumption of sudden deceleration maneuvers, an alternative method has been proposed, which presupposes a suitable approximation of the headway distribution function. The resulting macroscopic traffic equations are related to the hydrodynamic equations of ordinary fluids, but they contain a number of additional terms for three reasons:

1. Vehicles accelerate to a certain desired velocity.
2. A finite equilibrium variance of vehicle velocities is caused by imperfect driving.
3. Vehicle interactions are anisotropic and do not conserve energy or momentum.

The additional terms are responsible for certain instabilities of traffic flow, causing 'phantom traffic jams' or 'stop-and-go traffic'. They are also the origin of viscosity effects and of the divergence of 'traffic pressure' at maximum vehicle density. Here, it is essential that vehicular space requirements are taken into account [5], [13], [26]. Otherwise, the macroscopic traffic model would neglect certain characteristic terms, which would limit its validity to non-congested traffic situations.

For the purpose of computer simulations, it is advantageous to have the macroscopic traffic equations in flux representation. This has been analytically derived, but it contains the Gaussian error function. In contrast to previous results [26], the corresponding equations are not restricted to cases of small gradients.

Acknowledgments

The author wants to thank Martin Treiber, Tilo Schwarz, and Benno Tilch for providing Figs. 5, 6, and 7, respectively. He is also grateful to Henk Taale and the Dutch Ministry of Transport, Public Works and Water Management as well as to Thomas Bleile and the Robert Bosch GmbH for providing the empirical data shown in Figs. 1 to 3, and 7, respectively. The presented work has been financially supported by the DFG, Heisenberg scholarship He 2789/1-1, and by the BMBF, grant no. 13N7092 (collaborative research project "SANDY").

References

[1] Gazis D.C., Herman, R., Rothery R.W. (1961): Nonlinear Follow the Leader Models of Traffic Flow. Operations Research **9**, 545–567

[2] May A.D., Jr., Keller H.E.M. (1967): Non-Integer Car-Following Models. Highway Research Record **199**, 19–32

[3] Wiedemann R. (1974): *Simulation des Straßenverkehrsflusses* (Heft 8 der Schriftenreihe des IfV, Institut für Verkehrswesen, Universität Karlsruhe)

[4] Bando M., Hasebe K., Nakayama A., Shibata A., Sugiyama Y. (1995): Dynamical Model of Traffic Congestion and Numerical Simulation. Phys. Rev. E **51**, 1035–1042

[5] Helbing D. (1997): *Verkehrsdynamik: Neue physikalische Modellierungskonzepte* (Springer, Berlin)

[6] Schreckenberg M., Schadschneider A., Nagel K., Ito N. (1996): Discrete Stochastic Models for Traffic Flow. Phys. Rev. E **51**, 2939–2949.

[7] Nagel K., Paczuski M. (1995): Emergent Traffic Jams. Phys. Rev. E **51**, 2909–2918

[8] Nagatani T. (1995): Bunching of Cars in Asymmetric Exclusion Models for Freeway Traffic. Phys. Rev. E **51**, 922–928

[9] Krauss S., Wagner P., Gawron C. (1996): Continuous Limit of the Nagel-Schreckenberg Model. Phys. Rev. E **54**, 3707–3712

[10] Payne H.J. (1971): Models of Freeway Traffic and Control. In: Bekey G.A. (ed.) *Mathematical Models of Public Systems, Vol. 1* (Simulation Council, La Jolla, CA), 51–61

[11] Kerner B.S., Konhäuser P. (1994): Structure and Parameters of Clusters in Traffic Flow. Phys. Rev. E **50**, 54–83

[12] Kerner B.S., Konhäuser P., Schilke M. (1995): Deterministic Spontaneous Appearance of Traffic Jams in Slightly Inhomogeneous Traffic Flow. Phys. Rev. E **51**, 6243–6246

[13] Helbing D. (1996): Derivation and Empirical Validation of a Refined Traffic Flow Model. Physica A **233**, 253–282

[14] Helbing D. (1997): Empirical Traffic Data and their Implications for Traffic Modeling. Phys. Rev. E **55**, R25–R28

[15] Prigogine I., Herman R. (1971): Kinetic Theory of Vehicular Traffic (Elsevier, Amsterdam)

[16] Paveri-Fontana S.L. (1975): On Boltzmann-like Treatments for Traffic Flow. A Critical Review of the Basic Model and an Alternative Proposal for Dilute Traffic Analysis. Transportation Research **9**, 225–235

[17] Helbing D. (1996): Gas-Kinetic Derivation of Navier-Stokes-Like Traffic Equations. Phys. Rev. E **53**, 2366–2381

[18] Wagner C., Hoffmann C., Sollacher R., Wagenhuber J., Schürmann B. (1996): Second Order Continuum Traffic Flow Model. Phys. Rev. E **54**, 5073–5085

[19] Helbing D., Greiner A. (1997): Modelling and Simulation of Multilane Traffic Flow. Phys. Rev. E **55**, 5498–5508

[20] Helbing D. (1997): Modeling Multi-Lane Traffic Flow with Queuing Effects. Physica A **242**, 175–194

[21] Chapman S., Cowling T.G. (1970): *The Mathematical Theory of Nonuniform Gases* (3rd edition, Cambridge University Press, Cambridge)

[22] Jenkins J.T., Richman M.W. (1985): Kinetic Theory for Plane Flows of a Dense Gas of Identical, Rough, Inelastic, Circular Disks. Phys. of Fluids **28**, 3485–3494

[23] Lun C.K.K., Savage S.B., Jeffrey D.J., Chepurniy N. (1984): Kinetic Theories for Granular Flow: Inelastic Particles in Couette Flow and Slightly Inelastic Particles in a General Flowfield. J. Fluid. Mech. **140**, 223–256

[24] Goldshtein A., Shapiro M. (1995): Mechanics of Collisional Motion of Granular Materials. Part 1. General Hydrodynamic Equations. J. Fluid. Mech. **282**, 75–114

[25] Helbing D. (1997): Fundamentals of Traffic Flow. Phys. Rev. E **55**, 3735–3738

[26] Helbing D. (1997): Structure and Instability of Consistent High-Density Equations for Traffic Flow. Phys. Rev. Lett., submitted

[27] Helbing D. (1995): *Quantitative Sociodynamics. Stochastic Methods and Models of Social Interaction Processes* (Kluwer Academic, Dordrecht)

[28] Helbing D., Molnár P. (1995): Social Force Model for Pedestrian Dynamics. Phys. Rev. E **51**, 4282–4286

[29] Helbing D. (1993): Boltzmann-like and Boltzmann-Fokker-Planck Equations as a Foundation of Behavioral Models. Physica A **196**, 546–573

For further references cf. [5].

The Modelling Concept of Sociodynamics

Wolfgang Weidlich

Institut für Theoretische Physik der Universität Stuttgart

Abstract. A general concept is presented which allows of setting up mathematical models for stochastic and quasi deterministic dynamic processes in social systems. The basis of this concept is the master equation for the probability distribution over appropriately chosen personal and material macrovariables of the society. The probabilistic transition rates depend on motivation potentials governing the decisions and actions of the social agents. The transition from the probability distribution to quasi-meanvalues leads to in general nonlinear coupled differential equations for the macrovariables of the chosen social sector. Up to now several models about population dynamics, collective political opinion formation, dynamics of economic processes and the formation of settlements have been published.

1 Introduction

During the last three decades new branches of science of interdisciplinary scope have been developed starting from the conceptual framework of theoretical physics. One of the most prominent examples is *Synergetics* founded by H. Haken (1) and defined as the theory of spatial, temporal and functional macrostructures of multicomponent systems. The procedures and conclusions of synergetics have found many applications, in particular in the natural sciences.

Since the human society can *also* be considered as a complex multicomponent system consisting of individuals interacting with themselves and with their material environment it was a challenge to develop a strategy allowing of a general quantitative modelling procedure for collective dynamic macroprocesses in the society, too.

Sociodynamics, of which the present article gives a short introduction (see also (2) and (3)), is intended to tackle this ambitious program.

On the one hand, sociodynamics can be considered as part of synergetics, because it turns out, that there indeed exist many structural analogis on the macrolevel between social and natural multicomponent systems and that the general principles of synergetics apply to both kinds of systems.

On the other hand, the modelling procedures of sociodynamics could not be directly taken from synergetic algorithms primarily developed for systems in natural science for two reasons:

1. Although there exists a microlevel of interacting components (individuals, particles) and a macrolevel of collective phenomena in social and physical systems as well, in contrast to physics *no equations of motion*

are available on the microlevel for social systems. Nevertheless it is the modelling purpose of sociodynamics to derive dynamic equations for the macrovariables of the social system with a mathematical structure comparable to equations for macrovariables in physics.

2. The modelling procedure of sociodynamics should take into account by a decisive step *the specific structure of social systems* which differs from natural systems. The difference takes place at the microlevel: In physics and chemistry the interactions between the components (e.g. atoms or molecules) leading to changes of their states are rather *direct*, for instance by interaction forces or simple emission and absorption processes. Instead the interaction between members of the society is usually not a direct one like a Pawlow reflex between input stimulus and output reactions, but an *indirect one*. Indeed, in general there lie many estimations, considerations, deliberations and valuations between environmental input and individual decisions or reactions.

It is *not* the intention of sociodynamics to give a detailed description or even explanation of these considerations taking place in the brain of the individuals which mediate between their social situation and their decisions and actions. *However it is* the intention of sociodynamics to take into account the resultant effect of such considerations on the dynamics of the macrovariables of the society. The appropriate concepts for describing these macro-effects of individual social behaviour are found in Sect. 3.

Sometimes synergetics or sociodynamics were critizised to be "physicalistic" with the argument that because of the lack of isomorphy between microinteractions of physical and social systems no genuine structural analogies between both kinds of systems could exist.

This argument is however misleading because of the following reason: Although the components and their interactions of social and physical systems are indeed very different, both kinds of systems exhibit comparable structures (e.g. chaos, phase transitions etc.) on the macrolevel. This convergence to comparable macro-phenomena is possible because of the information compression wiping out many differences on the way from microlevel to macrolevel.

The organisation of the sections is the following: In Sect. 2 the intentions of the modelling concept are indicated and in Sect. 3 the main steps of the general modelling procedure are exhibited. Two applications of the modelling approach to very different processes follow in Sect. 4. The first example treats the interregional migration of interacting populations and the second example is a model of the dynamics of city evolution on the level of building sites.

2   The Intention of Sociodynamics

It is intended to build up a modelling strategy allowing in principle of an integrative quantitative description of dynamic macro-phenomena in the society in terms of coherent and sector-overlapping concepts. These concepts cannot make use of "micro-equations for individuals", because such equations are not available.

The modelling concept should establish the connection between the decisions and actions driven by motivations of individuals on the microlevel and the dynamics of collective material and abstract variables on the macrolevel.

In view of the complexity of human behaviour the starting point cannot be a deterministic but only a probabilistic treatment of individual decisions and actions, which however are guided by trends, utility considerations and resulting motivations.

From the stochastic microbehaviour there follows necessarily a stochastic description of the macrodynamics. The central equation for this probabilistic macrodynamics will be the master equation for the probability distribution of macrovariables, that means of the order parameters of the society or their sectors.

Approximate deterministic "quasi-meanvalue-equations" can thereupon be derived for the macrovariables. These coupled, in general nonlinear differential equations normally suffice for describing the essentials of social macrodynamics. In particular they are able to comprehend, due to their nonlinearity, phenomena like multiple equilibra, chaotic motion and social phase transitions.

3 The Main Steps of the Modelling Procedure

3.1 Step 1: The Configuration of Macrovariables

An appropriate choice of macrovariables playing the role of key-variables is essential for a sufficiently complete description of an approximately separable sector of the society. In the ideal case the chosen key-variables dominate ("slave") the sector-specific individual attitudes and activities (i.e. the microvariables) and among themselves they approximately obey a selfcontained subdynamics which can be considered as the relevant order parameter dynamics of that sector.

Two main kinds of macrovariables can be distinguished.

a) The Configuration of Collective Material Variables

Macrovariables of this kind are wellknown and have for long been used in particular in economics. One may distinguish, like in thermodynamics, between *intense* variables which are independent of the size of the system (examples: prices, productivity, density of commodities) and *extensive* variables proportional to the system's size (like extent of production, of investment, number of buildings etc.).

It is characteristic of collective material variables $m_1, \ldots, m_k, \ldots$ relevant for the society that their intensity or amount is influenced by individual actors or groups of actors (e.g. the management of firms), but that this influence is an indirect one, which means that the actions of individuals do not show up explicitly in m_k.

The set of material variables

$$\mathbf{m} = \{m_1, \ldots m_k, \ldots m_M\} \tag{1}$$

necessary to describe the social sector under consideration is denoted as *material configuration.*

b) Collective Personal Variables: the Socioconfiguration

In order to describe socio-psychological processes, too, material variables are not sufficient. Abstract collective phenomena like the social or political climate are more directly connected with attitudes, opinions or actions of individuals and their subgroups. In order to capture such phenomena the *socioconfiguration* is introduced:

In an abstract manner we distinguish a set of alternative states (attitudes or actions) $i, j, k, l, \ldots I$ which an individual may assume with respect to the considered social aspect or sector. Furthermore we may distinguish different subpopulations $p_\alpha, p_\beta, \ldots p_P$ discernible by constant characteristics, so that each individual is member of one subpopulation. Be n_i^α the number of members of p_α in state i. We denote the multiple

$$\mathbf{n} = \{n_1^1 \ldots n_I^1; \ldots ; n_1^\alpha \ldots n_i^\alpha \ldots n_j^\alpha \ldots n_I^\alpha; \ldots ; n_1^P \ldots n_I^P\} \tag{2}$$

as *socioconfiguration.* This multiple is a set of macrovariables and describes the (momentary) distribution of attitudes and actions among the subpopulations. Evidently the socioconfiguration is *directly* dependent on the decisions of individuals to retain or to change their attitudes or actions.

The total macroconfiguration is then given by the multiple $\{\mathbf{m}, \mathbf{n}\}$.

3.2 Step 2: Transition Rates Between Macroconfigurations and Their Interpretation

If all macrovariables $\{\mathbf{m}, \mathbf{n}\}$ remain constant with time, the society is in a stationary macroscopic equilibrium which may in some respects be compared to thermodynamic equilibrium.

In order to describe the dynamics of macrovariables we first introduce their elementary changes.

In the case of material macrovariables we may always choose appropriate units so that the elementary changes consist of the increase or decrease of one of the macrovariables, say m_k, by one unit. The material configuration $\mathbf{m}$ then makes the transition

$$\mathbf{m} \to \mathbf{m}_k^\pm = \{m_1, \ldots, (m_k \pm 1), \ldots m_M\} \tag{3}$$

The elementary change of the socioconfiguration $\mathbf{n}$ takes place if one individual of subpopulation p_α changes his/her state (attitude or action) from i to j. This leads to the transition

$$\mathbf{n} \rightarrow \mathbf{n}_{ji}^\alpha = \{n_1^1 \ldots n_I^1; n_1^\alpha \ldots (n_j^\alpha + 1), \ldots (n_i^\alpha - 1) \ldots n_I^\alpha; \ldots ; n_1^P \ldots n_I^P\} \tag{4}$$

The decisive dynamic quantities putting into effect a transition between the macroconfiguration $\{\mathbf{m}, \mathbf{n}\}$ and a neighboring configuration $\{\mathbf{m}_k^\pm, \mathbf{n}\}$ or $\{\mathbf{m}, \mathbf{n}_{ji}^\alpha\}$ are *transition rates* describing by definition the probability per unit of time that the respective transition takes place, given that initially the configuration $\{\mathbf{m}, \mathbf{n}\}$ is realised.

Since transition rates are positive (semi-)definite quantities they may always be written in exponential form!

For the transition between material variables from $\mathbf{m}$ to $\mathbf{m}_k^\pm$ and the inverse transition from $\mathbf{m}_k^\pm$ to $\mathbf{m}$ we thus obtain the form of transition rates:

$$w_k(\mathbf{m}_k^\pm, \mathbf{m}; \mathbf{n}, \boldsymbol{\kappa}) = w_k^\pm(\mathbf{m}; \mathbf{n}, \boldsymbol{\kappa}) = \mu_0 \exp\{M_k(\mathbf{m}_k^\pm, \mathbf{m}; \mathbf{n}, \boldsymbol{\kappa})\} \tag{3.5a}$$

$$w_k(\mathbf{m}, \mathbf{m}_k^\pm; \mathbf{n}, \boldsymbol{\kappa}) = w_k^\mp(\mathbf{m}_k^\pm; \mathbf{n}, \boldsymbol{\kappa}) = \mu_0 \exp\{M_k(\mathbf{m}, \mathbf{m}_k^\pm; \mathbf{n}, \boldsymbol{\kappa})\} \tag{3.5b}$$

Here we have introduced *motivation potentials* $M_k(\mathbf{m}_k^\pm, \mathbf{m}; \mathbf{n}, \boldsymbol{\kappa})$ being measures of the intensity of the transition from the initial state $\mathbf{m}$ to the final state $\mathbf{m}_k^\pm$. They are functions of the initial and final macroconfigurations, because these are compared and judged by the decision makers giving rise to that transition. Furthermore, they are also functions of a set of trend parameters $\boldsymbol{\kappa} = (\kappa_1 \ldots \kappa_T)$ describing the trends and the degree of the responsiveness of the decision makers to the numerical values of the initial and final values of the macrovariables $\{\mathbf{m}, \mathbf{n}\}$.

Similarly the transition rate for the transition from socioconfiguration $\mathbf{n}$ to $\mathbf{n}_{ji}^\alpha$ and the transition rate for the inverse transition from $\mathbf{n}_{ji}^\alpha$ to $\mathbf{n}$, respectively, read:

$$\begin{aligned} w_{ji}^\alpha(\mathbf{n}_{ji}^\alpha, \mathbf{n}; \mathbf{m}, \boldsymbol{\kappa}) &= \nu_0 n_i^\alpha p_{ji}^\alpha(\mathbf{n}_{ji}^\alpha, \mathbf{n}; \mathbf{m}, \boldsymbol{\kappa}) \\ &= \nu_0 n_i^\alpha \exp\{M_{ji}^\alpha(\mathbf{n}_{ji}^\alpha, \mathbf{n}; \mathbf{m}, \boldsymbol{\kappa})\} \end{aligned} \tag{3.6a}$$

$$\begin{aligned} w_{ij}^\alpha(\mathbf{n}, \mathbf{n}_{ji}^\alpha; \mathbf{m}, \boldsymbol{\kappa}) &= \nu_0(n_j^\alpha + 1) p_{ij}^\alpha(\mathbf{n}, \mathbf{n}_{ji}^\alpha; \mathbf{m}, \boldsymbol{\kappa}) \\ &= \nu_0(n_j^\alpha + 1) \exp\{M_{ij}^\alpha(\mathbf{n}, \mathbf{n}_{ji}^\alpha; \mathbf{m}, \boldsymbol{\kappa})\} \end{aligned} \tag{3.6b}$$

Here, p_{ji}^α and p_{ij}^α are individual transition rates from state i to j and state j to i, respectively and one has to take into account that the transition from $\mathbf{n}$ to $\mathbf{n}_{ji}^\alpha$ can be effected by n_i^α individuals in state i, and the transition from $\mathbf{n}_{ji}^\alpha$ to $\mathbf{n}$ by $(n_j^\alpha + 1)$ individuals in state j, respectively.

Since the motivations do not only depend on the initial and final macrovariables but also on the initial and final state of the individual effecting that transition, the motivation potential $M_{ji}^\alpha(\mathbf{n}_{ji}^\alpha, \mathbf{n}; \mathbf{m}, \kappa)\}$ now explicitly depends on the initial state i and final state j of the transition-making individual.

Let us now decompose the motivation potentials for the transitions $\mathbf{m} \to \mathbf{m}_k^\pm$; $\mathbf{n} \to \mathbf{n}_{ji}^\alpha$ and the inverse transitions $\mathbf{m}_k^\pm \to \mathbf{m}$; $\mathbf{n}_{ji}^\alpha \to \mathbf{n}$ into their symmetrical and antisymmetrical parts:

$$M_k(\mathbf{m}_k^\pm, \mathbf{m}; \mathbf{n}, \kappa) = M_k^{(s)}(\mathbf{m}_k^\pm, \mathbf{m}; \mathbf{n}, \kappa) + M_k^{(as)}(\mathbf{m}_k^\pm, \mathbf{m}; \mathbf{n}, \kappa) \qquad (7)$$

$$M_k(\mathbf{m}, \mathbf{m}_k^\pm; \mathbf{n}, \kappa) = M_k^{(s)}(\mathbf{m}_k^\pm, \mathbf{m}; \mathbf{n}, \kappa) - M_k^{(as)}(\mathbf{m}_k^\pm, \mathbf{m}; \mathbf{n}, \kappa) \qquad (8)$$

with:

$$M_k^{(s)}(\mathbf{m}_k^\pm, \mathbf{m}; \mathbf{n}, \kappa) = \frac{1}{2}\{M_k(\mathbf{m}_k^\pm, \mathbf{m}; \mathbf{n}, \kappa) + M_k(\mathbf{m}, \mathbf{m}_k^\pm; \mathbf{n}, \kappa)\} \qquad (9)$$

$$M_k^{(as)}(\mathbf{m}_k^\pm, \mathbf{m}; \mathbf{n}, \kappa) = \frac{1}{2}\{M_k(\mathbf{m}_k^\pm, \mathbf{m}; \mathbf{n}, \kappa) - M_k(\mathbf{m}, \mathbf{m}_k^\pm; \mathbf{n}, \kappa)\} \qquad (10)$$

and

$$M_{ji}^\alpha(\mathbf{n}_{ji}^\alpha, \mathbf{n}; \mathbf{m}, \kappa) = M_{ji}^{\alpha(s)}(\mathbf{n}_{ji}^\alpha, \mathbf{n}; \mathbf{m}, \kappa) + M_{ji}^{\alpha(as)}(\mathbf{n}_{ji}^\alpha, \mathbf{n}; \mathbf{m}, \kappa) \qquad (11)$$

$$M_{ij}^\alpha(\mathbf{n}, \mathbf{n}_{ji}^\alpha; \mathbf{m}, \kappa) = M_{ji}^{\alpha(s)}(\mathbf{n}_{ji}^\alpha, \mathbf{n}; \mathbf{m}, \kappa) - M_{ji}^{\alpha(as)}(\mathbf{n}_{ji}^\alpha, \mathbf{n}; \mathbf{m}, \kappa) \qquad (12)$$

with

$$M_{ji}^{\alpha(s)}(\mathbf{n}_{ji}^\alpha, \mathbf{n}; \mathbf{m}, \kappa) = \frac{1}{2}\{M_{ji}^\alpha(\mathbf{n}_{ji}^\alpha, \mathbf{n}; \mathbf{m}, \kappa) + M_{ij}^\alpha(\mathbf{n}, \mathbf{n}_{ji}^\alpha; \mathbf{m}, \kappa)\} \qquad (13)$$

$$M_{ji}^{\alpha(as)}(\mathbf{n}_{ji}^\alpha, \mathbf{n}; \mathbf{m}, \kappa) = \frac{1}{2}\{M_{ji}^\alpha(\mathbf{n}_{ji}^\alpha, \mathbf{n}; \mathbf{m}, \kappa) - M_{ij}^\alpha(\mathbf{n}, \mathbf{n}_{ji}^\alpha; \mathbf{m}, \kappa)\} \qquad (14)$$

Introducing now the notations for the symmetrical terms

$$\mu_0 \exp\{M_k^{(s)}(\mathbf{m}_k^\pm, \mathbf{m}; \mathbf{n}, \boldsymbol{\kappa})\} = \mu_k(\mathbf{m}_k^\pm, \mathbf{m}; \mathbf{n}, \boldsymbol{\kappa}) \tag{15}$$
$$= \mu_k(\mathbf{m}, \mathbf{m}_k^\pm; \mathbf{n}, \boldsymbol{\kappa})$$

and

$$\nu_0 \exp\{M_{ji}^{\alpha(s)}(\mathbf{n}_{ji}^\alpha, \mathbf{n}; \mathbf{m}, \boldsymbol{\kappa})\} = \nu_{ji}^\alpha(\mathbf{n}_{ji}^\alpha, \mathbf{n}; \mathbf{m}, \boldsymbol{\kappa}) \tag{16}$$
$$= \nu_{ij}^\alpha(\mathbf{n}, \mathbf{n}_{ji}^\alpha; \mathbf{m}, \boldsymbol{\kappa})$$

and assuming a slightly specialized form for the antisymmetrical parts of the motivation potentials:

$$M_k^{(as)}(\mathbf{m}_k^\pm, \mathbf{m}; \mathbf{n}, \boldsymbol{\kappa}) = u_k(\mathbf{m}_k^\pm; \mathbf{n}, \boldsymbol{\kappa}) - u_k(\mathbf{m}; \mathbf{n}, \boldsymbol{\kappa}) \tag{17}$$

$$M_{ji}^{\alpha(as)}(\mathbf{n}_{ji}^\alpha, \mathbf{n}; \mathbf{m}, \boldsymbol{\kappa}) = u_j(\mathbf{n}_{ji}^\alpha; \mathbf{m}, \boldsymbol{\kappa}) - u_i(\mathbf{n}; \mathbf{m}, \boldsymbol{\kappa}) \tag{18}$$

We may write the transition rates (3,5) and (3,6) in their final form:

$$w_k^\pm(\mathbf{m}; \mathbf{n}, \boldsymbol{\kappa}) = \mu_k(\mathbf{m}_k^\pm, \mathbf{m}; \mathbf{n}, \boldsymbol{\kappa}) \exp\{u_k(\mathbf{m}_k^\pm : \mathbf{n}, \boldsymbol{\kappa}) - u_k(\mathbf{m}; \mathbf{n}, \boldsymbol{\kappa})\} \tag{3.19a}$$

$$w_k^\mp(\mathbf{m}_k^\pm; \mathbf{n}, \boldsymbol{\kappa}) = \mu_k(\mathbf{m}, \mathbf{m}_k^\pm; \mathbf{n}, \boldsymbol{\kappa}) \exp\{u_k(\mathbf{m}; \mathbf{n}, \boldsymbol{\kappa})\} - u_k(\mathbf{m}_k^\pm; \mathbf{n}, \boldsymbol{\kappa})\} \tag{3.19b}$$

and

$$w_{ji}^\alpha(\mathbf{n}_{ji}^\alpha, \mathbf{n}; \mathbf{m}, \boldsymbol{\kappa}) = \nu_{ji}^\alpha(\mathbf{n}_{ji}^\alpha, \mathbf{n}; \mathbf{m}, \boldsymbol{\kappa}) \exp\{u_j^\alpha(\mathbf{n}_{ji}^\alpha; \mathbf{m}, \boldsymbol{\kappa}) - u_i^\alpha(\mathbf{n}; \mathbf{m}, \boldsymbol{\kappa})\} \tag{3.20a}$$

$$w_{ij}^\alpha(\mathbf{n}, \mathbf{n}_{ji}^\alpha; \mathbf{m}, \boldsymbol{\kappa}) = \nu_{ji}^\alpha(\mathbf{n}_{ji}^\alpha, \mathbf{n}; \mathbf{m}, \boldsymbol{\kappa}) \exp\{u_i^\alpha(\mathbf{n}; \mathbf{m}, \boldsymbol{\kappa}) - u_j^\alpha(\mathbf{n}_{ji}^\alpha; \mathbf{m}, \boldsymbol{\kappa})\} \tag{3.20b}$$

Interpretation of the Transition Rates

On the one side the form (3.19) and (3.20) of the transition rates for elementary changes of the values of macrovariables ist still almost fully general. On the other side it lends itself to a plausible interpretation:

The first factors μ_k and ν_{ji}^α of the rates (3.19) and (3.20) are symmetrical in the initial and final states and variables. They describe the flexibility or mobility of the change and give rise to the frequency by which the transitions take place.

The second factors depend exponentially on differences of functions u of the final states and variables after the transition and the initial states and variables before the transition. Therefore it is highly plausible to interpret the function $u_k(\mathbf{m}; \mathbf{n}, \kappa)$ as a measure for the utility of the configuration $(\mathbf{m}; \mathbf{n}, \kappa)$ seen by the decision makers in view of a transition $\mathbf{m} \rightarrow \mathbf{m}_k^{\pm}$. Similarly, $u_i^\alpha(\mathbf{n}, \mathbf{m}, \kappa)$ can be interpreted as a measure for the utility of the configuration $(\mathbf{n}; \mathbf{m})$ estimated by a member of subpopulation p_α being in state i. If this member of p_α considers a transition to another state j, this transition would lead not only to new macrovariables $(n_{ji}^\alpha; \mathbf{m})$ but also to a new utility $u_j^\alpha(n_{ji}^\alpha; \mathbf{m}, \kappa)$ for him/her.

Evidently, transitions are favoured (disfavoured) due to the form of the transition rates (3.19) and (3.20), if the utility of the final configuration and state is higher (lower) than the utility of the initial configuration and state.

The transition processes described by (3.20) between states of individuals - and simultaneously between different socioconfigurations - can be considered as generalized migration processes. Of course they include real migration processes, too, if the states $i, j, \ldots$ are identified as places of residence for a member of population p_α. In Sect. 4 this case will be explicitly discussed.

In view of this migratory interpretation, related but equivalent interpretations can be given to the mobility and utility terms of the transition rates (3.20). Thus, one may write the mobility $\nu_{ij}^\alpha = \nu_{ji}^\alpha$ in the form

$$\nu_{ij}^\alpha(n_{ji}^\alpha, \mathbf{n}; \mathbf{m}, \kappa) = \nu_0 \exp\{-d_{ij}(n_{ji}^\alpha, \mathbf{n}; \mathbf{m}, \kappa)\} \tag{21}$$

and interpret d_{ij}^α as a measure for the (generalised) distance between states i and j to be overcome in the intended transition from i to j or vice versa. Furthermore, the utility term in (3.20a) may be factorised into:

$$\mathrm{pull}_j^\alpha(n_{ji}^\alpha; \mathbf{m}, \kappa) = \exp\{u_j^\alpha(n_{ji}^\alpha; \mathbf{m}, \kappa)\} \tag{3.22a}$$

and

$$\mathrm{push}_i^\alpha(\mathbf{n}; \mathbf{m}, \kappa) = \exp\{-u_i^\alpha(\mathbf{n}; \mathbf{m}, \kappa)\} \tag{3.22b}$$

where evidently the pull-term pulls in case of large u_j^α the individual into the new state j and the push-term pushes in case of small or even negative u_i^α the individual away from the old state i.

Whereas we could give interpretations for the general terms of the transition rates it is not possible to "derive" the explicit form of mobilities and utilities. Appropriate forms can only be found for special models taking into account general trends and the mentality of the individuals. In this choice process the "feed back loop between qualitive and quantitive argumentation" plays a decisive role.

3.3   Step 3: Evolution Equations for the Macrovariables

The Master Equation

The transition rates (3.19) and (3.20) for the configuration of material and social macrovariables are sufficent for setting up the central evolution equation of sociodynamics, the master equation for the probability distribution over the macrovariables $\{\mathbf{m}, \mathbf{n}\}$.

Let the distribution function

$$P(\mathbf{m}, \mathbf{n}; t) \geq 0 \tag{23}$$

which is normalized by

$$\sum_{\mathbf{n}, \mathbf{m}} P(\mathbf{m}, \mathbf{n}; t) = 1 \tag{24}$$

be the probability to find the macroconfiguration $\{\mathbf{m}, \mathbf{n}\}$ at time t.

If the only nonvanishing transition rates are the next-neighbour rates (3.19) and (3.20), the *master equation* for $P(\mathbf{m}, \mathbf{n}; t)$ reads:

$$\frac{dP(\mathbf{m}, \mathbf{n}; t)}{dt} = -\sum_{k} \{w_k^+(\mathbf{m}, \mathbf{n}; t)P(\mathbf{m}, \mathbf{n}; t) + w_k^-(\mathbf{m}, \mathbf{n}; t)P(\mathbf{m}, \mathbf{n}; t)\} \tag{25}$$

$$+ \sum_{k} \{w_k^-(\mathbf{m}_k^+, \mathbf{n}; t)P(\mathbf{m}_k^+, \mathbf{n}; t) + w_k^+(\mathbf{m}_k^-, \mathbf{n}; t)P(\mathbf{m}_k^-, \mathbf{n}; t)\}$$

$$- \sum_{i,j,\alpha} \{w_{ji}^\alpha(\mathbf{n}_{ji}^\alpha, \mathbf{n}; \mathbf{m})P(\mathbf{m}, \mathbf{n}; t)\}$$

$$+ \sum_{j,i,\alpha} \{w_{ij}^\alpha(\mathbf{n}, \mathbf{n}_{ji}^\alpha; \mathbf{m})P(\mathbf{m}, \mathbf{n}_{ji}^\alpha; t)\}$$

It describes the change of the probability of configuration $\{\mathbf{m}, \mathbf{n}\}$ in terms of probability flows *from* $\{\mathbf{m}, \mathbf{n}\}$ *into* neighboring configurations $\{\mathbf{m}_k^\pm, \mathbf{n}\}$ and $\{\mathbf{m}, \mathbf{n}_{ji}^\alpha\}$ (see first and third line of the r.h.s. of (3.25) and from neighboring configurations $\{\mathbf{m}_k^\pm, \mathbf{n}\}$ and $\{\mathbf{m}, \mathbf{n}_{ji}^\alpha\}$ into $\{\mathbf{m}, \mathbf{n}\}$ (see second and fourth line of the r.h.s. of (3.25). The first two lines of the r.h.s. describe the probability changes due to transitions of the material configuration and the last two lines of the r.h.s. belong to probability changes due to transitions within the socioconfiguration.

Equations for Mean Values and Quasi-meanvalues

The meanvalues of functions of macrovariables are defined by

$$\langle f(\mathbf{m}, \mathbf{n}) \rangle_t = \sum_{\mathbf{m}, \mathbf{n}} f(\mathbf{m}, \mathbf{n}) P(\mathbf{m}, \mathbf{n}; t) \tag{26}$$

It is easy to derive equations of motion for the meanvalues $\langle m_k \rangle_t$ and $\langle n_i^\alpha \rangle_t$ of m_k and n_i^α from the master equation. They read

$$\frac{d\langle m_k \rangle_t}{dt} = \langle w_k^+(\mathbf{m}, \mathbf{n}) \rangle_t - \langle w_k^-(\mathbf{m}, \mathbf{n}) \rangle_t \tag{27}$$

$$\frac{d\langle n_i^\alpha \rangle_t}{dt} = \sum_j \langle w_{ij}^\alpha(\mathbf{m}, \mathbf{n}) \rangle_t - \sum_j \langle w_{ji}^\alpha(\mathbf{m}, \mathbf{n}) \rangle_t \tag{28}$$

These equations are exact but not closed, since the r.h.s. of (3.27) and (3.28) contains mean values of in general nonlinear functions of $\mathbf{m}$ and $\mathbf{n}$.

Only if the probability distribution $P(\mathbf{m}, \mathbf{n}; t)$ is unimodal and sharply peaked, the meanvalues of functions of $\mathbf{m}$ and $\mathbf{n}$ will approximately agree with the same functions of meanvalues $\langle \mathbf{m} \rangle_t$ and $\langle \mathbf{n} \rangle_t$. Then eqs. (3.27) and (3.28) are approximately closed. However, in case of multimodal probability distributions the meanvalues are no longer informative quantities because they lie between the peaks of the probability distribution and do not even approximately agree with characteristic trajectories $\{\mathbf{m}(t), \mathbf{n}(t)\}$ of the macrovariables.

On the other hand it can be shown, that *quasi-meanvalues* $\widehat{m}_k(t), \widehat{n}_i^\alpha(t)$, which by definition fulfil the closed autonomous and in general nonlinear differential equations

$$\frac{d\widehat{m}_k(t)}{dt} = w_k^+(\widehat{\mathbf{m}}, \widehat{\mathbf{n}}) - w_k^-(\widehat{\mathbf{m}}, \widehat{\mathbf{n}}) \tag{29}$$

$$\frac{d\widehat{n}_i^\alpha(t)}{dt} = \sum_j w_{ij}^\alpha(\widehat{\mathbf{m}}, \widehat{\mathbf{n}}) - \sum_j w_{ji}^\alpha(\widehat{\mathbf{m}}, \widehat{\mathbf{n}}) \tag{30}$$

represent the characteristics of the system-trajectories. Furthermore it can be shown, that the stable equilibrium points of (3.29), (3.30) coincide with the maxima of the stationary probability distribution $P_{st}(\mathbf{m}, \mathbf{n})$, if the transition rates fulfil the condition of detailed balance.

The fact, that the quasi-meanvalue equations (3.29), (3.30) and *not* the meanvalue equations (3.27), (3.28) describe in good (but smoothed out) approximation the true stochastic trajectories of which the ensemble consists, has a simple explanation: The true system point is hopping stochastically from cell $\{\mathbf{m}, \mathbf{n}\}$ to cell $\{\mathbf{m}', \mathbf{n}'\}$ in the configuration space via the probabilistic transition rates. The magnitude of these rates depends on the momentarily reached cell and its corresponding values $\{\mathbf{m}, \mathbf{n}\}$ of the macrovariables, but it does **not** depend on **meanvalues** of these rates taken over the whole - perhaps multimodal - momentary probability distribution. This fact is just captured by the quasi-meanvalue equations (3.29), (3.30) but it is missed by the (exact) meanvalue equations (3.27), (3.28) which instead describe the evolution of meanvalues taken over the whole — perhaps multimodal — probability distribution. In the case of multimodal probability distributions these meanvalues do not at all coincide with the true trajectories but lie inbetween them.

In general not enough empirical data are available to compare them with the full information contained in the probability distribution $P(\mathbf{m}, \mathbf{n}; t)$. Therefore the analysis of a model usually takes place in terms of the quasi-meanvalue equations associated with the master equation.

4 Examples of Sociodynamic Models

So far the modelling procedure sketched in the previous section has been applied to dynamic processes belonging to different sectors of the social sciences. We mention migration models [2] belonging to population dynamics, models of group dynamics [4] and of collective political opinion formation [5] belonging to sociology, models of competition between firms [6] and the dynamics of fashion demand [7] belonging to economics and models about settlement formation [8], [9] and urban dynamics [10], [11] belonging to regional science.

In the case of migration the modelling principles could be empirically validated by an extensive comparative investigation of interregional migration within six countries [12]. More recently complex processes of the interrelated evolution of traffic and cities have been investigated by integrated models on the basis of the same modelling principles [13].

Here we shall only exhibit the simplest, although nontrivial version of two models. The first model describes the migration of interacting populations. In this case the dynamic variables are only personal variables. The second model describes the evolution of cities on the level of sites on which different kinds of buildings can be erected. In this case only material variables are taken into account.

4.1 Example 1: Migration of Interacting Populations Between Regions

The masterequation and the quasi-meanvalue equations for the general case of P interacting populations between L regions can easily be formulated and has been extensively investigated. We treat the simplest nontrivial case of two populations p_μ, p_ν in two regions $i = 1, 2$ only, because already this model is able to describe three essentially different migratory situations and the phasetransitions between them, namely

1. the homogeneous mixing of p_μ and p_ν
2. the ghetto formation between p_μ and p_ν
3. the invasion-evasion migration between p_μ and p_ν

1.) Choice of Variables

Each individual is assumed to belong to one of two populations p_μ, p_ν with $2M$ and $2N$ members, respectively. The states or attitudes $i = 1, 2$ of the individuals are identified with "living in region i". The socio configuration consists of four variables

$$\mathbf{n} = \{n_1^\mu, n_2^\mu, n_1^\nu, n_2^\nu\} \tag{31}$$

where n_i^α is the number of members of p_α living in region i. Neglecting birth/death - processes, there hold the "conservation laws"

$$n_1^\mu + n_2^\mu = 2M \quad ; \quad n_1^\nu + n_2^\nu = 2N \tag{32}$$

so that only

$$m = \frac{1}{2}(n_1^\mu - n_2^\mu) \quad ; \quad -M \leq m \leq +M \tag{33}$$

$$n = \frac{1}{2}(n_1^\nu - n_2^\nu) \quad ; \quad -N \leq n \leq +N$$

are the relevant time-dependent variables.

2.) Transition Rates

$$w_{ji}^\mu(\mathbf{n}_{ji}^\mu; \mathbf{n}) \equiv w_{ji}^\mu(\mathbf{n}) = n_i^\mu p_{ji}^\mu \tag{34}$$
$$w_{ji}^\nu(\mathbf{n}_{ji}^\nu; \mathbf{n}) \equiv w_{ji}^\nu(\mathbf{n}) = n_i^\nu p_{ji}^\nu$$

with

$$p_{ji}^\mu = \tilde{\nu}_\mu \exp[u_j^\mu(n_j^\mu + 1, n_j^\nu) - u_i^\mu(n_i^\mu, n_i^\nu)] \tag{35}$$
$$p_{ji}^\nu = \tilde{\nu}_\nu \exp[u_j^\nu(n_j^\mu + 1, n_j^\nu) - u_i^\nu(n_i^\mu, n_i^\nu)]$$

3.) Choice of Utility Functions

We assume that the two regions $i = 1, 2$ are equivalent; that means that one region hat no natural preference before the other. Then the utility u_i^α of region i for a member of p_α can only depend on the number of members of the own population and the other population in region i. The simplest ansatz for this dependence is the linear one. Hence we put

$$u_i^\alpha(n_i^\mu, n_i^\nu) = \frac{1}{2}(\kappa_{\alpha\mu}n_i^\mu + \kappa_{\alpha\nu}n_i^\nu) \tag{36}$$

where the $\kappa_{\alpha\mu}$ and $\kappa_{\alpha\nu}$ ($\alpha = \mu, \nu$) are trend parameters describing the "social attitude" of the members of p_α concerning their relation to the members of p_μ and p_ν as far as this relation influences their migratory behaviour.

Inserting (4.6) into (4.5), and (4.5) into (4.4) and making use of the variables (4.3) one now obtains the explicit transition rates for the relevant variables (m, n):

$$w_{12}^\mu \equiv w_\uparrow^\mu(m, n) = \nu_\mu(M - m)\exp[\Delta u_\mu(m, n)] \tag{37}$$
$$\text{for}(m, n) \to (m + 1, n)$$
$$w_{21}^\mu \equiv w_\downarrow^\mu(m, n) = \nu_\mu(M + m)\exp[-\Delta u_\mu(m, n)]$$
$$\text{for}(m, n) \to (m - 1, n)$$
$$w_{12}^\nu \equiv w_\uparrow^\nu(m, n) = \nu_\nu(N - n)\exp[\Delta u_\nu(m, n)]$$
$$\text{for}(m, n) \to (m, n + 1)$$
$$w_{21}^\nu \equiv w_\downarrow^\nu(m, n) = \nu_\nu(N + n)\exp[-\Delta u_\nu(m, n)]$$
$$\text{for}(m, n) \to (m, n - 1)$$

with

$$\Delta u_\mu(m, n) = \kappa_{\mu\mu}m + \kappa_{\mu\nu}n \tag{38}$$
$$\Delta u_\nu(m, n) = \kappa_{\nu\mu}m + \kappa_{\nu\nu}n$$
$$\nu_\mu = \tilde{\nu}_\mu\exp(\frac{1}{2}\kappa_{\mu\mu}) \quad, \quad \nu_\nu = \tilde{\nu}_\nu\exp(\frac{1}{2}\kappa_{\nu\nu})$$

4.) Evolution Equations

Specialising eq. (3.25) to the present case which only contains personal variables, one obtains in straight forward manner the master equation

$$\frac{dP(m, n; t)}{dt} = -\{(w_\uparrow^\mu(m, n) + w_\downarrow^\mu(m, n) + w_\uparrow^\nu(m, n) + w_\downarrow^\nu(m, n))P(m, n; t)\}$$
$$+\{w_\uparrow^\mu(m - 1, n)P(m - 1, n; t) + w_\downarrow^\mu(m + 1, n)P(m + 1, n; t)$$
$$+w_\uparrow^\nu(m, n - 1)P(m, n - 1; t) + w_\downarrow^\nu(m, n + 1)P(m, n + 1; t)\}$$
$$\tag{39}$$

and by specialisation of (3.30) there follow the quasi-meanvalue equations

$$\frac{d\widehat{m}}{dt} = w_\uparrow^\mu(\widehat{m}, \widehat{n}) - w_\downarrow^\mu(\widehat{m}, \widehat{n}) \tag{40}$$
$$= 2\nu_\mu \{M \sinh[\Delta u_\mu(\widehat{m}, \widehat{n})] - \widehat{m} \cosh[\Delta u_\mu(\widehat{m}, \widehat{n})]\}$$
$$\frac{d\widehat{n}}{dt} = w_\uparrow^\nu(\widehat{m}, \widehat{n}) - w_\downarrow^\nu(\widehat{m}, \widehat{n})$$
$$= 2\nu_\nu \{N \sinh[\Delta u_\nu(\widehat{m}, \widehat{n})] - \widehat{n} \cosh[\Delta u_\nu(\widehat{m}, \widehat{n})]\}$$

Making the special assumptions

$$M = N \quad ; \quad \kappa_{\mu\mu} = \kappa_{\nu\nu} = \kappa \quad ; \quad \nu_\mu = \nu_\nu = \nu \tag{41}$$

and going over to scaled parameters and variables

$$\tilde{\kappa}_{\alpha\beta} = \kappa_{\alpha\beta} N \quad ; \quad \tau = 2\nu t \quad ; \quad x = \frac{\widehat{m}}{N} \quad ; \quad y = \frac{\widehat{n}}{N} \tag{42}$$

one obtains the scaled quasi-meanvalue equations:

$$\frac{dx}{d\tau} = \sinh(\tilde{\kappa}x + \tilde{\kappa}_{\mu\nu}y) - x \cosh(\tilde{\kappa}x + \tilde{\kappa}_{\mu\nu}y) \tag{43}$$
$$\frac{dy}{d\tau} = \sinh(\tilde{\kappa}y + \tilde{\kappa}_{\nu\mu}x) - y \cosh(\tilde{\kappa}y + \tilde{\kappa}_{\nu\mu}x)$$

5.) Selected Simulations of Scenarios

At first the effect of the parameters must be clear. This effect follows from the form (4.7) of the transitions rates. It turns out, that a large value of the "internal trend parameter" $\kappa = \kappa_{\mu\mu} = \kappa_{\nu\nu}$ favours the agglomeration of the members of one population in the same region; that means, $\kappa > 0$ is a measure for the trend of people to live together with people of their own. On the other hand, a positive "external trend parameter" $\kappa_{\alpha\beta}$ with $\alpha \neq \beta$ describes that people of p_α "like" to live together with people of p_β (and prefer to move to a region where more members of p_β live). A negative external trend parameter $\kappa_{\alpha\beta}$ describes the opposite trend, that members of p_α "dislike" to live together with members of p_β (and prefer to move to a region where less members of p_β live). Because of socio-cultural distances between different populations, negative values of $\kappa_{\alpha\beta}$, with $\beta \neq \alpha$, will be the more frequent case, whereas $\kappa = \kappa_{\mu\mu} = \kappa_{\nu\nu}$ will usually be positive.

In the following we depict the fluxlines of the scaled quasi-meanvalue equations (4.13) in four characterisitic cases:

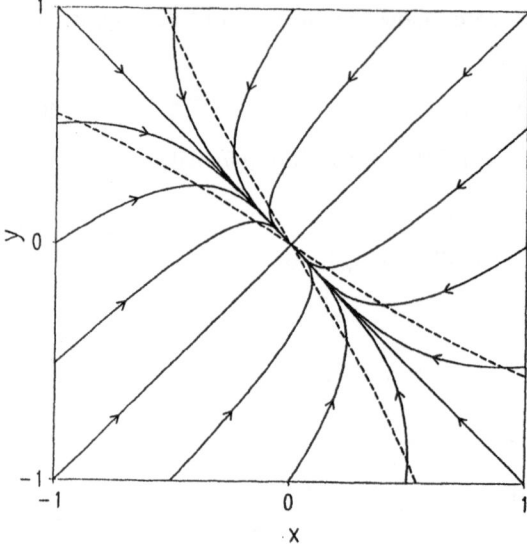

Fig. 1. Weak internal agglomeration trend $\tilde{\kappa} = \tilde{\kappa}_{\mu\mu} = \tilde{\kappa}_{\nu\nu} = 0.2$ and weak symmetrical external segragation trend $\tilde{\kappa}_{\mu\nu} = \tilde{\kappa}_{\nu\mu} = 0.5$. The result is one stable stationary point (0.0) describing the homogeneous mixture of populations p_μ and p_ν. All fluxlines approach the origin (0.0).

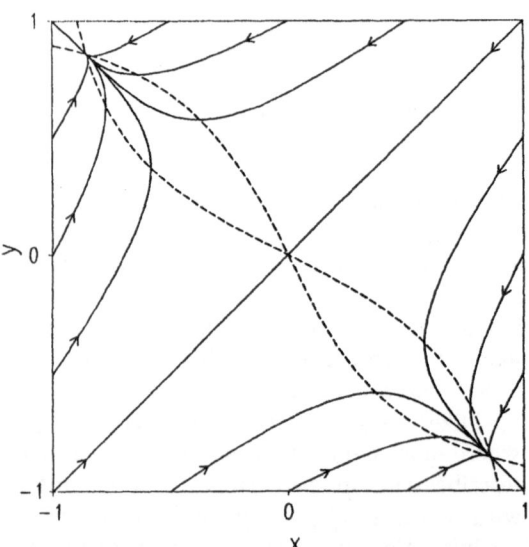

Fig. 2. Moderate internal agglomeration trend $\tilde{\kappa} = \tilde{\kappa}_{\mu\mu} = \tilde{\kappa}_{\nu\nu} = 0.5$ and strong symmetrial external segregation trend $\tilde{\kappa}_{\mu\nu} = \tilde{\kappa}_{\nu\mu} = -1.0$. Two stable stationary points in the second and fourth quadrants desribe the segregation of populations p_μ and p_ν in seperate "ghettos". The flux lines approach one of these stable equilibrium points.

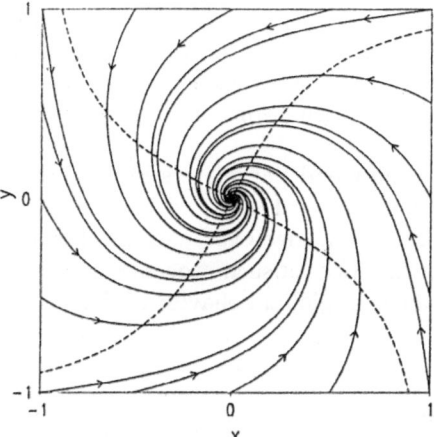

Fig. 3. Moderate internal agglomeration trend $\tilde{\kappa} = \tilde{\kappa}_{\mu\mu} = \tilde{\kappa}_{\nu\nu} = 0.5$ and strong asymmetric external trends $\tilde{\kappa}_{\nu\mu} = +1.0$; $\tilde{\kappa}_{\mu\nu} = -1.0$. There exists a stable focus (0.0) describing the homogeneous mixture of populations p_μ and p_ν. All flux lines spiral into the focus (0.0).

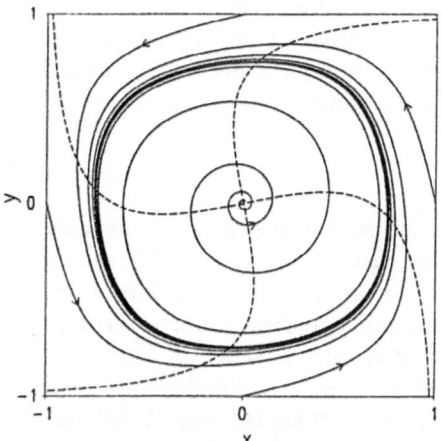

Fig. 4. Very strong internal agglomeration trend $\kappa = \kappa_{\mu\mu} = \kappa_{\nu\nu} = 1.2$ and strong asymmetric external trends $\tilde{\kappa}_{\nu\mu} = +1.0$; $\tilde{\kappa}_{\mu\nu} = -1.0$. The homogeneous population mixture (0.0) is now unstable. All fluxlines approach a limit cycle. It desribes a migration process never coming to rest and consisting of sequential evasion -invasion processes. This can be seen as follows: The limit cycle traverses all four quadrants. In quadrant 1 the majorities of populations p_ν and p_μ live together in region 1; the arrival at quadrant 2 means, that the majority of population p_μ has now settled in region 2 (evading population p_ν because of the segregation trend $\tilde{\kappa}_{\mu\nu} = -1.0$). Proceeding to quadrant 3 means, that population p_ν has invaded population p_μ (because of the external agglomeration trend $\tilde{\kappa}_{\nu\mu} = +1.0$) and that the majority of both populations is now settling in region 2; the approach of quadrant 4 means, that again population p_μ evades population p_ν by now settling in region 1.

Although the process described in Fig. 4 ist oversimplified due to only two available regions, it describes in principle correctly a process observed in some multicultural cities, namely the sequential erosion of quarters by migratory invasion-evasion processes between populations of different socio–cultural origin.

It is plausible that in the course of time trend parameters like the $\kappa_{\alpha\beta}$ undergo slow changes due to complex socio-cultural and political developments. Models of the kind presented here do not give explanations of these processes. They do however explain, that the effect of such processes can be a migratory phase transition from one to another global behavior of the social system.

4.2 Example 2: Evolution of Cities

Cities belong to the most complex systems observed in the human society. In order to obtain their simplified description by a mathematical model an appropriate "window of perception" must be chosen. The macrovariables providing this perspective will be material ones characterising the state of a city, whereas the activities of the decision makers (municipal authorities, house builders, architects) show up indirectly only in the evolution of this state.

However in setting up the model we follow the same general procedure composing the global evolution out of elementary trend–guided stochastic steps. This can be validated because also planning processes in general divide into several decision steps influenced by competing trends and the varying "Zeitgeist".

1.) Choice of Variables

On a sufficiently detailed level of description the state of a city can be characterised by the number, location and distribution of the different kinds of buildings fulfilling different purposes. In order to describe this state we introduce a certain kind of "occupation number representation" of the city.

At first the area of the city is tesselated into a lattice of "squares" or "sites" with discrete coordinates $i(i_1, i_2), j(j_1, j_2), \ldots$. For later purposes it is necessary to introduce the distance $d(i, j)$ between squares. A convenient measure for this distance is the "Manhattan metric", in particular in the case of a cartesian square lattice:

$$d(i,j) = |\, i_1 - j_1\,| + |\, i_2 - j_2\,| \tag{44}$$

Let now $x_i, y_i, \ldots$ denote the (integer) number of building units of kind x (lodgings), kind y (factories), and so on, on square $i(i_1, i_2)$. Further kinds of buildings could be service stations, schools, store houses, parks, etc. For the sake of simplicity, however, we here only consider lodgings (x) and factories (y).

The set of numbers of building units of all kinds on all squares

$$(\mathbf{x}, \mathbf{y}, \dots) = \{\dots, (x_i, y_i, \dots), \dots\} \tag{45}$$

is denoted as *city configuration*. The dynamics of this set of material variables must now be modelled.

2.) Transition Rates
The step by step evolution of the city configuration consists of transitions from $\{\mathbf{x}, \mathbf{y}\}$ to neighboring configurations:

$$\{\mathbf{x}, \mathbf{y}\} \Rightarrow \{\mathbf{x}_{j\pm}, \mathbf{y}\} = \{\dots (x_j \pm 1, y_j), \dots\} \tag{46}$$
$$\{\mathbf{x}, \mathbf{y}\} \Rightarrow \{\mathbf{x}, \mathbf{y}_{j\pm}\} = \{\dots (x_j, y_j \pm 1), \dots\}$$

and the corresponding transition rates for lodgings (x) and factories (y), respectively, have the general form introduced in (3.19) in the case of building up steps at square j:

$$w_{j\uparrow}^{(x)}(\mathbf{x}, \mathbf{y}) = \nu_\uparrow^{(x)} \exp\{\Delta_{j+}^{(x)} u(\mathbf{x}, \mathbf{y})\} \tag{47}$$
$$w_{j\uparrow}^{(y)}(\mathbf{x}, \mathbf{y}) = \nu_\uparrow^{(y)} \exp\{\Delta_{j+}^{(y)} u(\mathbf{x}, \mathbf{y})\}$$

and in the case of tearing down steps at square j:

$$w_{j\downarrow}^{(x)}(\mathbf{x}, \mathbf{y}) = \nu_\downarrow^{(x)} \exp\{\Delta_{j-}^{(x)} u(\mathbf{x}, \mathbf{y})\} \tag{48}$$
$$w_{j\downarrow}^{(y)}(\mathbf{x}, \mathbf{y}) = \nu_\downarrow^{(y)} \exp\{\Delta_{j-}^{(y)} u(\mathbf{x}, \mathbf{y})\}$$

where the expressions

$$\Delta_{j\pm}^{(x)} u(\mathbf{x}, \mathbf{y}) = u(\mathbf{x}_{j\pm}, \mathbf{y}) - u(\mathbf{x}, \mathbf{y}) \tag{49}$$
$$\Delta_{j\pm}^{(y)} u(\mathbf{x}, \mathbf{y}) = u(\mathbf{x}, \mathbf{y}_{j\pm}) - u(\mathbf{x}, \mathbf{y})$$

are differences of the utilities of the city configuration after and before the transition step.

3.) Choice of Utility Functions
The utility functions of city configurations entering the transition rates (4.17) and (4.18) must now be specified.

The ansatz for the utility $u(\mathbf{x}, \mathbf{y})$ comprises terms designed to describe the main effects influencing this utility. It is assumed to consist of two terms

$$u(\mathbf{x}, \mathbf{y}) = u_L(\mathbf{x}, \mathbf{y}) + u_I(\mathbf{x}, \mathbf{y}) \tag{50}$$

where the local term $u_L(\mathbf{x}, \mathbf{y})$ consists of a sum of utilities of the local configuration (x_j, y_j) on each site j:

$$u_L(\mathbf{x}, \mathbf{y}) = \sum_j u_j(x_j, y_j) \tag{51}$$

and the interaction term $u_I(\mathbf{x}, \mathbf{y})$ describes the –supportive or suppressive– effect of the distribution over the sites i, j, on the utility of the total configuration.

Its form is chosen as follows:

$$u_L(\mathbf{x}, \mathbf{y}) = \sum_{i,j} a_{ij}^{xx} x_i x_j + \sum_{i,j} a_{ij}^{xy} x_i y_j + \sum_{i,j} a_{ij}^{yy} y_i y_j \tag{52}$$

The case of negative a_{ij}^{xy} for $d(i, j) < d_0$ and positive a_{ij}^{xy} for $d(i, j) \geq d_0$ means, for instance, that it is unfavourable to build lodgings and factories in near neighborhood $d(i, j) < d_0$, but that it leads to a high utility to have them at distances $d(i, j) \geq d_0$.

The local utilities $u_j(x_j, y_j)$ are chosen to have the form

$$u_j(x_j, y_j) = p_j^{(x)} \ln(x_j + 1) + p_j^{(y)} \ln(y_j + 1) + p_j^{(z)} \ln(z_j) \tag{53}$$

where $p_j^{(k)}$ are measures for preferences to erect buildings of kind k on site j, and z_j is the empty disposable space on site j, if the opened up building capacity on site j is given by

$$c_j = x_j + y_j + z_j \tag{54}$$

This opened up capacity c_j on site j depends on the other hand on its situation relative to the center j_0 of the city. A simple assumption about c_j is a Gaussian capacity distribution:

$$c_j = c_0 \exp\left[-\frac{d^2(j, j_0)}{2\sigma^2(n_c)} \right] \tag{55}$$

where the width $\sigma(n_c)$ of this distribution - hence the size of the city - will depend on the city population number .

4.) Evolution Equations for City Configurations

The transition rates (4.17) and (4.18), which are explicit functions of the configuration $(\mathbf{x}, \mathbf{y})$ after inserting the utility functions (4.20), can now be used to set up in straight forward manner the master equation for the probability distribution $P(\mathbf{x}, \mathbf{y}; t)$ over the space of city configurations $(\mathbf{x}, \mathbf{y})$. It reads:

$$\frac{dP(\mathbf{x}, \mathbf{y}; t)}{dt} = \sum_j [w_{j\uparrow}^{(x)}(\mathbf{x}_{j-}, \mathbf{y})P(\mathbf{x}_{j-}, \mathbf{y}; t) - w_{j\uparrow}^{(x)}(\mathbf{x}, \mathbf{y})P(\mathbf{x}, \mathbf{y}; t)] \quad (56)$$

$$+ \sum_j [w_{j\downarrow}^{(x)}(\mathbf{x}_{j+}, \mathbf{y})P(\mathbf{x}_{j+}, \mathbf{y}; t) - w_{j\downarrow}^{(x)}(\mathbf{x}, \mathbf{y})P(\mathbf{x}, \mathbf{y}; t)]$$

$$+ \sum_j [w_{j\uparrow}^{(y)}(\mathbf{x}, \mathbf{y}_{j-})P(\mathbf{x}, \mathbf{y}_{j-}; t) - w_{j\uparrow}^{(y)}(\mathbf{x}, \mathbf{y})P(\mathbf{x}, \mathbf{y}; t)]$$

$$+ \sum_j [w_{j\downarrow}^{(y)}(\mathbf{x}, \mathbf{y}_{j+})P(\mathbf{x}, \mathbf{y}_{j+}; t) - w_{j\downarrow}^{(y)}(\mathbf{x}, \mathbf{y})P(\mathbf{x}, \mathbf{y}; t)]$$

proceeding as in Sect. 3), the closed, autonomous nonlinear system of quasi-meanvalue equations can be deduced:

$$\frac{d\widehat{x}_j(t)}{dt} = w_{j\uparrow}^{(x)}(\widehat{\mathbf{x}}(t), \widehat{\mathbf{y}}(t)) - w_{j\downarrow}^{(x)}(\widehat{\mathbf{x}}(t), \widehat{\mathbf{y}}(t)) \quad (57)$$

$$\frac{d\widehat{y}_j(t)}{dt} = w_{j\uparrow}^{(y)}(\widehat{\mathbf{x}}(t), \widehat{\mathbf{y}}(t)) - w_{j\downarrow}^{(y)}(\widehat{\mathbf{x}}(t), \widehat{\mathbf{y}}(t)) \quad (58)$$

5.) Stationary City Scenarios

For a given constant capacity distribution the evolution of the city configuration approaches a stationary state. In particular the quasi-meanvalue-equations reach one out of in general several possible stationary states. It depends on the initial conditions, which of the stationary states is eventually approached.

The interaction term in the utility functions has lead to a separation between residential and industrial districts within the urban area. (see Figs. 5-7)

The two examples show that the general modelling procedure may - after appropriate choice and implementation of the details - be applied to very different kinds of social dynamics and may perhaps provide the elements of a general modelling strategy in quantitative social science.

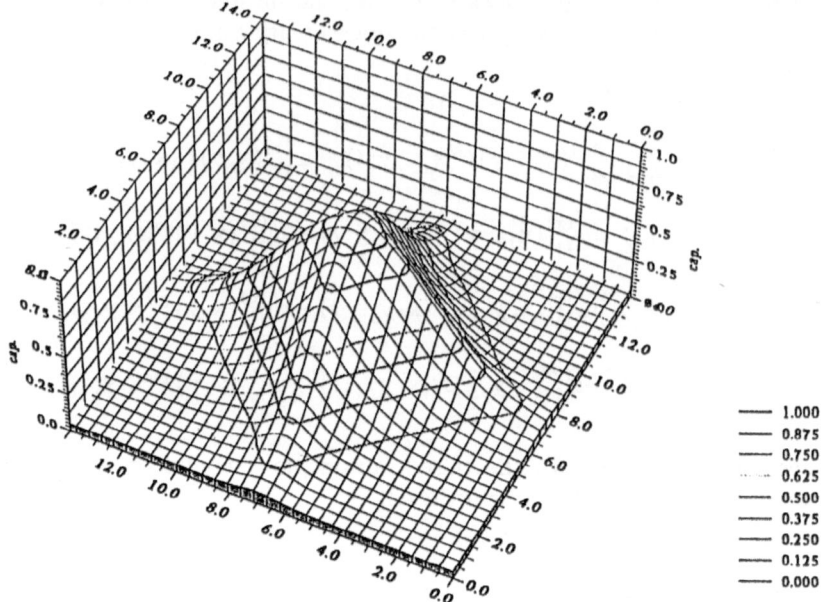

Fig. 5. Figure 5 shows a Gaussian capacity distribution over a city area tessellated into a lattice of 13 x 13 = 169 squares.

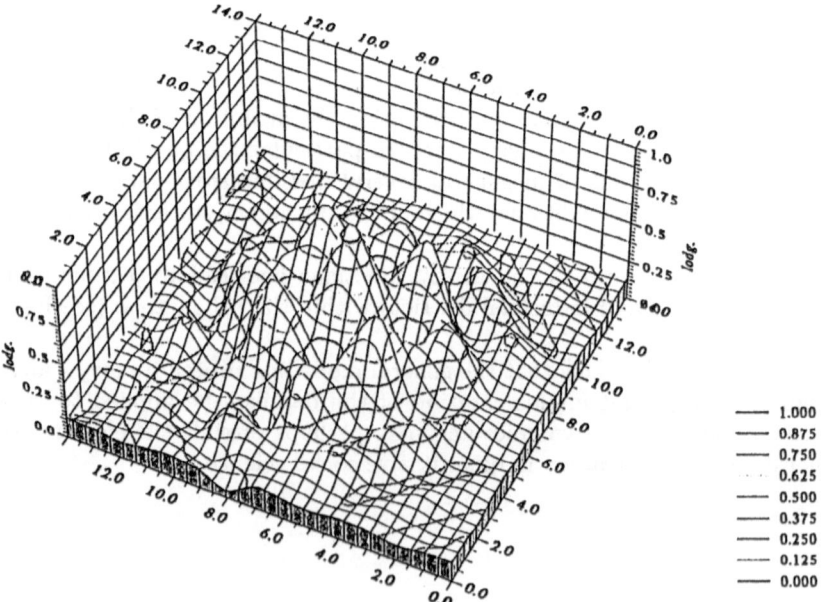

Fig. 6. Figure 6 exhibits - after appropriate calibrations of all parameters - a stationary solution of the quasi-meanvalue equations (4.27) for lodgings x_j on the same lattice of sites.

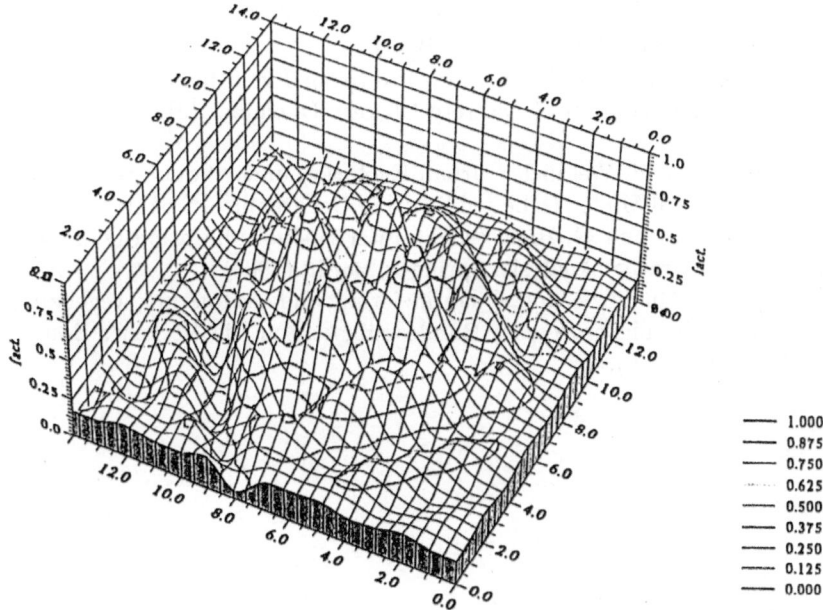

Fig. 7. Figure 7 exhibits the corresponding stationary solution of the quasi-meanvalue equations (4.28) for factories on this system of sites.

References

(1) H. Haken : Synergetics - An Introduction. Springer, Berlin (1978)

(2) W. Weidlich : Physics and Social Science - The Approach of Synergetics. Physics Reports 204, pp 1 - 163 (1991)

(3) D. Helbing : Quantitative Sociodynamics Kluver Academic Publishers (1995)

(4) W. Weidlich and D. Helbing : A Mathematical Model of Group Dynamics Including the Effects of Solidarity. Chapter 5 in "The problem of Solidarity: Theory and Models" Eds. T.J.Fararo and P. Doreias, Gordon and Breach, to appear 1997

(5) W. Weidlich : Synergetic Modelling Concepts for Sociodynamics with Applications to Collective Political Opion Formation Journal of Mathematical Sociology 18, pp 267-291 (1994)

(6) W. Weidlich and M. Braun : The Master Equation Approach to Nonlinear Economics Journal of Evolutionary Economics 2, pp 233-265 (1992)

(7) W. Weidlich and T. Brenner : Dynamics of Demand Including Fashion Effects for Interacting Consumer Groups, Studien zur Evolutorischen Ökonomik III, pp 79-115; A. Wagner and H.W. Lorenz (eds), Duncker & Humblat (Berlin) (1995)

(8) W. Weidlich and M. Munz : Settlement Formation I, A Dynamic Theory, Ann. Reg. Sci. 24, pp 83-106 (1990)

(9) M. Munz and W. Weidlich : Settlement Formation II, Numerical Simulation; Ann. Reg. Sci. 24, pp 177-196 (1990)

(10) W. Weidlich : Sociodynamics Applied to the Evolution of Urban and Regional Structures, Journal of Discrete Chaotic Dynamics (JDCD), Vol.1, (1997)

(11) T. Sigg and W. Weidlich : Urban Evolution under Population Pressure: A Mathematical Model, to be published in Geographical Systems (1997)

(12) W. Weidlich and G. Haag (eds) : Interregional Migration-Dynamic Theory and Comparative Analysis. Springer, Berlin (1988)

(13) G. Haag : "Qualifizierung, Quantifizierung und Evaluierung wegebau–induzierter Beförderungsprozesse"; Bundesministerium für Verkehr, Bonn, FE–Nr. 90436–95 (1996).

The Morphogenesis of Dictyostelium Discoideum – Pattern Formation in a Biological Excitable System

Florian Siegert[1], Bakhtier Vasiev[2] and Cornelis J. Weijer[2]

[1]Zoologisches Institut, Luisenstr. 14, 80333 München, Germany
[2]Department of Anatomy&Physiology, University of Dundee, Dundee, DD1 4HN, UK

1. Introduction

A major goal in the study of development of eukaryotic organisms is to understand the mechanisms of morphogenesis, i.e. how does a complex organism develop from a single cell, the fertilised egg and what determines its final shape. Mechanisms responsible for the development of multicellular organisms involve spatio-temporal control of cell proliferation, cell death and cell differentiation as well as differential cell movement. It is also clear that these processes all have to be precisely controlled in space and time and they have to be stable against external perturbations. This implies that these processes must be precisely regulated. This regulation is mediated via extensive cell-cell communication by extracellular signalling factors. These factors interact in characteristic positive and negative feedback loops to result in spatio-temporal regulation of cell division, cell death, differentiation and movement.

The mechanisms involved are complex and due to the multitude of cell types and signals often difficult to investigate in higher organisms. Since many of the basic properties of these mechanism are essentially conserved during evolution it makes sense to investigate them in simpler model organisms, containing fewer cell types. Furthermore it is advantageous to investigate organisms which are amenable to genetic analysis. The analysis of mutants in these mechanisms allows the study of perturbations on development.

For these reasons we have focused on the study of morphogenesis in a very simple organism the cellular slime mould *Dictyostelium discoideum*. Slime moulds are positioned between uni- and multi-cellular life in the evolutionary tree of life. Dictyostelium undergoes a starvation induced multicellular development (**Figure 1**) which shows many of the characteristic features of the development of higher organisms like controlled cell differentiation and differential chemotactic cell movement.

Normally slime moulds live as single amoebae in the soil. They feed on bacteria and divide by binary fission. Starvation induces the activation of a developmental program in which the cells aggregate chemotactically to form a multicellular mass of 10^4-10^5 cells. Since multicellular development occurs in the absence of food there is essentially no cell division, thus simplifying the analysis of morphogenesis. In the aggregate (mound) the cells start to differentiate into a number of different cell types, i.e. several prestalk types which will form the stalk, basal disk and upper and lower cup in the fruiting body as well as prespore cells which will continue to differentiate to form spores. The cells differentiate in random positions in the late aggregate [1]. The prestalk cells than sort out chemotactically to form the tip of the tipped mound

164

[2]. The mound erects and extends up in the air to form the standing slug which falls over and migrates away. The slug has a distinct polarity with a tip at the anterior end which guides all its movement. The slug is photo- and thermo-tactic which allows it to move up towards the soil surface. There it transforms into a small fruiting body (up to 4mm high) consisting of a stalk supporting a spore mass. The spores disperse and under suitable conditions germinate to release amoebae and the whole cycle can start all over again. In this article we will give an overview of the mechanisms that control aggregation, mound and slug formation and show that these processes can be viewed as pattern formation in a biological excitable system.

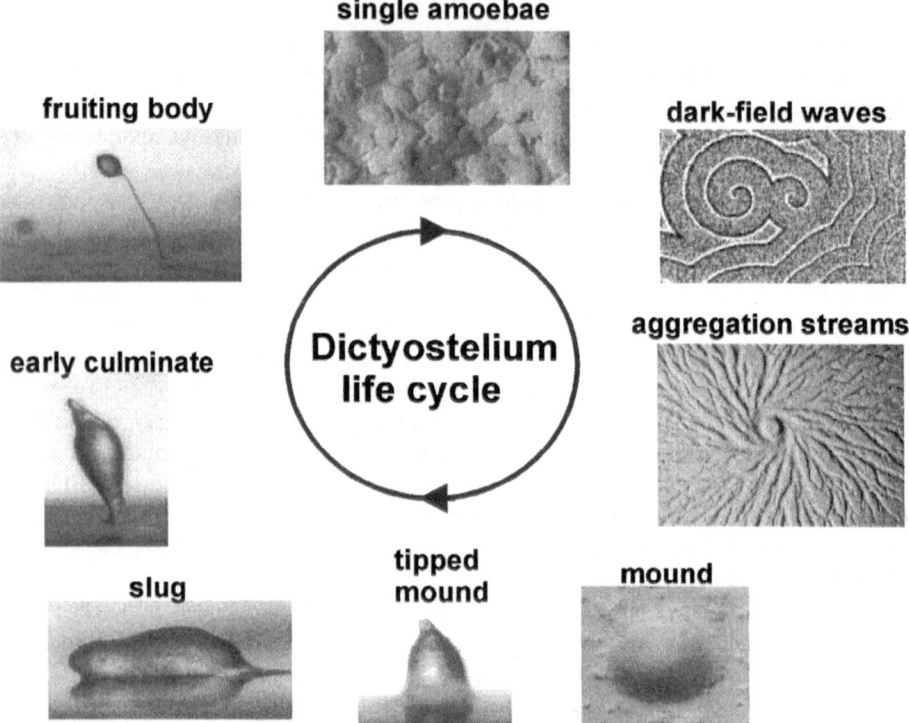

Figure 1: Dictyostelium life cycle. Shown are single amoebae, darkfield waves, aggregation streams, mounds, slug, an early culminate and a fruiting body. Development takes 24 hours at room temperature.

2. Aggregation

The principles that govern aggregation are now relatively well understood at the cellular level. Aggregation of individual *Dictyostelium* amoebae into multicellular aggregates occurs by chemotaxis to 3`-5`cyclic Adenosine monophosphate (cAMP) signals are released in a periodic fashion by cells in the aggregation centre. The cells in the aggregation centre periodically synthesise and secrete cAMP in the extracellular medium. Here it diffuses to neighbouring cells which detect this signal via highly specific and highly sensitive cAMP receptors. These receptors are transmembrane proteins with an extracellular cAMP binding domain and an intracellular effector domain. Binding of cAMP to the receptor leads to two competing processes, excitation and adaptation (reviewed in [3]). The excitation pathway leads to the activation of the enzyme adenylatecyclase, which produces cAMP from Adenosine-Tri-Phosphate (ATP). This intracellular cAMP is then secreted to the outside where it can bind to the receptors of the same cell leading to autocatalytic feedback and in addition it diffuses away to activate neighbouring cells. The adaptation pathway involves a desensitisation of the receptor, involving phosphorylation of its cytoplasmic tail, resulting in a termination of the autocatalytic relay response. The cells also secrete an enzyme cAMP phosphodiesterase which degrades cAMP. A fall in the extracellular concentration of cAMP then leads to a dephosphorylation and resensitsation of the receptor (**Figure 2A**). The adaptation process is responsible for the outward propagation of cAMP waves, since cells which have just relayed are refractory to further stimulation by cAMP. cAMP also leads to a chemotactic reaction in the direction of higher cAMP concentrations. The cells move up the gradient as long as the cAMP concentration is rising but stop to move as soon as the cAMP concentration starts to fall. This response leads to the periodic movement of the cells towards the aggregation centre guided by outward propagating waves of cAMP (**Figure 2B**). Superimposed on this system there is a complex feedback of the cAMP oscillations on the expression of various components of the signalling system. The cAMP pulses induce the synthesis of cAMP receptors, adenylatecyclase and phosphodiesterase resulting in an increase in excitability during aggregation [4].

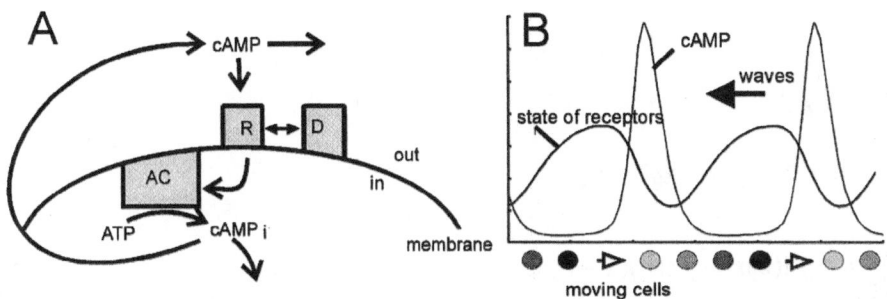

Figure 2: (A) Martiel-Goldbeter model for cAMP oscillations, R - active receptor, D-desensitised receptor, AC-Adenylatecyclase, cAMP and cAMPi - extra- and intra- cellular cAMP. (B) Scheme showing wave propagation and cell movement. The cAMP wave profile and fraction of active receptors are shown as calculated from equations (1-3). Waves propagate from right to left while cells (arrows and dots) move from left to right. Arrows represents moving cells, dark dot - non moving cell.

During early aggregation the cAMP waves can be seen as optical density waves using low-power darkfield optics [5, 6, 7]. These optical density waves are correlated with shape changes which cells undergo upon stimulation with cAMP.

Chemotactically moving cells are elongated and appear brighter than non-moving cells (**Figure 1,2B**). By correlating the cAMP signal via isotope dilution-fluorography with the waves, it was clearly demonstrated that the optical density waves observed during aggregation represent the propagating cAMP signal [8]. Most often waves appear as expanding spirals, in some strains also as concentric ring waves. Waves from neighbouring centres collide and annihilate each other leading to the formation of aggregation territories. Quantitative measurements showed that during aggregation the frequency of the waves increases while the wave propagation speed slows down. This is partly due to the cAMP dependent expression of components of the oscillatory system as well a to the dispersive properties of this excitable medium [9, 10].

There have been several attempts to model the early aggregation process. Two main questions need to be addressed: how do the cells produce cAMP waves and how do they move in response to these waves? Essentially two types of models were developed to describe mathematically the cAMP relay kinetics of Dictyostelium amoebae. The first model has been suggested by Martiel and Goldbeter [11, 12, 13]. It is based on the assumption that activation/inactivation of the cAMP receptors plays a key role in the response. A second model has been introduced by Tang and Othmer [14] in this model activation of activating and inhibitory G proteins control the periodic production of cAMP by the cells. Both models are able to describe oscillations in cAMP level in the cell suspensions as well as a cAMP wave propagation in a dispersed cell population [9, 15]. These models basically describe excitable and/or oscillatory media. For example the reduced Martiel-Goldbeter model [12] describing the cAMP relay system of individual cells, consists of three coupled non-linear equations which define the level of extra cellular and intracellular cAMP, and the activation state of the cAMP receptors (**Figure 2A**):

$$\frac{\partial c}{\partial t} = \left(\frac{k_t}{h}\right)\beta - k_e c \tag{1}$$

$$\frac{\partial \beta}{\partial t} = q'\Phi(r,c) - k_t \beta - k_i \beta \tag{2}$$

$$\frac{\partial r}{\partial t} = -f_1(c)r + f_2(c)(1-r) \tag{3}$$

Equation (1) describes the change in the level of extra-cellular cAMP (c), over time. These changes occur due to the secretion of intracellular cAMP β in the extracellular medium and hydrolysis by phosphosphodiesterase. Secretion is taken into account by the first term in right hand side of (1): cAMP is transported over the cell membrane with the rate defined by k_t. It is diluted in the extracellular medium by factor h which represents the ratio of extracellular to cell volume. Hydrolysis is taken into account by the second term and is assumed to be proportional (k_e) to extra cellular cAMP concentration. Equation (2) defines the level of intracellular cAMP (β). It takes into account the synthesis of intracellular cAMP by adenylatecyclase, its secretion and hydrolysis. The rate of cAMP synthesis (first term) is dependent on the level of extracellular cAMP, *c*, and the state of cAMP receptors (r) and represented by an allosteric non linear function (Φ(r,c)). Equation (3) reflects changes in the state of the receptors. It defines the relative number of activated (*r*) and inactivated (*1-r*) receptors whose (slow) interconversion is dependent on the level of extracellular cAMP. To describe cAMP waves propagating through a population of cells a diffusion term has to be included in (1). Finally the model becomes very similar to the prototype FitzHugh-Nagumo system.

cAMP waves not only propagate through the cell population but also co-ordinate their movement. cAMP orients the direction of otherwise randomly moving cells. There is strong evidence that the cells detect the gradient of cAMP over their length [16, 17, 18]. However there is also evidence that cells use the temporal derivative of cAMP and only move up the gradient as long as the cAMP level is rising [19, 20]. This allows cells to move chemotactically on the wave front rather than on the wave back. A number of mathematical models have been proposed for chemotactic cell movement. The best known one is the Keller-Segel model [21] describing a cell flux, J, as a function of cell density, ρ, and concentration of cAMP, c.

$$J = -D(c)\nabla\rho + \chi(c)\rho\nabla c \tag{4}$$

The first term on the right hand side describes random cell movement (the velocity can depend on the level of cAMP) and the second term describes the directed motion of the cells along the cAMP gradient. There are also a number of models where chemotactic cell motion is described in axiomatic way as rules for motion of units (cells) in a concentration field of cAMP [22, 23, 24, 25].

Much effort has been directed towards understanding the physical principles leading to the formation of streaming patterns by amoebae responding chemotactically to propagating cAMP waves. It has been proven analytically and numerically that streams form due to an instability caused by a coupling between the velocity of signal propagation and density of cells. Local accumulation of cells will, due to the dependence of the rate of cAMP accumulation on cell density, result in a local speeding up of the wave propagation at regions of high cell density. This local deformation of the wave front will lead to the attraction of even more cells to this region and finally to the formation of bifurcating aggregation streams [26, 25, 27, 28]. in which the cells move towards the aggregation centre. There they pile on top of each other to form a three dimensional hemispherical structure, the mound. In the streams the cells are elongated and connected by rather characteristic end to end contacts. The movement of individual cells however is still periodic, but somewhat faster than that of isolated cells, possibly suggesting a co-operative effect on their movement [29].

3. Wave propagation and cell movement in mounds

Special image processing techniques allowed to visualise faint optical density waves

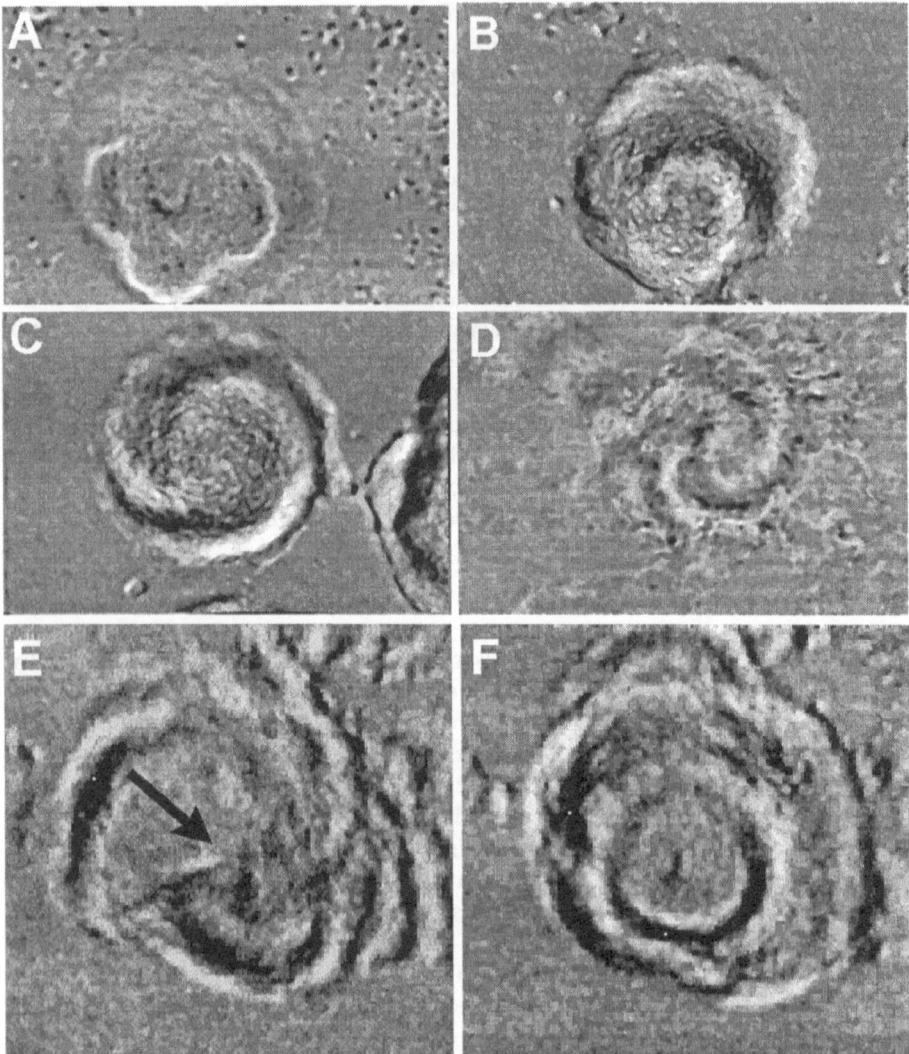

Figure 3: Different modes of signal propagation in mounds .(A) Centre organised by three pacemakers in strain XP55,(B) a single-, (C) Double- and (D) a triple- armed spiral centre in strain AX3. (E,F) Micro-injection of cAMP in mounds. Micro-injection of periodic cAMP pulses from a fine glass micro electrode can change the mode of signal propagation. A spiral centre (E) transforms to a concentric ring centre after periodic stimulation by exogenous cAMP through an electrode (F) The location of the tip of the needle is indicated by the arrow in (E).

in multicellular structures. These investigations showed that *Dictyostelium* mounds are organised by periodic signals displaying a whole range of different wave geometry's [30]. The exact pattern of the propagating cAMP signal seems to be strain specific. Some mounds are organised by single or by several coexisting pacemakers, others by single- or multi-armed spirals **(Figure 3A,B,C,D)**. The conversion of one type into another was also observed [30].

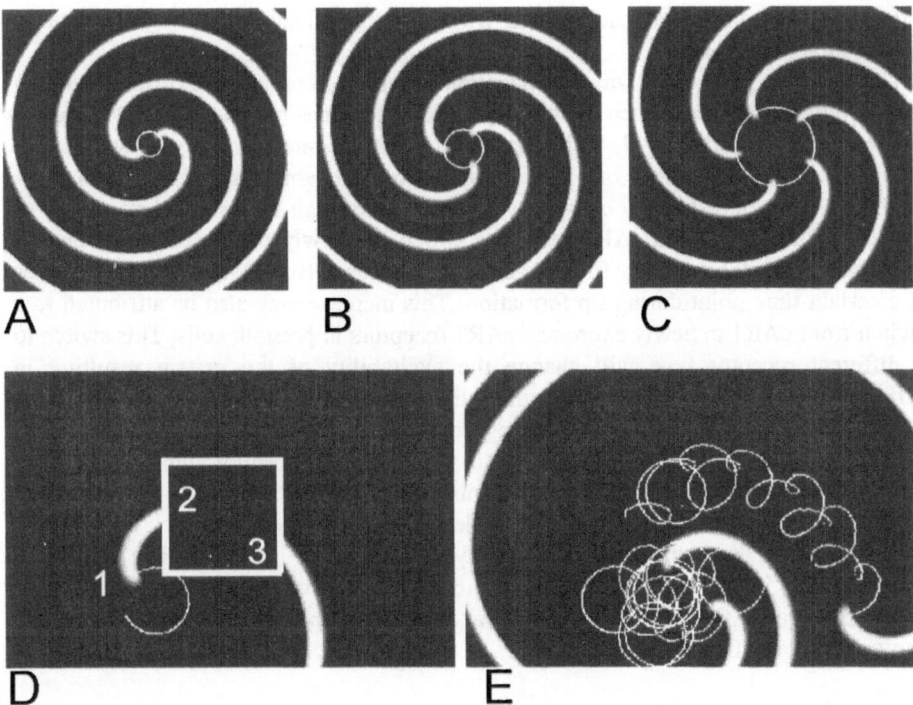

Figure 4: Formation of multi-armed spirals in a two-dimensional excitable medium described by the FitzHugh-Nagumo model. (A) Two- three- and five-armed spirals are shown in media with different excitability. The possible number of arms increases with a decrease in the excitability of the medium. Cycles in the centre of patterns indicate cores of the spirals. (B) A break of a single spiral can result in formation of a two-armed spiral. It happens when the break is located not too far from the spiral tip. We assume that multi-armed spirals in the mound occur by the same way.

Continuous measurements of the optical density waves from late aggregation until tip formation over a period of 3 hours demonstrated that there was a clear evolution in the dynamics of the waves [31]. Initially the waves propagate fast at low frequency but in the course of aggregation the wave frequency increased, while wave propagation speed decreased. Although we can observe a continuous succession of optical waves from aggregation to the mound stage it is not proven that the waves in the mound are caused by chemotaxis to propagating cAMP signals. In order to investigate this possibility we tested whether we could initiate optical density waves in mounds by periodic injection of cAMP in between the cells. Periodic micro-injection of 0.01nl pulses of 0.5mM cAMP into mounds initiated optical density waves which propagated from the electrode tip outwards and which interacted with endogenous waves **(Figure 3E,F)** shows, that periodic injection of cAMP pulses can override the endogenous multi-armed spirals and induce concentric ring waves emanating from the electrode tip. These observations clearly show, that optical density waves in mounds can be induced by cAMP oscillations and furthermore that induced waves annihilate endogenous waves upon collision, showing a common propagation mechanism.

The different patterns of wave propagation observed in mounds of different strains most likely are a consequence of strain specific differences in their ability to generate and relay the cAMP signal. This change in wave geometry observed within one strain, i.e. the change from one armed spirals to multi armed spirals in mounds of strain AX-3, are most likely caused by a switch from high affine cAR1 receptors to the less affine cAR2 and cAR3 receptors which are newly expressed at the end of aggregation [3]. Interestingly the period length increased from 2 minutes to 4 minutes at a certain time point during tip formation. This increase may also be attributed to a switch from cAR1 to newly expressed cAR2 receptors in prestalk cells. This switch to a different receptor type will change the excitability of the system resulting in different wave patterns and could explain the formation of multi-armed spirals from single armed spirals observed during aggregation [32]. Model calculations showed that this can happen in low excitable media via breaks in the spiral with one chemical wavelength from the core of the original spiral (**Figure 4**).

The diverse geometry of the signals leads to variety of complex cell motion patterns. Since cell movement is always opposite to the direction of signal propagation [5, 31] cell movement in mounds organised by concentric waves is directed towards the organising centre and slow. In the case of spiral wave cell movement is rotational and fast.

4. A model for mound formation

Observation of cells moving in mounds prompted theoretical investigations where the cell movement is considered as a flow of a viscous compressible liquid. This approach has been successfully used in a hydrodynamic model describing the whole process of aggregation until mound formation (**Figure 5**). In this model cell velocity has been defined by the Navier-Stokes equation:

$$\rho[\partial \mathbf{V} / \partial t + (\mathbf{V}\nabla)\mathbf{V}] = \mathbf{F}_{ch} + \mathbf{F}_{fr} + \eta\Delta\mathbf{V} + \xi\mathbf{grad}\,div\mathbf{V} + \mathbf{F}_{ad} - \mathbf{grad}p \qquad (5)$$

The left hand side of the equation describes the acceleration of cells under the influence of various forces described in the right hand side of the equation. $\mathbf{V}$ is the velocity of the cells. $\mathbf{F}_{ch}$ is the chemotactic force which is active on the front of cAMP waves, $\mathbf{F}_{fr}$ is a friction force responsible for slowing down cell movement, The third and fourth terms on the right hand side describe cell-cell friction: η and ζ are viscosity coefficients. $\mathbf{F}_{ad}$ takes into account cell-cell and cell-substrate adhesion forces, p is the pressure between the cells caused by the chemotactic accumulation of the cells (see [33] for details).

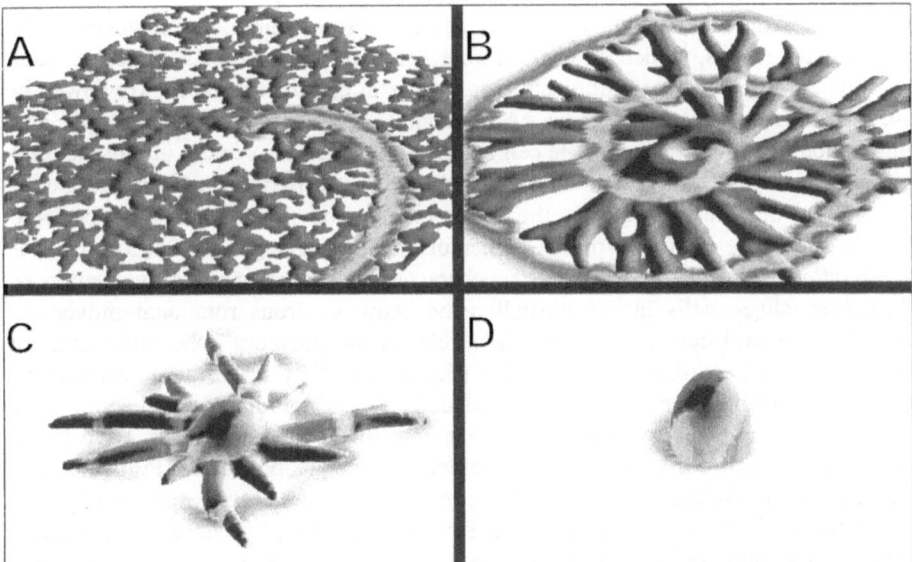

Figure 5: Simulation of aggregation and mound formation. Cell density is shown as a black iso-surface (ρ=0.5) and the cAMP spiral is mapped on this surface. The initial density of cells was zero everywhere in 3d-space except for the bottom plane (A). The cell density in each grid of this plane was represented by a random number varying between 0 and 1 so that average density in this plane was equal to 0.5. In response to cAMP spiral wave cells move and form aggregation streams (B-C) and then mound (D) which represents a stable solution of the system.

The evolution of the shape of the aggregate into a mound in this model has been obtained by solving an equation for the cell density field, i.e. by the equation for the conservation of mass:

$$\partial \rho \, / \, \partial t \, = \, D_\rho \Delta \rho \, - \, div(\rho \mathbf{V}) \tag{6}$$

The first term on the right hand side of the equation describes the random motion of the cells, while the second term describes co-ordinated chemotactic movement. Aggregation patterns found as solutions of these equations (5-6) in combination with a FitzHugh-Nagumo to describe the excitable cAMP kinetics are shown in **Figure 5**.

This model although rather qualitative shows some remarkable similarities with the formation and appearance of aggregates observed in real life (**Figure 1,5**). Furthermore it is able to describe some mutant phenotypes frequently observed [33].

5. Wave propagation in slugs

Up to the mound stage cAMP wave propagation can be seen as optical density wave using darkfield optics and digital image processing techniques [6, 30]. During slug migration and culmination an extra cellular slime sheet, which is secreted continuously, surrounds the slug. This gives the slug some mechanical stability but also impedes the observation of optical density waves. There are however many experimental results, which indicate a role for extra cellular cAMP during later development in cellular communication as well as in cell differentiation (reviewed in [34, 35]. To find out, if periodic cAMP signals also control slug migration and culmination, we investigated single cell behaviour and cell movement in multicellular structures, assuming, that periodic signals should cause periodic cell movement. It was shown, that indeed cells in the prespore zone of slugs go through periodic velocity and shape changes typical for chemotactically moving cells [36, 37]. Further investigations showed, that there exists a characteristic pattern of cell movement in *D. discoideum* slugs: cells in the prestalk zone show vigorous rotational movement around the central core of the tip, while cells in the prespore zone move straight forward in the direction of slug migration (**Figure 6A**) [38]. From these observations the geometry of the propagating signal was deduced: a three-dimensional scroll wave (spiral wave) produces rotational cell movement in the tip and planar wave fronts produce the straight forward movement observed in the prespore zone (**Figure 6B**).

Computer simulations showed, that the conversion of a scroll into a series of planar waves occurs, if there is a substantial difference in excitability between the prestalk and prespore cell population or if not all but only a part of the prespore population is actively relaying the cAMP signal (**Figure 6B**) [39, 40].

During the slug stage there is a further specification of cell types. pstA cells are formed in the anterior outer part of the prestalk zone. pstO cells are found at the boundary between prestalk and prespore cells, while pstB cells, which will form the stalk, are found in the central core of the prestalk zone. The prespore cells are located in the back two thirds of the slug [41]. Prespore genes need cAMP for their induction and stabilisation. Expression of the prestalk specific ecmB gene by the pstB cells is inhibited by high concentrations of extra cellular cAMP while ecmA expression by pstA cells requires high concentrations of cAMP [42, 43, 44]. Computer simulations using the Martiel-Goldbeter model for cAMP oscillations showed, that the core of the scroll wave in the prestalk zone is a region of low average extracellular cAMP, exactly the condition which facilitates the expression of the stalkspecific ecmB gene in the central core of the prestalk zone (**Figure 6C**) [40]. Despite the complex mode of wave propagation it gives rise to a relatively simple spatial pattern of average cAMP, which can be read out by the cells in different positions in the slug to keep the differentiation state of the cells in the slug stable (**Figure 6D**). These simulations suggest that the wave propagation pattern not only is responsible for the control of cell movement but also might be involved in the differentiation of the prestalk cell types [40].

6. Tip formation, cell differentiation and sorting

One of the most interesting but also most complicated phases of development is slug formation. During aggregation the cell density increases dramatically and cells start to move up on top of each other in the third dimension. In the mound cells begin to differentiate in prestalk and prespore cells. The prestalk cells differentiate at random positions, but then sort towards the top of the mound to form a tip [45]. The mound then contracts at the base while extending up in the air to form a standing slug. Slug formation can be seen as a two step process, i.e. sorting of prestalk cells to form a tip followed by a tip induced contraction and elongation of the mound to form a standing slug (**Figure. 1**).

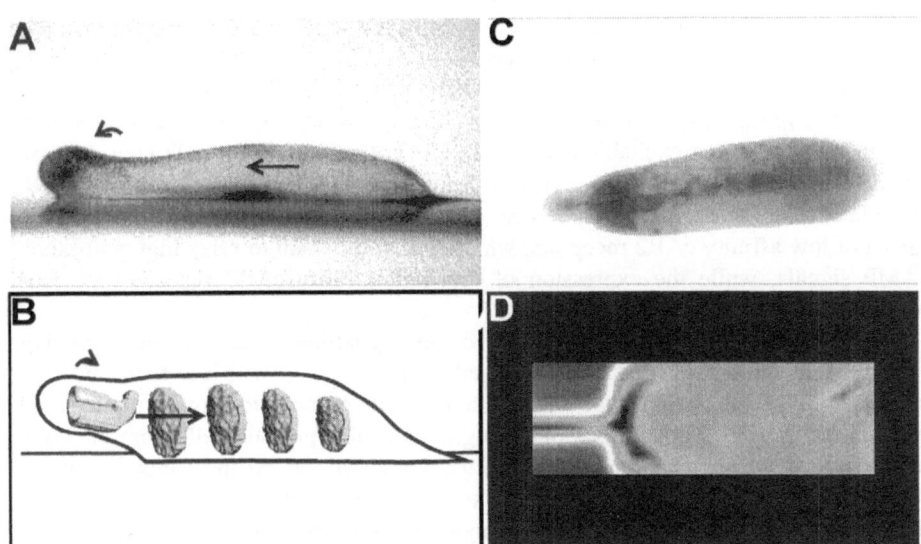

Figure 6: Wave propagation, cell movement and differentiation in slugs. (A) Photograph of a neutral red stained slug. The dark stained region on the left hand side is the prestalk region. The arrows indicate the direction of cell movement. (B) model for waves in the slug. The arrows indicate direction of cAMP wave propagation. (C) Photograph of slug showing expression of the prestalk specific gene ecmB (dark region in slug), note the expression in the slug middle and at the prestalk-prespore boundary. (D) average cAMP levels after integration over 10 periods of wave rotation. Note the close correspondence between average cAMP (D) and cell type differentiation (A,B).

Cell sorting most likely results form differential chemotactic cell movement towards cAMP. Experiments showed that prestalk cells preferentially sort towards an artificial cAMP source [46]. Furthermore mutants which overexpress cAMP phosphodiesterase are blocked at the mound stage of development and defective in cell sorting [47]. The difference in effective movement speed towards a cAMP source could be caused by cell type specific differences in the motive force generated by prestalk and prespore cells. Differences in motive force could result from differences in the cytoskeleton or cell-cell adhesion, i.e. prespore cells being more adhesive than prestalk cells. There is experimental evidence for both types of mechanisms: Isolated pstA cells, which will sort to the top of the aggregate, are able to move faster to an artificial cAMP source as isolated prespore cells [45]. Furthermore several mutants with defects in components of the cytoskeleton are arrested at the mound stage [48, 49]. Using a cold sensitive myosin mutant it has been shown, that there are two stages in development where myosin II is absolutely required for morphogenesis, at the mound stage during tip formation and during culmination [50]. Secondly in multi cellular tissues cell-cell interactions are likely to play an important role in the control of cell movement. It is known that prespore cells are more adhesive as prestalk cells [51]. Prestalk cells may therefore move more efficiently in a multicellular aggregate consisting of prespore and prestalk cells. We therefore suspect that cell sorting involves all these mechanisms, i.e. differential chemotaxis towards cAMP, cell type specific differences in the generation of motive force as well as cell type specific differences in cell-cell interactions and cell-substrate interactions.

Cell sorting will feedback on the signalling patterns in the tipped mound since prestalk and prespore cells differ in their excitability. Many experiments suggest that prestalk cells are more excitable than other cells in the mound. Prestalk cells express higher amounts of the enzymes involved in the synthesis and degradation of cAMP, adenylatecyclase and phosphodiesterase [52, 53, 54] and they express a specific subset of low affinity cAR2 receptors, which will allow them to relay high amplitude cAMP signals, while the expression of the high affinity cAR3 receptor becomes restricted to prespore cells [55, 56].

Taken together cell sorting should affect the signalling system in the following way: the collection of fast oscillating prestalk cells in the tip will lead to an increase in excitability in the tip. This will result in a loss of spiral arms to form a simple scroll wave in the tip [32]. The removal of the highly excitable prestalk cells from the body of the mound will result in a decrease in local excitability and to conversion of the scroll wave in the tip to a twisted scroll wave in the mound [39, 40]. This will lead to a twisted rotational cell movement in the mound. As a result the mound contracts and extends up into the air.

We are now testing this possibility by an extension of the model from mound formation by incorporation of different cell types (fluids) with different chemotactic and relay properties. We consider the mound to be a drop of liquid consisting of two kinds of fluids and use a two-field description of this drop to model cell sorting. The velocity of the liquid, V defined in (5), is assumed to have two components corresponding to velocities of prestalk, V_1, and prespore, V_2, cells:

$$V = (\rho_1 V_1 + \rho_2 V_2) / (\rho_1 + \rho_2) \tag{7}$$

Since the liquid is incompressible: $\rho_1 + \rho_2 = 1$ in the mound. The chemotactic forcing, F_{ch}, in (5) is also assumed to consist of two components corresponding to chemotactic forcing of prestalk and prespore cells.

$$\mathbf{F}_{ch} = (\rho_1 \varphi_1 (\partial g / \partial t) + \rho_2 \varphi_2 (\partial g / \partial t)) \mathbf{grad} g \tag{8}$$

The difference in cAMP signalling between prestalk and prespore cells is also taken into account by two different sets of parameters for the FitzHugh-Nagumo model. To find the velocities of prestalk and prespore cells we put expressions (7-8) for velocity and chemotactic force into equation (3), put coefficient $\rho_1 + \rho_2$ to the last term in (3) and by separating terms consisting ρ_1 and ρ_2 we get two equations for V_1 and V_2:

$$\partial V_i / \partial t = V_i \nabla (V_i \rho_i) / \rho_i - (V_i \nabla) V + F_i + \eta \Delta (V_i \rho_i) / \rho_i - \mathbf{grad} p \quad \text{where i=1,2} \tag{9}$$

The two equations (9) each coupled with the equation of conservation of mass (5) (used to define the densities of both fluids) give the evolution of the cell density fields over time. Our preliminary computations show that starting from a random distribution of cell types in the mound one can obtain spatially separated patterns of cell types as well as tip formation. In response to scroll waves rotating along the axis of hemispherical mound

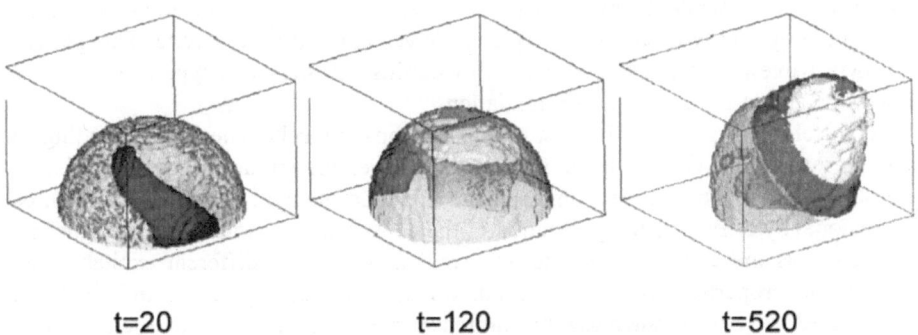

| t=20 | t=120 | t=520 |

Figure 7: Simulation of cell sorting and tip formation. To describe cell movement we used equations (5,7,8,9) and to describe cAMP signalling we used the FitzHugh-Nagumo equations. A sequence of images is shown depicting sorting of highly excitable fast moving prestalk cells (white) from less excitable slow moving prespore cells (light grey(transparant)) as well the as the cAMP signal (dark grey). A scroll wave was initiated in the mound with an initially random distribution of prestalk and prespore cells. In the course of time the cells sort which result in a twisting of the scroll wave. The period of the scroll wave decreases from 38 to 21 time units.

the liquid begins to rotate in the opposite direction. The faster moving fluid (prestalk cells) accumulates in the centre and top of the mound. Since the faster fluid is more excitable their separation leads to the mound becoming inhomogeneous with respect to its excitability. This in turn results in a change of the cAMP wave shape. Since the density of the excitable cells is higher on the top of the mound the velocity of wave propagation there is higher as in the base of the mound and the scroll becomes twisted. As a consequence the cells experience a vertical chemotactic force which results in the further collection of faster moving cells on the top. Finally all the faster cells collect at the top of the mound and form a tip, exactly as it happens in real mounds **(Figure 7)**.

7. Conclusions

It is now becoming clear that periodic signals not only control aggregation but also all later stages of morphogenesis. Multicellular mounds are organised by either concentric ring waves emanating from one or more centres or by spiral waves with up to 10 arms. These signals are used to regulate the process of cell sorting, in which the most excitable cells move on top of the aggregate. This sorting process leads to a highly excitable tip and a less excitable main body. This spatial separation feeds back on the signal geometry. We propose that the cells in the tip are organised by a rotating scroll wave of cAMP with the core of the scroll wave coinciding with the long axis of the tip. The scroll wave converts into a twisted scroll wave and planar waves in the body of the mound. This pattern of wave propagation leads to a rotational movement of the prestalk cells in the tip and a periodic upward movement of the cells in the base of the mound. Furthermore it results in a contraction of the tip and an elongation of the mound into a standing slug which becomes unstable and topples over. The slug now moves away, while the movement of the cells is still being controlled by a scroll wave in the tip and twisted scroll or planar waves in the prespore zone. This pattern of cAMP wave propagation is also used to stabilise prestalk cell type specific gene expression and to initiate stalk differentiation.

Dictyostelium is possibly the first organism whose morphogenesis is beginning to be understood at the cellular level. During early development morphogenesis is based on wave propagation in a two dimensional excitable medium which becomes three dimensional by chemotactic aggregation of the cells. More complexity is brought into the system as the cells differentiate into several types with different excitable and chemotactic properties. Due to these additional levels of regulation and feedback complicated wave forms such as multi-armed spirals, twisted scroll waves etc. arise.

Acknowledgements. We like to thank Dirk Dormann, Jens Rietdorf and Till Bretschneider for performing some of the experiments and calculations and the BBSRC and Deutsche Forschungsgemeinschaft for financial support.

8. References

[1]   J. Williams, Current Opinion in Genetics & Development, **5**, 426-431, 1995.

[2]   D. Traynor, M. Tasaka, I. Takeuchi, and J. Williams, Development, **120**, 591-601, 1994.

[3]   C. A. Parent and P. N. Devreotes, Annu Rev Biochem, **65**, 411-440, 1996.

[4]   R. A. Firtel, Current Opinion in Genetics & Development, **6**, 545-554, 1996.

[5]   F. Alcantara and M. Monk, J Gen Microbiol, **85**, 321-334, 1974.

[6]   J. D. Gross, M. J. Peacey, and D. J. Trevan, J Cell Sci, **22**, 645-656, 1976.

[7]   F. Siegert and C. Weijer, J Cell Sci, **93**, 325-335, 1989.

[8]   K. J. Tomchik and P. N. Devreotes, Science, **212**, 443-446, 1981.

[9]   J. J. Tyson and J. D. Murray, Development, **106**, 421-426, 1989.

[10]   P. Foerster, S. Muller, and B. Hess, Development, **109**, 11-16, 1990.

[11]   A. Goldbeter, Biochemical oscillations and cellular rhythms. The molecular bases of periodic and chaotic behaviour.: Cambridge Univ. Press, 1996.

[12]   J.-L. Martiel and A. Goldbeter, Biophys J, **52**, 807-828, 1987.

[13]   J. L. Martiel and A. Goldbeter, Lect Notes Biomath, **71**, 244-255, 1987.

[14]   Y. H. Tang and H. G. Othmer, Math Biosci, **120**, 25-76, 1994.

[15]   Y. H. Tang and H. G. Othmer, Phil Trans R Soc Lond B, **349**, 179-195, 1995.

[16]   J. M. Mato, A. Losada, V. Nanjundiah, and T. M. Konijn, Proc Natl Acad Sci USA, **72**, 4991-4993, 1975.

[17]   G. Gerisch, Biol Cellulaire, **32**, 61-68, 1978.

[18]   M. J. Caterina and P. N. Devreotes, FASEB J, **5**, 3078-3085, 1991.

[19]   D. Wessels, D. Shutt, E. Voss, and D. R. Soll, Molecular Biology of the Cell, **7**, 1349-1349, 1996.

[20]   D. Wessels, J. Murray, and D. R. Soll, Cell Motil Cytoskel, **23**, 145-156, 1992.

[21]   E. F. Keller and L. A. Segel, J Theor Biol, **26**, 399-415, 1970.

[22]   B. Novak and F. F. Seelig, J Theor Biol, **56**, 301-327, 1976.

[23]   S. A. Mackay, J Cell Sci, **33**, 1-16, 1978.

[24]   O. O. Vasieva, B. N. Vasiev, V. A. Karpov, and A. N. Zaikin, J Theor Biol, **171**, 361-368, 1994.

[25]   C. van Oss, A. V. Panfilov, P. Hogeweg, F. Siegert, and C. J. Weijer, J Theor Biol, **181**, 203-213, 1996.

[26]   H. Levine and W. Reynolds, Phys Rev lett, **66**, 2400-2403, 1991.

[27]   B. N. Vasiev, P. Hogeweg, and A. V. Panfilov, Phys Rev Lett, **73**, 3173-3176, 1994.

[28]   T. Hofer, J. A. Sherratt, and P. K. Maini, Proc. R. Soc. London B, **259**, 249-257, 1995.

[29]   N. J. Savill and P. Hogeweg, J Theor Biol., **184**, 229-235, 1997.

[30]   F. Siegert and C. J. Weijer, Curr Biol, **5**, 937-943, 1995.

[31]   J. Rietdorf, F. Siegert, and C. J. Weijer, Dev Biol., **177**, 427-438, 1996.

[32]   B. N. Vasiev, Siegert F., Weijer C.J., Phys Rev lett, **78**, 2489-2492, 1997.

[33]   B. Vasiev, F. Siegert, and C. J. Weijer, J of Theor Biol, **184**, 441, 1997.

[34]   C. D. Reymond, P. Schaap, M. Veron, and J. G. Williams, Experientia, **51**, 1166-1174, 1995.

[35]   M. Y. Chen, R. H. Insall, and P. N. Devreotes, Trends in Genetics, **12**, 52-57, 1996.

[36]   A. J. Durston, F. Vork, and C. Weinberger, , J. G. Vassileva-Popova and E. V. Jensen, Eds. New York: Plenum, 1979, pp. 693-708.

[37]   F. Siegert and C. J. Weijer, Physica D, **49**, 224-232, 1991.

[38]   F. Siegert and C. J. Weijer, Proc Natl Acad Sci USA, **89**, 6433-6437, 1992.

[39]   O. Steinbock, F. Siegert, S. C. Muller, and C. J. Weijer, Proc Natl Acad Sci USA, **90**, 7332-7335, 1993.

[40]   T. Bretschneider, F. Siegert, and C. J. Weijer, Proc Natl Acad Sci USA, **92**, 4387-4391, 1995.

[41]   K. A. Jermyn, K. T. Duffy, and J. G. Williams, Nature, **340**, 144-146, 1989.

[42]   M. Berks and R. R. Kay, Development, **110**, 977-984, 1990.

[43]   N. A. Hopper, C. Anjard, C. D. Reymond, and J. G. Williams, Development, **119**, 147-154, 1993.

[44]   R. D. M. Soede, N. A. Hopper, J. G. Williams, and P. Schaap, Dev Biol, **177**, 152-159, 1996.

[45]   A. Early, T. Abe, and J. Williams, Cell, **83**, 91-99, 1995.

[46]   S. Matsukuma and A. J. Durston, J Embryol Exp Morphol, **50**, 243-251, 1979.

[47]   D. Traynor, R. H. Kessin, and J. G. Williams, Proc Natl Acad Sci USA, **89**, 8303-8307, 1992.

[48]   A. De Lozanne and J. A. Spudich, Science, **236**, 1086-1091, 1987.

[49]   F. Rivero, R. Furukawa, A. A. Noegel, and M. Fechheimer, Molecular Biology of the Cell, **7**, 3141-3141, 1996.

[50]   M. L. Springer, B. Patterson, and J. A. Spudich, Development, **120**, 2651-2660, 1994.

[51]   C. H. Siu and R. K. Kamboj, Dev Genet, **11**, 377-387, 1990.

[52]   S. S. Brown and C. L. Rutherford, Differentiation, **16**, 173-183, 1980.

[53]   M. Pahlic and C. L. Rutherford, J Biol Chem, **254**, 9703-9707, 1979.

[54]   C. Rutherford, S. Brown, and D. Armant, "Enzymatic potential for establishing gradients of cyclic AMP during cell determination in Dictyostelium discoideum.," in *Adv. Cycl. Nucl. Res.*, vol. 14, 1981, pp. 705.

[55]   C. L. Saxe III, Y. M. Yu, C. Jones, A. Bauman, and C. Haynes, Dev Biol, **174**, 202-213, 1996.

[56]   Y. M. Yu and C. L. Saxe, Dev Biol, **173**, 353-356, 1996.

Path Optimization in Chemical and Biological Systems on the Basis of Excitation Waves

Oliver Steinbock

Otto-von-Guericke-Universität Magdeburg
Institut für Experimentelle Physik, Abteilung Biophysik
Universitätsplatz 2, D-39106 Magdeburg

Keywords. Reaction-diffusion coupling, path optimization, chemical waves

1 Introduction

Optimization strategies play a crucial role for the survival of living species. It is therefore no surprise, that biological evolution has developed a broad spectrum of optimized mechanisms, structures, and behavior. Today certain examples appear to be obvious, such as the specialization of organisms to an aerobic or an anerobic environment [1]. On the other hand, there are still many optimization strategies to be unraveled. Some of them might teach us useful knowledge on how general optimization problems can be solved or approximation of optimized solutions can be obtained without the use of modern computers and numerical algorithms.

In this contribution the construction of shortest paths in systems with labyrinthine obstacles is discussed. The key idea is to exploit the features of excitation waves to accumulate information that is relevant for optimized motion in the system [2,3]. The concept is demonstrated for the chemical Belousov-Zhabotinsky (BZ) reaction [3-5]. Based on these results, the approach is compared to phenomena found in aggregating slime mold colonies [6].

The features of propagating excitation waves form the backbone of our pathfinding 'algorithm'. Excitation waves can be observed in a wide variety of chemical reaction systems, such as the oxidation of carbon monoxide on Pt(110) surfaces [7] or the chlorite-iodide reaction [8]. Also certain biological media are known to support the propagation of these waves (e.g., cardiac and neural tissue [9-11]). An intriguing example is also the occurrence of Ca^{2+} waves in human oocytes shortly after fertilization [12]. Regardless of the particularities of the system, excitation waves can arise from the spatial coupling of nonlinear temporal dynamics by short-range transport processes. Reaction-diffusion systems [13], such as the BZ reaction, are a prominent example for this coupling: Initially the entire system is in its chemical steady-state. A local perturbation that exceeds a certain threshold triggers an autocatalytic reaction leading to the formation of steep concentration gradients. The resulting diffusional flux now triggers the autocatalytic reaction to start in the surrounding area. Consequently, a circular front of chemical activity propagates through the entire system. For one-dimensional media or planar fronts in higher dimensions the front velocity v is constant and obeys the equation

$$v \propto (k\,D)^{1/2} \qquad , \quad (1)$$

where k and D are the rate constant of the autocatalytic reaction and the diffusion coefficient of the autocatalytic species, respectively [9,14]. The concentration profiles of the involved compounds remain also constant in a coordinate system moving with the front. The velocity of a curved front, however, depends on the local front curvature K as expressed in the eikonal equation

$$N = v - D\,K \qquad , \quad (2)$$

with N denoting the normal front velocity [15]. Hence, an expanding circular front ($K > 0$) propagates slower than a planar one. The difference $v - N$, however, rapidly decreases with the decreasing curvature of the expanding wave. It should be noted, that the curvature-velocity dependence [2] is a potential source of errors in our path-finding approach.

The local dynamics behind the front determine whether the system is capable of supporting one solitary front only or allows the propagation of periodic wave trains. In bistable reactions, such as the iodate-arsenous-acid reaction [16] or the acid catalyzed chlorite oxidation of tetrathionate [17], the front consumes at least one of the reactants completely. Truly excitable or oscillatory reactions (e.g., the BZ reaction) recover after the passage of a reaction front and allow the formation and the propagation of periodic wave structures. An important feature of these systems is that the recovery process requires a certain time in which the reaction is refractory. A direct consequence of this behavior is the existence of a dispersion relation $v(f)$ and a limiting frequency above which no stable wave trains exist [18].

One of the most striking phenomenological differences between excitation waves and electromagnetic or acoustic waves is the mutual anihilation of excitation fronts upon collision or, in other words, the absence of interference [19]. During a collision of two excitation fronts the excitation is captured in between the refractory zones of the waves, thus, preventing further propagation. For the same reason, reflection of chemical waves is impossible at most boundaries (e.g., no-flux boundaries). Nevertheless, excitation waves obey *Fermat's principle* to a good approximation. Accordingly, effects such as wave refraction obeying Snell's law can be observed along a line of velocity jump [20].

Fermat's principle, the absence of interference and reflection, and the constant propagation velocity and concentration profiles characterizing excitation waves define the basic ingredients of chemical pathfinding as described in the following.

2 Chemical Pathfinding

2.1 The Belousov-Zhabotinsky Reaction

The Belousov-Zhabotinsky (BZ) reaction is the oxidation of malonic acid by bromate in sulfuric acid [4,5,21]. The reaction is catalyzed by redox couples such as Ce(III)/Ce(IV) or ferroin/ferriin [22]. Most studies of spatio-temporal phenomena are carried out in gel systems or membranes to avoid hydrodynamic perturbations [22,23]. In the ferroin-catalyzed medium the unexcited state is red according to the color of the reduced catalyst, while excitation waves appear as blue bands due to the absorption spectrum of ferriin. The wave propagation is controlled by the autocatalytic production and the diffusion of $HBrO_2$. Typical propagation velocities are of the order of 1 mm/min. In the wave back the system shows high concentrations of bromide and ferriin. The decay of these species is controlling the recovery process of the system. This process is mainly responsible for the magnitude of frequencies observed in periodic BZ wave patterns (0.1-0.01 Hz).

2.2 Experimental

Labyrinth-like structures were prepared from vinyl-acrylic membranes (Gelman Metricel; thickness 140 μm; pore size, 0.45 μm) by carefully cutting out rectangular or circular regions [3]. The membranes were then soaked with a BZ reaction mixture of the following composition: $[NaBrO_3]$ = 0.2 M, $[H_2SO_4]$ = 0.4 M, $[CH_2(COOH)_2]$ = 0.17 M, $[NaBr]$ = 0.1 M, $[Fe(phen)_3^{2+}]$ = 1.0 mM. All values correspond to initial concentrations. Finally, membranes were covered with silicone oil to prevent changes of concentration due to evaporation. The composition of the BZ solution was selected so that no spontaneous wave initiations occurred; however waves could be reproducibly initiated by allowing a silver wire to contact the reagent-loaded membrane. Temperature was kept constant at 25.0° ± 0.2°C. The wave propagation was monitored by image analysis of monochromatic light (λ = 500 nm) reflected from the medium. The video signals were recorded and then digitized into image sequences, which were enhanced and analyzed with standard imaging routines.

2.3 Distance analysis

The simple procedure for the preparation of labyrinth-like reaction matrices (described above) allows the realization of a variety of geometries and provides an effectively two-dimensional system without hydrodynamic perturbations. The rectangular or circular holes cut out of the membrane act as obstacles for wave propagation. Figure 1A shows a 50-frame composite image of a chemical wave traveling through a BZ membrane 'labyrinth'. The wave was initiated at the lower left corner. During its journey through the system it splits at each junction and therefore reaches all sites of the labyrinth that are connected to the starting point. Note, that split segments traveling along different routes can meet again, thus, leading to wave collisions. Two of these collisions can be seen in the depicted example (Fig. 1; see white arrows). The creation of periodically orbiting waves is impossible as long as the wave segments stay in contact with the boundaries. For different experimental conditions Agladze et al. [24] reported the detachment of high frequency wave trains from a curved boundary giving rise to spiral waves. In our experiments, however, no detachment of waves and, therefore, no periodic phenomena were observed.

The data shown in Figure 1 allow the measurement of shortest distances between the site of wave initiation and the points located along the wave fronts. The excitation propagates with a constant velocity of $v = 2.41 \pm 0.18$ mm/min through the system obeying Fermat's principle to a good approximation. Hence, the front recorded at the time t consists of points having a distance of $s = vt$ to the initiation site. Image sequences with higher temporal resolution allow the generation of maps that give the path length from any location in the maze to a given target point, within the resolution of the composite image. The major source of errors in this approach arises from the curvature-velocity dependence [15]. While a temporally changing propagation velocity $v(t)$ could be easily accounted for in our analysis, it is much harder to incorporate spatial dependences $v(x,y)$ that arise from the eikonal equation. However, curvature-induced changes of propagation velocity should be tolarable in our experiments, since the channel width is large (2-3 mm) compared to the expected critical radius ($r_{crit} = D/v \approx 50$ μm). To further minimize this error it is necessary to increase the length scale of the labyrinth and/or to increase wave velocity by adjustments of the chemical composition.

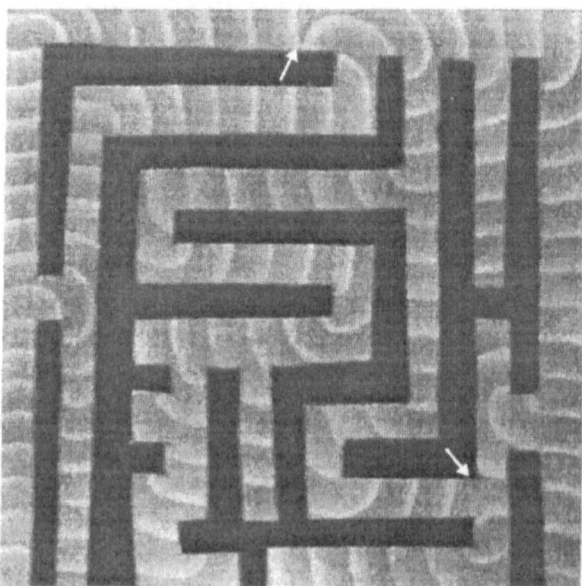

Figure 1 Composite image showing 50 snapshots ($\Delta t = 50$ s) of a solitary wave propagating through a BZ membrane labyrinth. Obstacles appear as black rectangles. Image area: 3.2 cm by 3.2 cm. (Figure taken from ref. 3).

2.4 Construction of Shortest Paths

Chemical distance analysis yields the optimal transit time and distance for every location in the excitable system by monitoring the propagation of a *single* wave till complete extinction. The data accumulated provide the basis for the determination of detailed minimum-length paths [3]. The optimal pathway from an arbitrary point to the starting point is easily determined according to the vector field of normal wave velocities. A typical example of such a vector field is shown Figure 2. Alternatively to the direct analysis of normal wave velocities, one can also calculate the gradient vector field $V = \nabla(t(x,y))$, where t denotes the time when the site (x,y) becomes excited. The vector field obtained transforms the global problem of path optimization to the local task of following the $-V$ vectors. Optimal trajectories can also be found by searching for the the nearest point indexed $t\text{-}\tau$, where t is the elapsed time from the wave initiation to the current position and τ is the time increment between successive images.

Five examples of shortest pathways are shown in Figure 2. The figure reveals that the path optimization includes local features such as diagonal trajectories through the rectangular segments of the labyrinth. Note, that the neighboring points at approximately $(x,y) = (18,30)$ have quite different pathways back to the starting point. These points are separated by a line along which a wave collision has occurred (compare, Fig. 1). Since the detection of wave collisions is simple, the experiment yields directly this useful information on the location of separatrices segregating domains of significantly different pathways. Similar 'watersheds' have been observed in experiments with synchronous initiation of several waves, yielding information on optimal transits back to a particular initiation site [3].

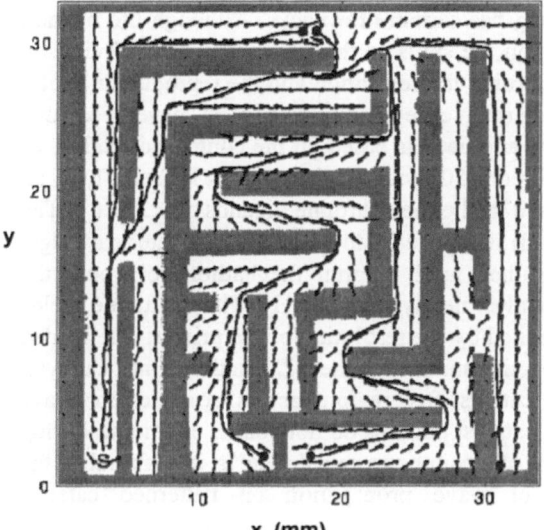

Figure 2 Vector field of normal wave velocities obtained from the data shown in Fig. 1. S denotes the point of wave initiation. The solid lines give five examples of shortest paths.
(Figure taken from ref. 3).

Figure 3 Excitation waves on BZ membranes with printed catalyst patterns. Waves appear as white segments. The checkerboard arrangement (dark: catalyst-loaded; medium gray: catalyst free) gives rise to hexagonal wave geometries. Image area: 14.5 cm^2; side length of triangular cells $\approx$ 1.65 mm. (Figure taken from ref. 25).

2.5 An Alternative Technique for Creating Patterned Reaction Matrices

The preparation of active systems with unexcitable obstacles as used for the demonstration of chemical pathfinding has certain drawbacks. Although the method is simple, the quality obtained relies on the skills of the experimentalist. Obviously, there are also heavy restrictions concerning the resolution and the reproducibilty of a certain geometry. Therefore, an alternative technique has been developed that overcomes these drawbacks [25]. The approach is based on the controlled loading of polysulphone membranes (Gelman Supor-450) with the BZ catalyst Fe(batho)$_3^{2+}$ - an octahedral iron-complex having three bathophenathroline ligands [22,25,26]. The catalyst solution (in glacial acetic acid) is pipetted into a clean, dry printer cartridge. A Hewlett Packard DeskJet 520 is then used to print the bathoferroin onto the membranes which immobilize the catalyst nearly immediately. The technique yields high-quality catalyst patterns only limited by the resolution of the printer. The loaded membranes can then be used as a reaction matrix for the BZ reaction. Only catalyst-loaded regions are excitable, while catalyst-free regions act as obstacles for wave propagation. Notice, that the printed catalyst patterns do not vanish or smooth during the course of the reaction, since the polysulphone membrane is immobilizing the bathoferroin. A typical example of wave propagation on patterned catalyst membranes is shown in Figure 3.

Additional experiments have been performed to investigate possibilities of using photosensitive ruthenium-complexes for the described printing technique. The classic Ru-catalyst, the water soluble complex $Ru(bpy)_3^{2+}$ [27], does not immobilize in polysulphone membranes. Therefore, we have synthesized $Ru(batho)_3SO_4$ which could be readily immobilized with our technique. Unfortunately, this complex turned out to be a poor catalyst for the BZ reaction. Combinations of bathoferroin and $Ru(batho)_3^{2+}$, however, are creating a photosensitive medium, in which wave propagation occurs. The use of two different catalysts in the presented printing technique allows the controlled generation of (a) unexcitable (catalyst-free) areas, (b) excitable areas that are not influenced by light (bathoferroin-loaded), and (c) photosensitive domains where excitability can be externally controlled (mixed-catalysts).

3 Optimization in Slime Mold Aggregation

Pathfinding and path optimization are important tasks for many living species. A male moth can find a 'calling' female moth by sensing smallest amounts of volatile pheromones (some moths can detect this scent trail up to 3 km) [28]. Other studies have reported path optimization performed by ants commuting between a food source and their nest [29]. One of the most fascinating examples, however, is the control of wiring of the brain, that is how neurons find their correct targets. Our knowledge on this 'axon pathfinding' and the target recognition in often complex surroundings is still relatively small, although several of the involved compounds, such as repulsive guidance molecules, have already been identified [30].

An intruiging question is whether certain living systems are exploiting a pathfinding technique similar to the approach discussed above for the chemical Belousov-Zhabotinsky reaction. A promising candidate for related investigations is the cellular slime mold *Dictyostelium discoideum*.

3.1 The Cellular Slime Mold *Dictyostelium discoideum*

Dictyostelium discoideum is well suited for the study of cellular communication and a widely investigated example of excitable systems in biology [31]. *Dictyostelium* cells are amoebae that can easily change their shape by forming pseudopodia exploring their surrounding. The typical size of a cell is about 10 μm. *Dictyostelium discoideum* can be found in forrest soils where it feeds on bacteria and grows by cell devision. Certain factors, such as starvation, can trigger a fascinating developmental cycle that causes 10^4-10^5 single amoebae to aggregate and form a multicellular fruiting body bearing spores [6,31].

The aggregation is controlled by intercellular propagation of excitation waves [31,32]. Extracellular 3',5'-cyclic adenosine monophosphate (cAMP) acts as a chemoattractant and is the propagator species of the underlying reaction-diffusion mechanism [9,32,33]. Cells in the aggregation center periodically produce cAMP. The cAMP is secreted into the extracellular medium, where it diffuses away.

Neighboring cells detect this chemoattractant via cell surface receptors [34]. The stimulated cells then produce huge amounts of cAMP, which they in turn secrete. This feedback process results in a wave-like propagation of the cAMP signal from cell to cell and from the aggregation center outwards. Binding of cAMP to the receptor induces a phosphorylation of the receptor leading to desensitization and the shut down of the cAMP production. An extracellular enzyme (phosphodiesterase) degrades cAMP allowing the system to recover. Stimulated cells use the information encoded in the cAMP concentration wave to move chemotactically toward the aggregation center.

A typical experimental procedure for measuring excitation patterns is briefly explained in the following: *Dictyostelium* cells (here: strain AX2) are harvested from nutrient solution at a cell density of 5×10^6 cells/ml. They are carefully washed with phosphate buffer to remove all nutrients and then spread uniformly on the surface of agar plates. An useful parameter controlling the dynamics of wave patterns and aggregation is caffeine that is added to the agar gel at concentrations of about 0 to 5 mM. The classic methodology for the detection of wave patterns is darkfield-photography [31,32]. Typical wave lengths and periods found during the early *Dictyostelium* aggregation are approximately 1-5 mm and 5-10 min, respectively. Cell migration is easily monitored by recording microscopic images (bright-field illumination) with a time-lapse video recorder [35].

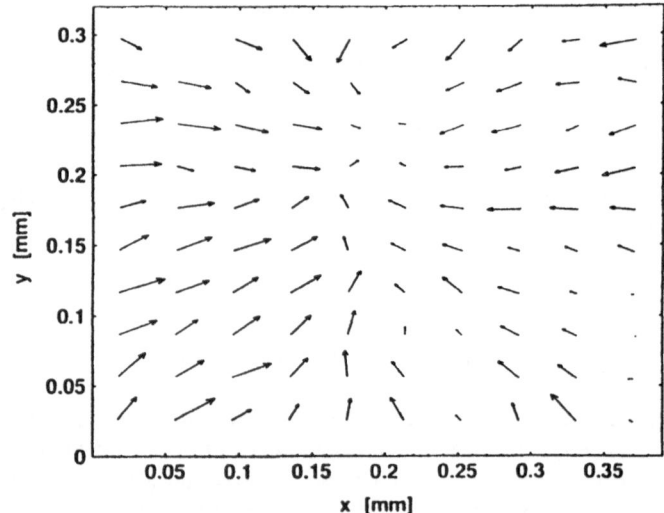

Figure 4 Velocity-field of aggregating *Dictyostelium* cells controlled by a target pattern of propagating cAMP waves. The pacemaker of the concentric waves is located close to (0.2 mm; 0.25 mm). The amoebae describe a motion along radial lines, since chemotactic cell motion in this system is always perpendicular to the chemical wave front (i.e.; parallel to the concentration gradient). Aggregation takes place on the surface of a thin agar gel. (Figure taken from ref. 36).

3.2 Chemical Waves Guide Amoebae Along Shortest Paths

Chemotactic cell migration is always oriented perpendicular to the propagating fronts of cAMP concentration. Notice, that in contrast to the erratic chemotaxis in bacterial systems, which is reminiscent of random walks, slime mold amoebae move in a fairly homogenous fashion. In earlier studies we have investigated microscopic areas of aggregation patterns by velocity-flow analysis [35,36]. The results showed that the migration velocity of amoebae changes rhythmically between approximately 0 and 30 µm/min with periods (5-10 min) that coincide with the periods of macroscopic excitation patterns monitored by dark-field photography. Most of these patterns are spirals or expanding concentric waves (target patterns). An example of a vector field of chemotatic cell motion close to the center of a target pattern is shown Figure 4. The velocity field is extracted from a 120 s sequence by mutual-correlation analysis [35]. The pacemaker of a target pattern controls cell aggregation like a star-shaped node attracting points in phase space.

The chemotactic migration behavior of *Dictyostelium* amoebae has a fascinating connection to the path-finding procedure described in Chapter 2. In our chemical demonstration, a computer was used for the construction of the vector-fields of normal wave velocity which allowed the determination of shortest paths from any point in the system back to the site of wave initiation. The chemical wave was exploited to transform the global problem of path optimization into a local one. In other words, the local geometry and the local propagation direction of the wave carries crucial information on how to leave the labyrinth along the shortest path. Obviously, this information is highly useful for an observer-entity that wants to optimize its motion, but is not aware of the global structure of its surrounding. This is exactly the situation that slime mold amoebae face during aggregation. The cells are capable of detecting normal wave velocities via concentration gradients of cAMP. This detection mechanism enables the slime mold to perform optimized navigation in complex surroundings such as given by their natural habitats (soil etc.).

While it has been known for a long time, that chemical waves of attractant concentration are used by the amoebae to find the aggregation center [6,31], it is surprising that the aggregation occurs in an optimized fashion. The main features of excitation waves, including constant profiles and the absence of interference phenomena, are crucial components that allow this optimization. If the cells in the aggregation center would constantly produce the attractant, large amounts of cAMP would be necessary to establish concentration gradients that are steep enough for reliable orientation. In particular, oversaturation of receptor molecules might be a serious problem in such an aggregation process and cells would be confronted with significant changes of attractant concentration while approaching the center. The time required by diffusion to build a detectable gradient in a distance of about 1 cm from the aggregation center might be an additional factor, favoring the use of excitation waves. Nevertheless, a recent study of Wang and Kuspa [37] suggests, that certain mutants of Dicyostelium that are not able to perform extracellular cAMP signaling can aggregate, if the initial cell density is higher than approximately 14×10^5 cells/cm^2. This value has to be compared to a critical cell density of about 1×10^5

cells/cm^2 for the wild type. It will be interesting to study the aggregation mechanism of the mutant and to compare it with the mechanisms discussed above.

3.3 Wave Collisions, Aggregation Territories, and Voronoi Diagrams

So far, we have discussed the possibility that the aggregation of slime mold amoebae occurs along shortest pathways. In other words, cells are approaching their aggregation center in a nearly optimal time. Another point of interest, however, is the competition and the separation between adjacent aggregation territories. Figure 5 shows an numerical simulation of collisions between six solitary waves that are initiated from the sites $S = (p_1, p_2, ..., p_n)$ synchronously. The colliding waves mutually annihilate along straight lines that form an array of polygons. This array is identical to the well-known Voronoi diagrams $Vor(S)$:

$$Vor(p_i) = \{r \in \mathbf{R}^2 \mid dist(r, p_i) \leq dist(r, p_j); j \neq i\} \qquad (4).$$

The collision lines demarcate boundaries between these basins which consist of points with shortest distance to one particular initiation site. It seems reasonable to assume that *Dictyostelium* might exploit chemical waves to estimate these Voronoi diagrams for different aggregation centers. Consequently, cells would migrate to the nearest wave source, thus adding a second level of optimization to their aggregation process.

Figure 5 Superposition of six circular excitation waves simulated with a generic reaction-diffusion model (cf., ref. 39). The lines definded by the locations of wave collisions coincide with the Voronoi diagram (black lines) of the six initiation sites. They form watersheds between basins of points which have their shortest distance to one particular wave source (Voronoi polygons). Waves were initiated synchronously.

What are the limits of this approach? A major limitation is that the exact construction of a Voronoi diagram requires the synchronous initiation of waves: Two circular waves initiated at the sites (0, 0) and (1,0) synchronously will collide along the line $x = 0.5$. A time lag $\Delta\tau$ in wave initiation, however, causes the fronts to collide along the curve

$$r = [(v\,\Delta\tau)^2 - 1]\,[2\,(v\,\Delta\tau - \cos\phi)]^{-1} \qquad\qquad , \quad (3)$$

where r and ϕ are the polar coordinates and v denotes the wave velocity (disregarding curvature effects). Further deviations from the simple Voronoi diagrams appear for periodic wave trains having different frequencies. Due to the master-slave dynamics of wave patterns in excitable systems, the curve of wave collisions would slowly shift toward the low frequency pacemaker until the high frequency source would remove its competitor completely. In the case of *Dictyostelium discoideum*, however, cell motion is partially compensating for this effect: As soon as wave collisions occur, cell density decreases in the corresponding areas due to the migration of cells and the adjacent aggregations territories start to decouple. After approximately 30 min a nearly cell-free and therefore unexcitable corridor has formed between the territories.

4 Conclusions

Excitation fronts allow the accumulation of data for complete Euclidean distance maps in systems with or without complex geometry. These fronts can be also used to transform the global task of finding shortest paths to a problem that can be solved on the basis of *local* information. We have demonstrated this approach experimentally for the BZ reaction. In this context, an efficient technique for the preparation of chemical systems with complex geometries has been discussed. The slime mold *Dictyostelium discoideum* seems to be an excellent example of a living system exploiting pathfinding on the basis of excitation waves. However, further experiments are necessary to support this hypothesis.

The use of excitation waves has certain fascinating features that are hard to realize with numerical approaches. One of the most striking characteristics is that an excitation front analyses systems in a highly parallel fashion: The time required for performing the experiment depends only on the longest distance between the point(s) of wave initiation and the set of locations connected to the starting point(s). It is neither influenced by the total area nor by the number of branches.

Curvature effects on propagation velocity can be a source of errors. Since our major motivation, however, is to investigate non-numerical and non-electronic approaches for path optimization that can be exploited by living systems, these minor deviations appear to be less relevant. Nevertheless, it will be useful to further quantify these factors. There might be even interesting applications of curvature-induced deviations such as the shut-down of channels having a half-width below the critical radius $r_{crit} = D / v$ [3].

5 Acknowledgment

Experiments on chemical pathfinding were carried out in collaboration with K. Showalter (West Virginia University) and Á. Tóth (Jozsef Attila University, Szeged, Hungary). Their help is gratefully acknowledged. I also thank P. Kettunen (University of Oulu, Finland), P. Palmer (West Virgina University), and S. C. Müller (Magdeburg University). The present study was partly supported by the Fonds der Chemischen Industrie.

6 References

1. B. Alberts, D. Bray, J. Lewis, M. Raff, K. Roberts, and J. D. Watson, Molecular Biology of the Cell (Garland Publishing Inc., New York, 1994).
2. J. A. Sepulchre, A. Babloyantz, and L. Steels, in Proceedings of the International Conference on Artificial Neural Networks, (T. Kohonen, K. Makisara, O. Simula, and J. Kangas; Eds.) (Elsevier, Amsterdam, 1991), pp. 1265-1268.
3. O. Steinbock, Á. Tóth, and K. Showalter, Science 267, 868 (1995).
4. R. J. Field and M. Burger (eds.), Oscillations and Traveling Waves in Chemical Systems (Wiley, New York, 1985).
5. A. N. Zaikin and A. M. Zhabotinsky, Nature 225, 535 (1970).
6. W. F. Loomis, Development of Disctyostelium (Academic Press, San Diego, 1982).
7. G. Ertl, Physikalische Blätter 46, 339 (1990); see also: S. Jakubith et al., Phys. Rev. Lett., 65, 3013 (1990); G. Ertl, Science 254, 1750 (1991).
8. P. De Kepper, J. Boissonade, and I. R. Epstein, J. Phys. Chem. 94, 6525 (1990).
9. J. D. Murray, Mathematical Biology (Springer, Heidelberg, 1988).
10. A. Gorelova and J. Bures, J. Neurobiol. 14, 353 (1983).
11. J. M. Davidenko et al., Nature 355, 349 (1992).
12. M. Sousa, A. Barros, and J. Tesarik, Mol. Hum. Reprod. 2, 265 (1996); see also: A. Galione et al., Science 261, 348 (1993); J. Parrington et al., Nature 379, 364 (1996).
13. T. S. Akhromeyeva, S. P. Kurdyumov, G. G. Malinetskii, and A. A. Samarski, Phys. Reports 176, 189 (1989).
14. R.-L. Luther, Zeitschrift für Elektrochemie und angewandte physikalische Chemie 12, 596 (1906); R. A. Fisher, Ann. Eugenics 7, 353 (1937).
15. J. J. Tyson and J. P. Keener, Physica D32, 327 (1988); P. Foerster, S. C. Müller, and B. Hess, Science 241, 685 (1988).
16. D. Horváth and K. Showalter, J. Chem. Phys. 102, 2471 (1995).
17. I. Nagypál and I. R. Epstein, J. Phys. Chem. 90, 6285 (1986); I. Lengyel and I. R. Epstein, Proc. Natl. Acad. Sci. U.S.A. 89, 3977 (1992).
18. J. D. Dockery, J. P. Keener, and J. J. Tyson, Physica D30, 177 (1988); H. Sevcikova and M. Marek, Physica D39, 15 (1989).

19. V. A. Vasilev, Y. M. Romanovskii, D. S. Chernavskii, and V. G. Yakhno, Autowave Processes in Kinetic Systems (VEB Deutscher Verlag der Wissenschaften, Berlin, 1986).
20. A. M. Zhabotinsky, M. D. Eager, and I. R. Epstein, Phys. Rev. Lett. 71, 1526 (1993).
21. F. W. Schneider and A. F. Münster, Nichtlineare Dynamik in der Chemie (Spektrum Akademischer Verlag, Heidelberg 1996).
22. T. Yamaguchi et al., J. Phys. Chem. 95, 5831 (1991).
23. B. Neumann, Zs. Nagy-Ungvarai, and S. C. Müller, Chem. Phys. Lett. 211, 36 (1993).
24. K. Agladze et al., Science, 264, 1746 (1994).
25. O. Steinbock, P. Kettunen, and K. Showalter, Science 269, 1857 (1995).
26. A. Lazar, Z. Nosziticzius, H. Farkas, and H.-D. Försterling, Chaos 5, 443 (1995).
27. V. Gáspár, G. Bazsa, and M. Beck, Z. Phys. Chem. 264, 43 (1983); L. Kuhnert, K. I. Agladze, and V. I. Krinsky, Nature 337, 244 (1989); O. Steinbock and S. C. Müller, Physica A188, 61 (1992).
28. J. S. Kennedy, Science 184, 999 (1974); J. S. Kennedy, Physiol. Ent. 8, 109 (1983).
29. R. Beckers, J. L. Deneubourg, and S. Goss, J. Theor. Biol. 159, 397 (1992); S. Goss, S. Aron, J. L. Deneubourg, and J. M. Pasteels, Naturwissenschaften 76, 579 (1989).
30. M. Tessier-Lavigne and C. S. Goodman, Science 274, 1123 (1996).
31. G. Gerisch, Naturwissenschaften 58, 430 (1971).
32. F. Siegert and C. J. Weijer, J. Cell Sci. 93, 315 (1989); J. Rietdorf, F. Siegert, and C. J. Weijer, Develop. Biol. 177, 427 (1996).
33. J.-L. Martiel and A. Goldbeter, Biophys. J. 52, 807 (1987); J. J. Tyson, K. A. Alexander, V. S. Manoranjan, and J. D. Murray, Physica D34, 193 (1988); B. Vasiev, F. Siegert, and C. J. Weijer, J. Theor. Biol. 184, 441 (1997); a direct measurement of cAMP waves is reported in: K. J. Tomchik and P. N. Devreotes, Science 212, 443 (1981).
34. See e.g., Y. Yu and C.L. Saxe III, Develop. Biol., 173, 353 (1995).
35. O. Steinbock, H. Hashimoto, and S. C. Müller, Physica D49, 233 (1991).
36. O. Steinbock and S. C. Müller, Zeitschr. f. Naturforsch. 50c, 275 (1995).
37. B. Wang and A. Kuspa, Science 277, 251 (1997).
38. D. T. Lee, IEEE Trans. Comput., C-33, 1072 (1984).
39. D. Barkley, Physica D49, 61 (1991).

BSE Viewed Dynamically:
A Possible Early Cure Based on Passive
Immunization Against PrP^{Sc}

Otto E. Rössler[1], John L. Hudson[2], Reimara Rössler[3], and Jürgen Parisi[4]

[1] Division of Theoretical Chemistry, University of Tübingen, D-72076 Tübingen, Germany
[2] Department of Chemical Engineering, University of Virginia, Charlottesville, Virginia 22901, USA
[3] Medical Policlinic, University of Tübingen, D-72076 Tübingen, Germany
[4] Faculty of Physics, Department of Energy and Semiconductor Research, University of Oldenburg, D-26111 Oldenburg, Germany

Abstract. Bovine spongiform encephalopathy (BSE) is proposed to be a "dynamical disease" in the sense of Mackey and Glass. It according to the Prusiner–Gajduesek model involves an autocatalysis with an exceedingly low threshold for "going off." A few trigger molecules are believed to be sufficient. This extreme sensitivity – combined with the fact that only natural body constituents are involved – paradoxically makes intervention rather easy: "Increase the wall reactions". Hence increasing the scavenging of the autocatalytic bodily constituent may suffice. Homologous antibodies against the spontaneously occurring "folded-sheet" isoform (PrP^{Sc}) of the natural prion protein (PrP^{C}) should therefore be effective. A single injection may abolish the disease in an earliest stage.

The dynamical behavior of BSE may be analogous to a well-known chemical reaction, the oxygen–hydrogen reaction ("popping gas"). This reaction does not go off spontaneously because of the presence of wall reactions [1]. These infrequent, easily saturable spontaneous side reactions prevent the autocatalysis from going off through the scavenging of the autocatalytic species at low concentrations. Thus the self-acceleration is halted.

Current knowlege postulates an exceedingly low infectious dosis for BSE-inducing prions taken up orally and passing the gut barrier and the blood–brain barrier [2]. Then the infectious molecules are believed to trigger an autocatalytic process – just like UV radiation would trigger the oxygen hydrogen mixture in a closed glass capillary. This means that in BSE, the scavenging, natural "wall reaction" must be exceedingly weak.

The infectious molecular specimens can therefore be compared with – and are intuitively perceived as – whole invading parasitic organisms or parasitic molecules (like viruses), since in this latter case also a single specimen suffices in theory (although not in practice). This biological assumption is responsible for the fact that reaction-kinetic thinking is not usually brought to bear when it comes to understanding the dynamics of BSE.

Although the idea of a threshold is not uncommon in reaction kinetics, nevertheless it is unusual for the threshold of a reacting system to be as low as it apparently is in the case of BSE. Because the threshold is so very low, the "therapy" could be quite easy. It suffices to raise the threshold up to more usual values. This is easy to do in the case of a very low-threshold (easily going-off) autocatalysis. The introduction of a few scavenger molecules suffices to accomplish this increase in threshold.

The autocatalytically active species are responsible for the disease as far as one knows today. They are the PrP^{Sc} molecules [3]. These molecules are conformal isomers – "beta-pleaded sheet configurations" [3] – of naturally occurring globular (fully 3-dimensional, non-flat) molecules. These normal molecules are the PrP^{C} molecules. They are produced by the brain cells for an unknown purpose. They are not necessary for survival since animals that are no longer able to make them through genetic manipulation live normal lives (apart from being immune to BSE) [4].

The isoforms – the pathological PrP^{Sc} molecules – are naturally generated from their globular precursors at an exceedingly low rate (otherwise the spontaneously occurring cases would be more frequent). For it is believed that they trigger the disease through autocatalysis in the spontaneously occurring cases of Creutzfeld–Jacob disease. It is also believed that they do the same thing – multiply autocatalytically – after a sufficient number of "sibling molecules" have been introduced from the outside. These helper molecules (in overcoming the threshold for autocatalysis) either are identical in type when coming from the same animal species or are not quite identical from the molecular point of view when coming from a different animal species. They are believed to be introduced through feeding (or mite bites, respectively), in the case of BSE or its analogues [2,3].

To interrupt the autocatalysis, it suffices to increase the outflux rate of the autocatalytic species. To this end, antibodies can be used in principle as mentioned. Homologous (samer-species) antibodies against PrP^{Sc} should be easy to obtain by using adjuvans-type immunization methods in animals. The animals to be used to obtain the antiserums from them would thus have to be injected with viable PrP^{Sc} molecules. Nevertheless, they will presumably not develop the disease themselves in spite of having been injected with large doses of PrP^{Sc}. For the antibodies formed would protect them through active immunization. Monoclonal antibodies would be desirable to be obtained eventually in order to yield an optimal passive immunization. A single injection with such an antiserum should be sufficient to generate a "flooding" of all the dangerous sites in the organism where the autocatalysis can go off (in the brain and the lymphatic system), by a sufficient amount of scavenger molecules for a sufficiently long time. The disease (the autocatalysis) would then be prevented from "going off" for a sufficiently long time that afterwards, the risk is no higher any longer than it is for any "non-infected" organism.

How can one arrive at such an optimistic interpretation? The reason is, as mentioned, reaction kinetics. The underlying reaction-kinetic scheme is that of a two-variable oscillator, as proposed some time ago for a "temperature compensated chemical clock" [5]. There is a slow "influx reaction" to the first variable present (corresponding to PrP^C in the present case), and there is a first-order decay reaction present for the same first variable (natural degradation of PrP^C in the present case). In addition, there exists a very weak side reaction in which the first variable is transformed into the second variable (PrP^{Sc} in the present case). This second species is autocatalytically active. It is also subject to a scavenging reaction analogous to a wall reaction that is "saturable" – like a Langmuir adsorption isotherm or a Michaelis–Menten type outflux. This is the case in all low-threshold autocatalytic reactions [5]. Thus the same functional elements as discussed above form a reaction-kinetic relaxation oscillator.

This relaxation-type oscillator can also be run as a bistable system as we shall discuss below. Before doing this, let us briefly mention in passing that the oscillatory potential of the full two-variable scheme can also be taking advantage of in principle. In this more complicated case, one would interfere with the spontaneous production of the precursor (PrP^C) or else would increase its outflux reaction. The latter aim could be accomplished by introducing antibodies once more, but this time to PrP^C itself, the precursor to the second autocatalytic variable. Hereby, one would need fairly high concentrations, however. This is because unlike PrP^{Sc}, PrP^C is not present only in tracer amounts.

This method could even be tried much later – after the autocatalysis has already set in. The on-going high-level autocatalysis would then come to a halt since the substrate – needed by any autocatalytic "fire" – would have fallen below the threshold value that admits self-sustenance. This second method would also amount to a sort of "cure" at first. However, eventually the natural production of PrP^C would, after the gradual disappearance of the antibodies, be responsible that the concentration of PrP^C would go up again to reach its former steady-state value. Shortly before that, another autocatalytic event (and hence relapse of the disease) would be bound to set in, triggered by rest material of the second variable from last time. In this case we would thus indeed have an "oscillator".

This proposal amounts to a therapeutic scheme of last resort, so to speak, even after the manifest onset of the disease – interference with the PrP^C molecules in their native form through passive (or active) immunization. Many problems would have to be effectively overcome in this case, however, starting with the blood–brain barrier which is hard-to-overcome by large (non-tracer-doses) of protein molecules (antibodies), and ending with induced up-regulation of the production of the precursor molecule. Also an induced inflammatory response would have to be coped with. We therefore return to our main proposal – removal of the second variable by tracer amounts of a reaction partner (antibodies).

This main proposal is much more limited dynamically speaking since it looks only at the bistable nature of the second variable itself. To prevent the latter from going off, it suffices to introduce (by passive immunization) tracer amounts of antibodies to this variable (PrP^{Sc}). This appears to be feasible in principle – in analogy to the fate of the ingested (rather than injected) molecules which trigger the disease in reality.

In spite of its straightforwardness and simplicity, the proposal made contains a surprising feature. The specific nature of the infectious agent itself (the exact type of prion [6]) is not crucial for the outcome. The "infection" becomes conceptually reduced to a threshold-surpassing "trigger event" in a far-from-equilibrium reaction system (cf. [7]). This change of philosophy – from thinking in terms of autonomous infective agents to thinking in terms of ordinary dynamical variables like an influx concentration – may turn out to be crucial for the interventional strategies to be adopted regarding BSE.

The proposal made above consists in the introduction of antiserums against PrP^{Sc} – passive immunization. Alternatively, active immunization (which would be less expensive) could be used as well in animals. Both methods would be much more economical than disposing of the live animals themselves. In view of the inherent economic potential of our proposal, it appears worthwhile to try and find out soon whether the immunization method works out on smaller animals like hamsters (which generally allow one to arrive at definitive results much faster, cf. [2]).

The proposal at the same time is of potential interest also in the context of how to treat human suspects who have been freshly exposed – like a surgeon who inadvertently cut himself in the presence of contaminated material. The same monoclonal antibodies could be used for trying to prevent the disease from going off in humans. The method if it works would act as cure here, too.

Once early diagnosis of the first manifest stages becomes possible – in lymphatic tissue, for example –, the same method may still work, although only in its active-immunization variant. Note that at that stage, there would be no need for concern any more as far as the artificial introduction of infectious molecules along with an immune-stimulating adjuvans to generate an active antibody response is concerned.

The second method mentioned above (stimulation of an immune response to the precursor protein PrP^{C}) may also become of interest in this case. However, this is a much less encouraging option as we saw. It may even contribute to the active mechanism of the disease itself since deposited immune complexes involving the precipitated precursor protein may be formed. Formation of such immune complexes may play a role in the natural course of the disease, by contributing to the massive increase in the production of PrP^{C} in an up-regulated form (which may cause the cells to be turned into single-purpose factories of this protein before they die). Nevertheless even this second method in its desperateness still illustrates the potential advantages to be gained from thinking, not in terms of an "infectious parasite" but in terms of ordinary dynamical systems.

The fact that on the level of the population of the host, a "selection for infectiosity" in the Darwinian sense takes place will have to be taken into account in any future complete theory of the disease. However, this should not detract from the potential usefulness of the purely dynamical paradigm considered here.

To conclude, we have proposed to view BSE as a triggerable "dynamical disease" in the sense of Mackey and Glass [8] (whereby the triggerability would be a new feature), rather than as an infectious disease. In this view, no self-replicating infectious agent is required on the level of the individual. All pathological substances – fibrils of beta-sheets and globular amyloid – would be entirely host-made. The Creutzfeld–Jacob disease would be more like a "dyscrasia" – a wrong mixture of natural body constituents – than like an infectious disease in the traditional sense. Much as in the early times of tuberculosis, where "most" of the damage would be caused by the defense mechanisms, including amyloidosis [9], most or – perhaps even all – features of BSE would be nonspecific. This change of emphasis nourishes hopes that rather simple strategies of intervention may indeed be effective.

Acknowledgments We thank Peter Wills, Theo Dingermann, Gerold Baier, Achim Kittel, Martin Bünner, John McCaskill and Dieter Fröhlich for discussions. For J.O.R.

References

[1] Semenov, N., Chemical Kinetics and Chain Reactions, Oxford University Press, London, 1935.
[2] Dingermann, T., Bovine Spongiforme Enzephalopathie. Deutsche Apothekerzeitung 136, 165-174 (1996).
[3] Gajdusek, D.C., J. Neuroimmunol. 20, 95-100 (1988).
[4] Büeler, H. et al., Cell 73, 1339-1347 (1993).
[5] Rössler, O.E., San Diego Biomedical Symposium 14, 99-104 (1975).
[6] Wills, P.R., J. Theoret. Biol. 122, 157-178 (1986).
[7] Nicolis, G. and Prigogine, I., Self-Organization in Nonequilibrium Systems, Wiley, New York 1977.
[8] Mackey, M. and Glass, L., Science 197, 287-289 (1977).
[9] Letterer, E., Allgemeine Pathologie (General Pathology), Thieme, Stuttgart 1959.

Molecular Semiotic Structures in the Cellular Immune System: Key to Dynamics and Spatial Patterning ?

Walter Schubert
Inst. f. Medical Neurobiology, Neuroimmunology and Molecular Pattern Recognition Research Group, Otto-von-Guericke-University of Magdeburg

1. Introduction

The organization of complex unicellular and higher organisms involves the proper spatial distribution of macromolecules, ions and cell organells, the interplay of these components and their individual regulation due to functional requirements. A major goal of biological research is the understanding of these complex cellular mechanisms and spatial arrangements at the molecular level. Molecular strategies, such as biochemical approaches and molecular genetic techniques have significantly enhanced our knowledge on the structure and function of single molecular classes or monomolecular interactions, such as the interaction of hormones with their specific receptors. However, we are not aware of higher complex combinatorial molecular specification of functions in biological systems. The latter refers, for example, to the cell surface membrane whose specialized functions must require distinct protein compositions. Whilst it is elegantly possible by molecular methods to identify and analyze the single proteins of such higher complex protein systems, it remains a major challenge (and hitherto unresolved problem) to identify the particular compositions of proteins in cells (at the single cell level) that may encode a particular function. Given that such molecular organization exists at cell surfaces, the combinations of proteins, which are specific for given biological contexts (or functions), might be visualized by fluorescence methods as combinatorial fluorescence signals or "signs" (semiotic elements) within a setting of syntactical rules set by the cells of a given system. Semiotic has at its goal a general theory of signs at all their forms and manifestations (8). It is postulated here, that molecular semiotic as a science of the combination of molecules (single components or signs) within a biological system will be critical for understanding complex cellular interactions in health and disease. Approaching that problem requires significant technological improvements to overcome the pesent limitations of fluorescence microscopy.

Based upon observations in skeletal muscle (11) a new fluorescence labeling methodology was developed (12,14,15). It provides a tool to identify protein compositions at the single cell level with high power of combinatorial molecular discrimination. Applied to the cell surface organization in the cellular immune system, the methodology provides combinatorial molecular data on the immune cell surface in vivo. The data hitherto obtained by the methodology indicate that a large, but relatively limited number of different receptors, adhesion and signaling molecules are grouped into an enormous number of different combinations that fulfill the formal criteria of a semiotic system. The data give rise to a concept suggesting that these combinations are part of a molecular network (cooperative units of a battery of different gene products) that encodes specificity of migratory pathways in the cellular immune system, and thus may profoundly contribute to the organization of basic immune functions.

2. The Problem:
Selectivity of Migration Patterns in the Cellular Immune System

Unlike cells of solid organs, the cells of the immune system (T- and B -lymphocytes, natural killer cells, monocytes/macrophages, granulocytes) must be able to both traffic thoughout the organism and to localize sites where protecting immune functions must be exerted (immunosurveillance). When infections are detected, lymphocytes rapidly accumulate in the relevant site. This is brought about by mechanisms which mediate the interaction of circulating lymphocytes (lymphocytes circulating in the blood) with the affected endothelium lining the wall of small blood vessels at this site. Migration of circulating lymphocytes to secondary immune organs involves similar directed and highly selective trans-endothelial migration. Characteristically, the lymphocyte-endothelial interaction leads to extravasation (transendothelial migration) of the lymphocyte to accumulate and/or further migrate within the affected site or secondary immune organ. Similarly, lymphocytes in certain organ-specific autoimmune diseases accumulate within the intact tissue, organize themselves as densly packed lymphocyte networks and thereby contribute to the tissue destruction. Selective migration of lymphocytes to sites of bacterial or viral infection and the transmigration of lymphocytes through the diverse solid organs and their recirculation to secondary immune organs are physiological homing processes that are implicated in immuno-surveillance and homeostasis in the cellular immune system. In addition, organ-specific lymphocyte invasion in autoimmune diseases may be mediated by similar, albeit pathological homing mechanisms exerted at the cell surface of invasive lymphocytes. Hence, molecular specification of the cell surface is critical for the dynamics of the cellular immune system.

Recently there has been a remarkable increase in our understanding of the molecules that participate in the lymphocyte/granulocyte homing processes. The rapid pace of discovery is underscored by the burgeoning number of publications in the field (for review see ref. 1). It is now clear that particular members of the selectin-family, the integrin-family, and the immunoglobulin-superfamily of adhesion molecules are important components of the lymphocyte/granulocyte homing machinery (3,7,9,10), and thus play a role in the overall processes of cellular immunodynamics and spatial patterning. For example, it was shown that L-selectin (CD62L), expressed at the surface of circulating lymphocytes, mediates the selective trans-endothelial migration of lymphocytes to lymph nodes. Consequently a concept of molecular `addressins´ emerged rapidly refering to the early assumption that lymphocytes are capable of expressing highly specific single molecular classes that serve to direct lymphocytes to particular organ `addresses´. However, it became evident that many adhesion molecules or their ligands are not only expressed by lymphocytes, but also by granulocytes and may serve different adhesive functions. L-selectin, for example, mediates lymphocyte recirculation through peripheral lymph nodes as well as granulocyte extravasation (transendothelial migration) at sites of acute inflammation. How, then, can the same molecules be used for such diverse processes and `addresses´ ? The answer may come from use of various combinations of adhesion and signalling molecules to recruit particular leukocyte subsets into the appropriate tissue. Furthermore, a combinatorial receptor machinery may also serve to organize both migratory pathways and lymphoid network formation within particular tissues. Hence, there are important questions to be answered. Is there a combinatorial molecular coding system at the cell surface of migratory immune cells? Which molecules or molecular classes belong to this system? How can a suface code be identified?

3. How to Recognize Cell Surface Associated Protein Networks: the Concept of Molecular Semiotics

To establish whether there is a combinatorial cell surface protein code for homing and spatial patterning in the cellular immune system in vivo (in situ) requires new concepts and technological developments.

First, it is infered that a protein network encoding a larger number of different cell surface functions by differential combinations of its components (proteins) will be recognizable by its specific syntax of combinations. For example, a combination A expressed during invasion of immune cells (IC) in the brain will be inherently different from a combination B during invasion of the muscle etc. Hence, if combinatorial coding systems exist, it is expected that (i) the selectivity-coding combinations expressed at the cell surface of invasive IC are always specific for a given biological context and (ii), in principle, those combinations can be detected by appropriate imaging methodes with a high power of combinatorial molecular discrimination, when all components of the system are examined in parallel. It is further infered, that these combinations must be quantified within the intact biological system, because a disruption of the biological context (isolation of IC from organ, homogenization of tissues) would lead to a loss of the specific syntactical information, although the single components of the system will still be detectable (proteins or their corresponding RNA species). Similarly, molecular genetic approaches to simultaneously monitor large numbers of RNA´s will lead to gene expression profiles in given samples (6,16), but will not unravel combinatorial specificities wthin network structures of cells or tissues. The above described requirement for network structure detection, i.e. the syntax of it, may be comparable with those related to the constraints of a language: letters laid down on a table (symbolizing the biological structure: cell or tissue) will only encode a meaning if they are assembled according to the correct syntactical rules of their combinations (words, or semantic combinations of words forming sentences). Understanding the code (meaning) requires that (i) the correct follow-up of letters and words is not destroyed, and (ii) the reader is able to recognize all possible letter combinations (symbolizing the imaging method with unlimited power of combinatorial molecular discrimination). If the same letters (forming semantic structures, i.e. sentences) are then assembled in a dish, the syntactical information will be lost (the meaning is destroyed), although it will still be possible to determine which letters are present in the system and at which number (symbolizing molecular genetic approach to quantify a given gene expression profile from the source of a cell homogenate).

Second, methods for the simultaneous localization of random numbers of proteins to identify specificitic combinations (syntax within a molecular network) must be developed, because the power of combinatorial molecular discrimination of the conventional or present fluorescence systems is too low.

Third, if a syntax of molecular combinations is identified in vivo, the next necessary step will be to identify its function, i.e. the information which is encoded by given protein combinations. Such analyses will strictly rely on organismic model systems, i.e. immunological mouse models combined with new microscopic devices for combinatorial molecular detection, and molecular biological techniques.

4. Methodological Remarks

In an attempt to analyze the combinatorial protein sets at the cell surface of muscle-invasive T lymphocytes a new approach was developed that overcomes the limitations of present combinatorial fluorescence methodologies (12, 14, 15). It was hypothesized that a principle operating in the intact immune system might be used to generate an in-situ labeling technology with unlimited numbers of different specificies at the level of a single cell: The ability of the immune system to produce hundreds of different antibodies (with different protein specificities) against a battery of different cell surface proteins of one and the same bacterium may be the basis of such approach. These antibodies will all specifically bind to these proteins, because the underlying antibody-specific kinetics include that the antibody can move across the barrier of the dense carbohydrate layer at the very bacterial surface in order to reach the protein binding sites located beneath these structures near the plasma membrane. Hence, it can be expected, that antibodies moving towards their target are not significantly influenced by sterical hindrances. Therefore it might be expected, that a given cell immobilized at a surface (i.e. object slide) can be specifically labeled or loaded with a large number of different antibodies applied sequentially, independent of whether these antibodies are conjugated or unconjugated to a dye (12). It turned out that indeed, at least 20 different dye-conjugated monoclonal antibodies can be applied to successively and specifically tag the corresponding epitopes of a given cell surface without a significant loss of fluorescence signal intensity. By using repetitive incubation -imaging -bleaching cycles it was possible to show specific labeling patterns for each given antibody within a battery of up to 18 different proteins (12). A disadvantage of the technique is that it is very time consuming to collect an appropriate number of combinatorial patterns . Within the last 8 years, using analyzes (cell per cell) by hand, it was possibe to collect not more than 300.000 cell patterns. These data provide new insights into the molecular organization of the immune cell surface (i) during muscle invasion, (ii) within T cell areas of the tonsil (a secondary immune organ), and (iii) in the blood (manuscript in preparation). The data have prompted us to develop a new microscope type (multi-epitope-ligand-"Kartierungs" microscope, MELK) that allows for automatic and higly standardized detection and mathematical analyzis of combinatorial protein patterns (manuscript in preparation). By this technology it is now possible to capture combinatorial molecular structures in cells within a much shorter time schedule leading to data sets which are the basis for consecutive approaches including mathematical models. It will also provide the possibility to combine multi-epitope-mapping procedures with time-resolved dynamic studies of molecular interactions at the cell surface (5).

5. Polymyositis, a Human Disease as a Model to Analyze Semiotic Structures on Invasive T Lymphocytes

Polymyositis (PM) is a human inflammatory skeletal muscle disease of unknown etiology. The characteristic pathogenic feature of PM is muscle invasion by T lymphocytes. These cells entering the muscle tissue from the circulation transmigrate across the endothelial cells of small blood vessels (extravasation), then migrate towards intact muscle fibers and penetrate the basal lamina cylinders (endomysial tubes), which surround these fibers (Fig. 1).

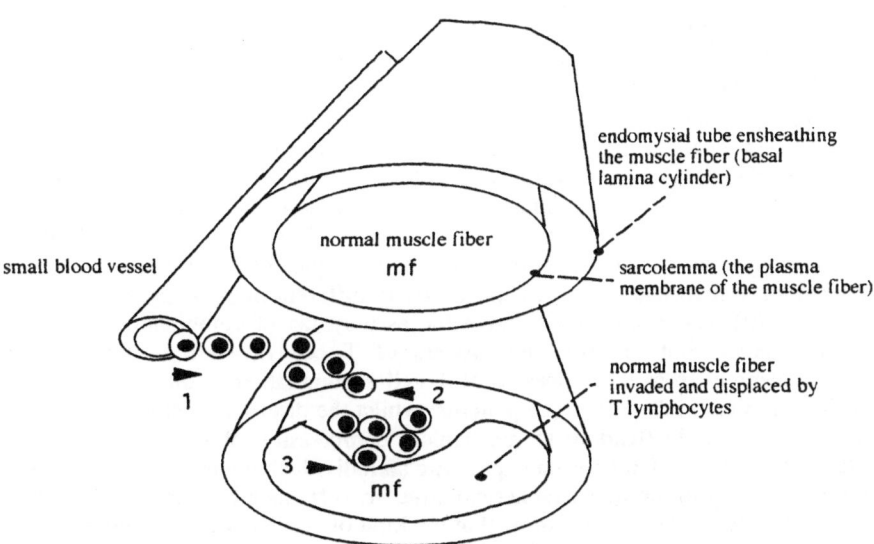

small blood vessel

normal muscle fiber
m f

endomysial tube ensheathing
the muscle fiber (basal
lamina cylinder)

sarcolemma (the plasma
membrane of the muscle fiber)

normal muscle fiber
invaded and displaced by
T lymphocytes

m f

Fig. 1. Schematic illustration of the main morphological features of polymyositis. (1) T lymphocytes (morphologically homogeneous mononuclear cells) circulating in the blood enter the muscle tissue (pathological homing, transendothelial migration across the endothelial wall of small blood vessels); (2) These invasive T lymphocytes have a high migratory potential and migrate towards single intact muscle fibers (directed migration). Leading front T lymphocytes (2) penetrate the endomysial tube and thereby enter the space between the basal lamina cylinder and the plasma membrane of the muscle fiber. (3) In later stages of migration both the leading front T lymphocytes and the T lymphocytes following the invasive front cells have displaced and compressed the muscle fiber. Hence, there is no cytolytic cell death of the fiber but rather a mechanically mediated compression and injury.

This initial process is then followed by a progessive migration of these T cells, which leads to a compression and displacement of the plasma membrane of the muscle fiber. By using MELK localization procedures it was found, that 18 different characteristic cell surface recepetor and adhesion molecules are expressed by these T cells. However, expression of the same molecules is also found in many other compartments of the immune system, indicating that the expression of each molecular class itself is unspecific related to the diverse anatomical sites or processes. However, overlay of the fluorescence distribution patterns of each given cell surface protein at the level of each given T cell reveals several interesting new informations suggesting that there are rules for combination of receptor proteins. First, nearly each given T cell within a T cell infiltrate migrating towards a muscle fiber shows an individual combinatorial receptor pattern (a pattern that is not expressed by the surrounding T cells). Second, the T cell initially penetrating the basal lamina of intact muscle fibers (leading front T lymphocyte) expresses a combinatorial pattern, which is inherently different from that in the cells behind the invading front (Fig. 2). Third, the comparative investigations of the combinations expressed by the same 18 cell surface receptor proteins in other immune compartments revealed that the pattern in muscle are clearly distinct from those expressed by immune cells in the blood and in secondary immune organs (tonsil). Interestingly, within the total possible combinatorial spectrum of the 18 proteins (2^{18}) 149 different combinations were found, 45 were specific for muslce invasion, roughly 100 were specific for PMBL in the blood and only 2 to 4 were specific for T cell areas in the tonsil. Hence there is a clear cut association of given combinatorial forms with defined biological contexts: processes, anatomical sites or immune compartments. These data give rise to the following conclusions: (i) the 18 proteins detected are part of a combinatorial cell surface receptor system that expresses specificity in different immune processes and compartments, (ii) this specificity is detectable as a syntax of combinations within defined anatomical contexts, (iii) the process of T cell invasion in muscle is intrinsically different from the processes of T cell accumulation in the tonsil at the level of the combinatorial cell surface patterns: whilst the directed (vectorial) type of migration in the muscle (leading to mechanical compression of muscle fibers) is associated with arrays of T cells each expressing an individual combinatorial pattern, T cells accumulating in the tonsil (spherical structures) frequently repeat 4 different combinatorial patterns. These data show that the kind of combination and the type of expression within accumulations of T cells is clearly related to the organization and the destination of T cell network structures and thus to the dynamics of immune cells. Hence, formally there is structural evidence for a molecular semiotic system at the cell surface of migrating immune cells.

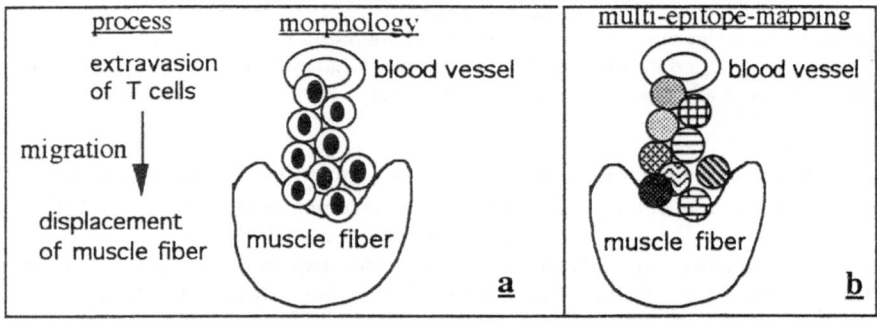

Fig. 2. Simplified schematic illustration of differential combinatorial cell surface receptor patterns on muscle invasive T lymphocytes. The cominatorial patterns are directly related to the normal morphological stain of the invasive cells by the method of multi-epitope-ligand-"Kartierungs" microscopy (MELK). T cells are morphologically homogeneous cells which express "individual" (muscle-invasion-specific) combinatorial receptor patterns at the cell surface during invasion (simultaneous detection of 18 different cell surface epitopes, resp. receptor proteins). The patterns are undetectable by conventional fluorescence microscopy, which is limited to the simultaneous observation of not more than 4 to 5 channels. Note that the individual protein combinations in muscle-invasive T cells are different from those in the blood and the tonsil and that the heterogeneity of combinations within the inflammatory infiltrate signifies the vectorial type of migration of the T cells towards intact muslce fibers (see also Fig. 3).

6. Possible Mechanism of Spatial Migration and Patterning

The above described results give rise to a concept which is schematically summarized in Fig. 3. Briefly, within the immune system there might be a priming pool that generates combinatorial protein sets at the cell surface of given immune cell types (IC). These combinatorial sets of proteins are chosen out of a limited number of at least 20 different cell surface proteins (most of which are known as CD markers). This pool system can be regarded as a repertoire or "lexical store house", that engineers combinatorial molecular addresses at the cell surface of IC. IC are then released into the blood circulation. Their addresses may be stably expressed during circulation.

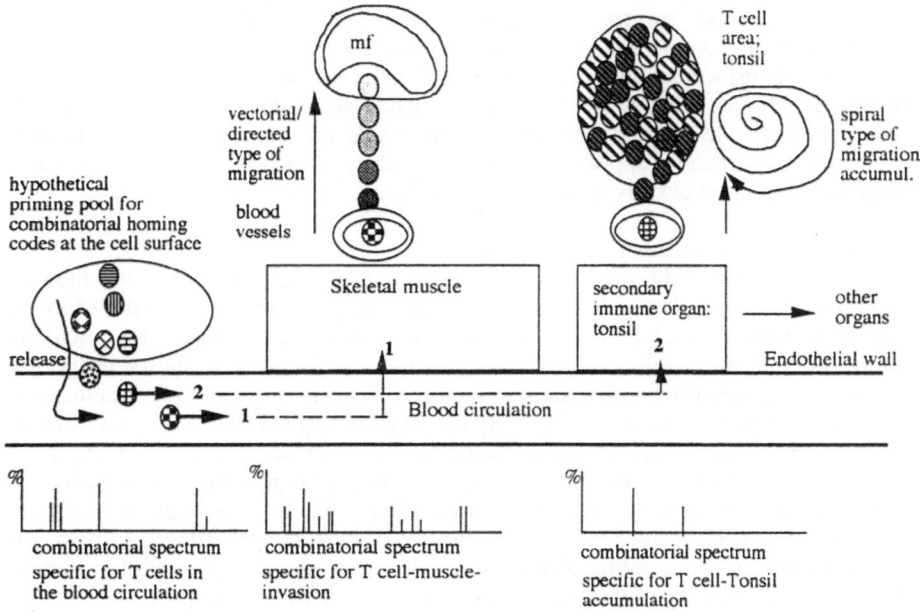

Fig. 3. Possible mechanism organizing homing, dynamics and organ-specific spatial patterning of lymphoid cells. The model suggested here is based upon combinatorial cell surface receptor patterns which express context-specific structures (identified by MELK). These structures can be described as a semiotic system of receptors and adhesion molecules likely to encode the spatial organization of the cellular immune system in health and disease. For explanations see text. Note that arrays of T cells each carrying an individual combinatorial receptor pattern at the cell surface are related to a vectorial type of directed migration in muscle, whilst the use of not more than 2 (to 4) different combinatorial cell surface patterns (repeats of tonsil-specific combinations) on the T cell surface may be related to a spiral type of migration that finally may lead to spherical T lymphocyte accumulations characteristic for T cell areas in secondary immune organs. The combinatorial protein spectra occuring in the different immune compartments are demonstrated schematically in the lower part of the figure: each vertical line represents the frequency of cells expressing a particular cell surface protein combination within the total possible spectrum of 2^{18} different combinations. Note that the spectra are different in each compartment.

There may be a twofold character of the combinatorial molecular system, analogous to other sign systems (4). Both the IC carrying the combinatorial address code and the adressee (interacting endothelial cell type foming the inner layer of the blood vessel wall) must have at their disposal more or less the same "filing cabinet of prefabricated" combinatorial molecular representations. In fact, endothelial cells express at their cell surfaces molecular classes, which are also common to the IC surface (2,13). Once a circulating IC with a given combinatorial address has bound to the corresponding endothelium, it transmigrates into the corresponding organ. During this extravasation process the invasive IC switches the combinations at the cell surface to a combinatorial program that is specifically designed for an interaction with the organ-specific microenvironment (Fig. 3, different patterns indicating the different combinatorial sets of cell surface proteins). It appears that the different combinations of cell surface proteins generate differential adhesion. Hence, it is possible that the arrays of T cells in muscle, within which each T cell expresses a different cell surface combination pattern of adhesive proteins, generates a gradient of adhesive forces that may contribute to a directed movement of these T cells towards the muscle fiber. By contrast, in the tonsil the frequent use of 4 different combinations (repeats) appears to generate spherical accumulations of T cells, possibly by addition of cells in a spiral-like manner.

7. Conclusions, Questions and Perspectives

Together, the data hitherto obtained by multi-epitope imaging microscopy indicate that there are highly ordered combinatorial protein structures at the cell surface of migrating immune cells. These expression patterns are clearly not random: localization of the single components (proteins) of the system shows unspecific ubiquitous expression in the cellular immune system, whereas the simultaneous combinatorial patterns express selectivity for given biological functions, processes, or compartments. This indicates that combinatorial cell surface patterns may express a new level of function, because they show a highly regulated syntax within an enormous number of possible combinations. These results may suggest that these combinatorial cell surface structures encode migratory patterns throughout the organism and may thus be involved in the self-organization of the cellular immune system. The specificity of the patterns are recognizable by MELK procedures in situ rather than by conventional microscopy. They can be described as a molecular semiotic system at each level (syntax, semantic, pragmatic) and will therefore imply the existence of molecular semantic rules. The combinatorial structures hitherto identified appear to be part of a probably more complex system of protein components. In previous MELK analyzes, it appeared however, that roughly 20 different proteins of the immune cell surface (out of approximately 180 different known receptor and adhesion molecules) may belong to the coding system. Interestingly, as a methaphor, this number nearly corresponds to the number of letters -for example - of the english language. Is it that each protein of this system represents a sign that can be assembled with other protein signs to semantic sructures at the cell surface ? Then it must be postulated that - independent of how the system is regulated - each immune cell must have at its disposal a non-random repertoire of combinations. For example, it appears that out of the possible number of combinations of 20 different proteins (2^{20}) only a limited number will be expressed in a given biological system. At the biological level of complex molecular architectures on cell surfaces, these restricted combinations of cell surface proteins may respresent mirror images of all possible microenvironments, which can be potentially faced by a cell during cell-to-cell or cell-to-matrix interactions. The present data suggest that the microscopic screening for syntactical protein structures at the IC surface by localization of all proteins of a system in parallel, and within the same intact biological tissue, is a prerequisite for the detection of combinations which are relevant for migratory specificity and intercellular structural organization.

Many questions remain to be answered. For example, are there inter-individual differences in the usage of combinations? Is there an information flow (inside-out or outside-in, at the cellular level) on the basis of given combinations, for example from cell to cell within a lymphoid structure (cell accumulation)? How is the pattern of migration or accumulation modified, when the combinatorial system is altered? Given that the degree of freedom in the genetic system coding the single proteins used for combinations in IC is practically zero (freedom to differentially combine DNA base pairs in a gene), the immune cell may be less constrained to combine proteins to new contact structures at the cell surface. Hence, it is possible that the immune system - at this level of combinatorial molecular organization - can create new combinatorial structures and contexts, which may significantly feed back to the spatial migration - adhesion pattern, or, the evolution of the system. It seems a necessary conclusion, that a disturbance of the system (i.e. co-expression of proteins that are usually not combined in a normal immune system) may lead to new selectivities and/or adhesiveness: one possible consequence may be, that abnormal invasion patterns or cell accumulations within particular organs will occur, as for example in autoimmune diseases (i.e. multiple sclerosis, inflammatory bowel diseases, Hashimoto thyroiditis, etc) or neurodegenerative diseases with low degree of cell invasion. In this context, abnormal cell surface protein sets in circulting immune cells may be a proximal cause of pathological autoinvasion rather than expression of autoantigen(s) by the target cells (15).

Elucidation of combinatorial molecular organization in the cellular immune system (or other cellular systems) requires the combination of new microscopic techniques, pattern recognition algorithms, mathematical methods, biophysical and appropriate biological model systems. It may be a new interdiscplinary challenge. It is conceivable that - if the combinatorial addresses of circulating immune cells are identified by these approaches- strategies will be developed to selectively block cells primed for invasion of particular organs during disease.

Acknowledgments: This work was supported by DFG (SFB 387; INK15, TpA1; 627/-2-2, 8-1); BMBF 07NBL04 Tp B10; and Ministery of Culture and Education SA.

References

1. Butcher, E.C., Picker, L.J.: Lymphocyte homing and homeostasis. Science 272, 60-66, 1996.
2. Defilippi, P., Silengo, L., Tarone, G.: Regulation of adhesion receptor expression in endothelial cells. In: Dunon, D., Mackay, C.R., Imhof, B.A. (eds): Adhesion in leukocyte homing and differentiation. Springer, Berlin. pp 79-87 (1993).

3. Hauzenberger, D., Klominek, J., Bergstrom, S.E., Sundqvist, K.G.: T lymphocyte migration: the   influence of interactions via adhesion molecules, the T cell receptor, and cytokines. Crit. Rev. Immunol. 15, 285-316 (1995).

4. Jakobson, R.: Selected writings. Vol. II. Word and language. The Hague. 1991.

5. Lendeckel, U., Wex, T., Ittenson, A., Arndt, M., Frank, K., Maiboroda, O., Schubert, W., Ansorge, S.: Rapid mitogen-induced aminopeptidase N surface expresion in human T cells is dominated by mechanisms independent of de novo protein biosynthesis. Immunobiol. 197, 55-69 (1997).

6. Lockhart, D., Dong, H., Byrne, M.C., Follettie, T., Gallo, M.V., Chee, M.S., Mittmann, M., Wang, Ch., Kobayashi, M., Horton, H., Brown, E.L.: Expression monitoring by hybridization to high-density oligonucleotide arrays. Nature Biotechnology 14, 1675-1680 (1996).

7. Mackay, C.R., Andrew, D.P., Briskin, M., Ringler, D.J., Butcher, E.C.: Phenotype, and migration properties of three major subsets of tissue homing T cells in sheep. Eur. J. Immunol. 26, 2433-2439   (1996).

8. Morris, Ch.: Signification and Significance. A study of the relations of signs and values. The M.I.T. Press, Massachusetts Institute of Technology, Massachusetts. 1964.

9. Picker, L.J., Terstappen, LWMM, Rott, L.S., Streeter, P.R., Stein, H., Butcher, E.C.: Differential expression of homing - associated adhesion molecues by T cell subsets in man. J. Immunol. 145, 3247-3255 (1990).

10. Picker, L.J., Warnock, R.A., Burns, A.R., Doerschuk, C.M., Berg, E.L., Butcher, E.C.: The neutrophil selectin LECAM-1 presents carbohydrate ligands to the vascular selectin ELAM-1 and GMP-140. Cell 66, 921-933 (1991).

11. Schubert, W., Zimmermann, K., Cramer, M., Starzinski-Powitz, A.: Lymphocyte antigen Leu19 as a molecular marker of regeneration in human skeletal muscle. Proc. Natl. Acad. Sci. USA 86: 307 - 311 (1989).

12. Schubert, W.: Multiple antigen - mapping microscopy of human tissue. In: G. Burger, M. Oberholzer, G.P. Vooijs (eds.): Advances in analytical cellular pathology. pp 97 - 98. Excerpta medica.Elsevier, Amsterdam, 97-98 (1990).

13. Schubert, W.: Triple immunofluorescence confocal laser scanning microscopy: spatial correlation of novel cellular differentiation markers in human muscle biopsies. Eur. J. Cell Biol. 55: 272 - 285 (1991).

14. Schubert, W.: Antigenic determinants of T lymphocyte $\alpha\beta$ receptor and other leukocyte surface proteins as differential markers of skeletal muscle regeneration: detection of spatially and timely restricted patterns by MAM microscopy. Eur. J. Cell Biol. 58, 395 - 410 (1992)

15. Schubert, W., Masters, C.L., Beyreuther, K.: APP[+] T lymphocytes selectively sorted to endomysial tubes in polymyositis displace NCAM-expressing muscle fibers. Eur. J. Cell Biol. 62, 333-342 (1993).

16. Zhang, L., Zhou, W., Velculescu, V.E., Kern, S.E., Hruban, R.H., Hamilton, S.R., Vogelstein, B., Kinzler, K.W.: Gene expression profiles in normal and cancer cells. Science 276, 1268-1272 (1997).

Molecular Evolutionary Dynamics

Christian V. Forst[1,2]

[1] Inst. für Molekulare Biotechnologie, Beutenbergstr. 11, D-07745 Jena, Germany[**]
[2] Santa Fe Institute, 1399 Hyde Park Road, Santa Fe, NM 87501, U.S.A

Abstract. Evolutionary dynamics, especially dynamics on bio-molecular scale, bear inherent nonlinear properties. We analyze these dynamics of evolution by subdividing it in less sophisticated processes: population dynamics, population-support dynamics, and genotype-phenotype mapping. Molecular evolutionary biology provides a sufficiently simple experimental setup for a quantitative analysis of these subsystems. RNA secondary structures serve as model of reasonable phenotype mapping. Preimages of these mappings reveal neutral behavior and percolate genotype-space as so called *neutral networks*. The spatial organization of these networks significantly determine evolutionary dynamics: By stochastic flow transitions between two networks (two phenotypes) take place, by neutral drift better phenotypes are explored and evolutionary optimization towards a global maximum is possible, and by diffusion complex dynamical units (such as hypercycles) are competitive against parasites.

1 Introduction

Information and its development and processing in biology has properties that distinguishes it from information in chemistry and physics. Information in biology is encoded and is processed similar as in information technology and computer science. It is stored in genotypes and transferred to offspring through inheritance and less directly through epigenic processes. Dynamics of nonlinear bio-chemical reaction networks, such as cellular metabolisms, are easily interpreted as information processing. But any comprehensive understanding of biological phenomena requires an interpretation in evolutionary terms as Theodosius Dobzhansky (1973) formulated: *"Nothing in biology makes sense expect in the light of evolution"*. This sentence, rephrasing Galilei's famous quote (Galilei, A. Favaro ed., 1968; Park, 1988), is much stronger since it postulates the existence of a formal language to describe and explain observations in nature.

Initialized by the epochal discoveries made by Francis Crick and James Watson in the year 1953 *molecular biology* came into being . First steps in the direction of a *molecular evolutionary biology* were already taken in the late sixties by the pioneer work of Sol Spiegelman (1971) who developed *serial transfer experiments* as a new method of molecular evolution in the test tube. Manfred Eigen (1971) at about the same time formulated a kinetic

---

[**] address to be used for correspondence

theory of molecular evolution. Since then studying evolution of molecules in laboratory systems has become a research area of its own. This approach simplifies evolutionary systems as much as possible and makes them accessible to an analysis by the conventional methods of physics and chemistry.

Evolutionary dynamics itself is a highly complex process. Therefore we omit additional difficulties in considering spatiotemporal patterns and introduce a comprehensive model which tries to account for most of the relevant features of molecular evolution. Peter Schuster proposed an interaction of three processes described in three different abstract metric spaces (Schuster, 1995):

- the *sequence space* of genotypes being DNA or RNA sequences,
- the *shape space* of phenotypes, and
- the *concentration space* of biochemical reaction kinetics.

The sequence space $\mathcal{Q}_\alpha^n$ is a metric space containing all sequences of chain-length n with alphabet-length α (i.e. the number of letters used in an alphabet $\mathcal{A}$). For DNA or RNA sequences α is 4, for peptides α would be 20 due to 4 distinct nucleotides common in DNA or RNA and 20 aminoacids in peptides and proteins resp. As a metric serves the Hamming metric (Hamming, 1986). The shape space covers all possible structures considered. The definition of structures itself is highly context dependent. Clearly two biopolymers with different sequences will form different spatial conformations at atomic resolution. This descriptions is, no question, adequate for comparison of active sites of enzymes (or ribozymes). In contrast, stating that all tRNAs have the same shape refers to a different understanding of structure on a coarse-grained level. Secondary structures may serve as one example which is suitable for our purpose of developing a mathematical model of molecular evolution. Similarities and dissimilarities of RNA structures can be expressed by means of mathematical measures with metric properties. Finally concentration space is the conventional space in which chemical reaction kinetics is described. Martin Feinberg formalized this concept in mathematical terms (Feinberg, 1977). The conceptual cycle which is formed by a projection of evolutionary dynamics upon the three abstract spaces is sketched in Fig. 1.

Interactions through catalysis, predator-prey or host-parasite behavior which are inherent in biological systems suppress optimization of individual fitness. Formulating molecular evolution in terms of evolutionary dynamics has to take into consideration not only approaching a steady state but also nonlinear dynamical phenomena like oscillations or chaotic behavior in space and time. The key towards comprehensive understanding of biological systems is the mapping of genotypes into phenotypes. Indeed, all biological functions are properties of phenotypes. Thus in the next section let us introduce a mathematical framework of such relationships.

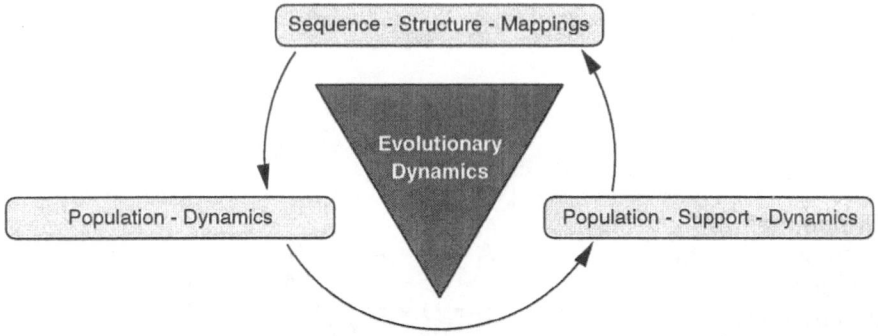

Fig. 1. Evolutionary dynamics in sequence-, shape-, and concentration-space.

2 Topology of RNA Sequence – Structure Maps

2.1 Secondary Structures and Compatible Sets

RNA secondary structures and the induced sequence – structure relationship are a suitable and generic description for genotype – phenotype mapping which is important in molecular evolutionary biology. RNA secondary structures represent a type of coarse-graining of biopolymer structures. They commonly are understood as list of Watson-Crick (**A=U** and **G≡C**) and Wobble (**G−U**) base-pairs which are compatible with unknotted and pseudo-knot-free two-dimensional graphs (for a precise formal definition we refer to Waterman (1978)).

Defining secondary structures independently of chemical or physical restrictions yields a general description based on contacts with respect to arbitrary alphabets $\mathcal{A}$ with arbitrary pairing rules Π. A *pairing rule* Π on $\mathcal{A}$ is given as a set of pairs of letters from the given alphabet (i.e. **AU, UA, GC, CG, GU, UG** for natural RNA-molecules with alphabet **A, U, G, C**). This concept can easily be extended to a general description of biopolymer structures via contact maps (Kopp *et al.*, 1996). Similar to secondary structures a general *contact structure* c is determined by a *set of contacts* of c omitting the trivial contacts due to adjacent letters in the succession of the sequence.

A relevant concept in studying sequence – structure relation is how sequences have to be composed to fulfill necessary conditions for folding into a desired structure. In the following we define *compatibility* of a sequence to a given structure: A sequence x is said to be *compatible* to a structure s if all base-pairs required by s can be provided by x_i and $x_j \in x$ with respect to the pairing rule Π for each base pair (Fig. 2). $\mathbf{C}(s)$ is the set of all sequences which are compatible to structure s. The number of compatible sequences is readily computed for secondary structures (with n_u unpaired bases and n_p base pairs this evaluates to $4^{n_u} \cdot 6^{n_p}$).

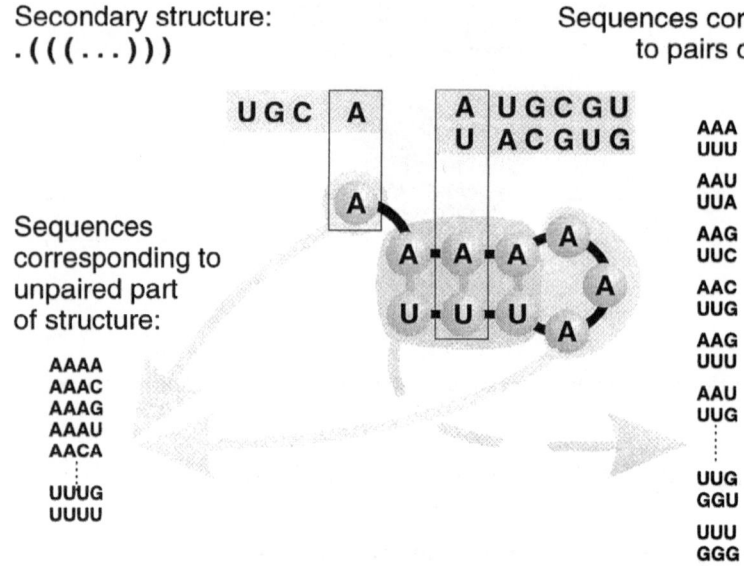

Secondary structure:
.(((...)))

Sequences corresponding to pairs of structure:

Sequences corresponding to unpaired part of structure:

Fig. 2. Compatible sequences with respect to a fixed secondary structure. A sequence is called *compatible* with a given secondary structure if for all base pairs in the structure there are pairs of matching bases in the sequence. Sequences compatible to a structure *do not* fold in general into this structure. However the structure will always be found as result of suboptimal folding.

2.2 Preimages and Complete Mappings

The relation between RNA sequences and secondary structures is understood as a (non-necessarily invertible) mapping f_n from sequence space $\mathcal{Q}_\alpha^n$ into shape space $\mathcal{S}_n$. This section presents some results on *Generic Properties of Combinatory Maps* by C. Reidys et al.. A mathematical framework with proofs can be found there (Reidys et al., 1997). Essential insights are as follows:

The set of all sequences folding into a given structure is denoted as *neutral network* $\Gamma_n(s)$ with respect to s. $\mathcal{Q}_\alpha^n$ denotes the generalized hypercube of dimension n over an alphabet $\mathcal{A}$ of size α (i.e. the number of letters in $\mathcal{A}$ is α), and $s \in \mathcal{S}_n$ is a fixed secondary structure. Mathematically $\Gamma_n(s)$ refers to the induced subgraph of $f_n^{-1}(s)$ in $\mathbf{C}(s)$ ($f_n^{-1}(s)$ indicates the *preimage* of a fixed structure s w.r.t the mapping f_n). A sketch of these embeddings is shown in Fig. 3.

Remark 1 *The graph of compatible sequences $C(s)$ to a fixed secondary structure s is*

$$C(s) = \mathcal{Q}_\alpha^{n_u} \times \mathcal{Q}_\beta^{n_p} \tag{1}$$

α *is the number of different nucleotides, and* β *is the number of different types of base pairs that can be formed by* α *different nucleotides.*

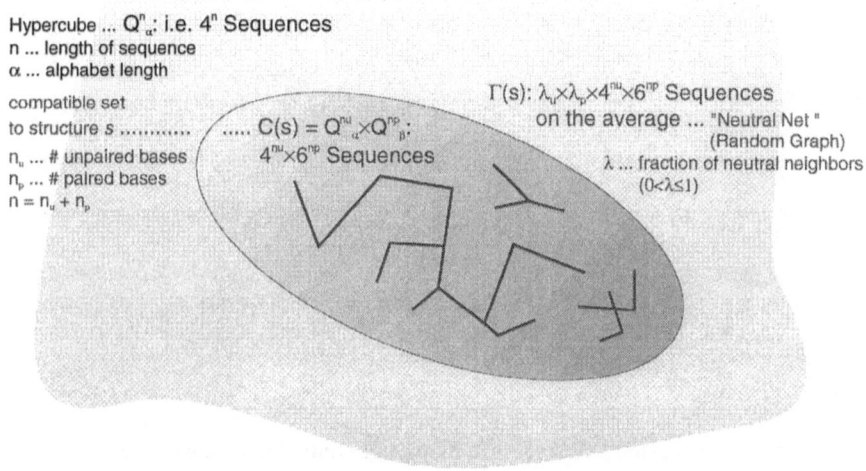

Fig. 3. Sketch of a neutral network $\Gamma(s)$ (shown as solid line graph) embedded in the set of sequences compatible to structure s (i.e. $\mathbf{C}(s)$ – indicated as oval) which itself is embedded in sequence space $\mathcal{Q}_\alpha^n$ (realized as shaded background).

Sequences folding into the same structure are thus represented by a subgraph that is *randomly* induced on the set of compatible sequences corresponding to this structure (Reidys *et al.*, 1997). Accordingly this model is of probabilistic nature and properties of random subgraphs are studied as functions of a single parameter – a probability measure over all possible induced subgraphs in a given sub-hypercube with a *choosing parameter* λ. This parameter represents the mean fraction of neighbors that are neutral with respect to the structure. – the *fraction of neutrality* with $0 \leq \lambda \leq 1$. Considering paired and unpaired regions we are dealing with two independent assignments for each corresponding choosing probability λ_p and λ_u resp. are used. Now vertices in both sub-cubes $\mathcal{Q}_\alpha^{n_u}$ and $\mathcal{Q}_\beta^{n_p}$ are chosen independently with these probabilities λ_u and λ_p resp. This is equivalent to choosing pairs of vertices in $\mathcal{C}(s)$ and yields exactly the desired neutral network $\Gamma_n(s)$ as probability space with its corresponding probability measure.

Two remarkable results are assertions about *density* and *connectivity* of subgraphs. In analogy to percolation theory these properties are fulfilled almost sure if λ exceeds the threshold value

$$\lambda^* = 1 - \sqrt[1-\alpha]{\alpha}$$

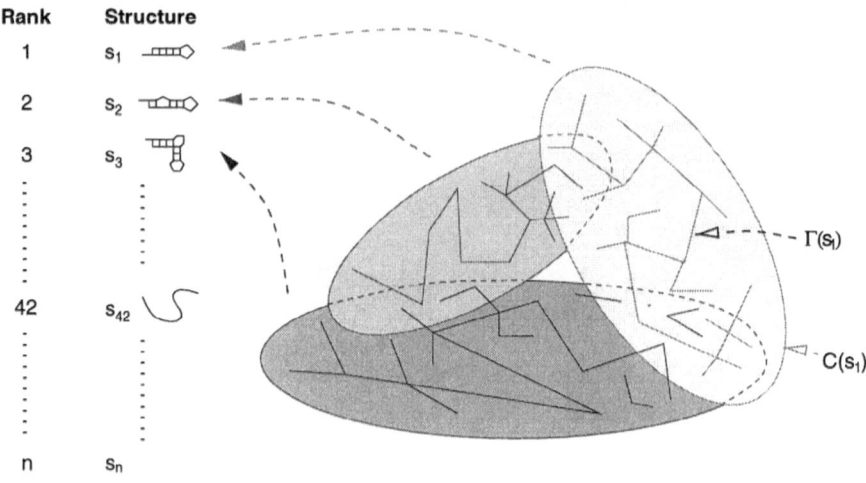

Fig. 4. Complete sequence to structure mapping: neutral networks are constructed by subsequent (random) mapping of sequences to an ordered set of structures. Note that any pair of compatible sets have a non-empty intersection.

Once we know how to construct a neutral network $\Gamma_n(s)$ we order the set of secondary structures $\mathcal{S}_n$ and define a complete mapping by iterating the construction process of the corresponding neutral network w.r.t. the ordering. The preimage for the structure with highest rank s_1 is assigned independently. For all other structures s_i, $i > 1$ the mapping depends on all previous assignments. A visualization of this process is shown in Fig. 4.

3 Evolutionary Dynamics

Our restricted view on evolutionary dynamics is the formalization of Schuster's suggestion outlined in the introduction. The dynamics of a population of individuals with genotypes and phenotypes (and a distinct mapping between these two properties) living in an artificial world are studied. Error-prone autocatalytic and catalyzed replication, unspecific dilution due to limited recourses, specific dilution due to predation, alteration due to reaction between individuals are taking place and change the composition of the population in time. We first give an introduction and present an adequate method to simulate a set of interacting units by stochastic means. We show implications of the partly neutral genotype – phenotype relationship on dynamics. And we present some insights of Darwinian evolution as a "simple" hill climbing scenario.

3.1 Dynamics of Finite Populations

Motivated by simulating chemical reactions by stochastic means, Gillespie (1977) developed an algorithm which performs this task. He describes a "population" $\mathbf{P}$ of size N of chemical agents $(P_i \,|\, i \in N)$ performing reaction in a chemical reaction network in terms provided by the theory of point processes:

Each chemical agent (or species, manifested in one or multiple entities) is identified by an integer valued measure ϕ. The time evolution of ϕ is obtained by a mapping from $(P_i \,|\, i \in N)$ to the family $(P_i' \,|\, i \in N)$ as follows: we select an ordered pair (P_k, P_l) where $P_k, P_l \in \{\, P_i \,|\, i \in N \,\}$. For this purpose let $\mathcal{F}_c$ be a "fitness" of a chemical agent in a generalized sense – i.e. the sum over all reaction rates for all reaction which "produces" agent s weighted by concentrations of the involved species. Accordingly the average "fitness" of ϕ reads

$$\mathcal{F}_\phi = \sum_s \phi(\text{Population of agent } s)\mathcal{F}_s.$$

Now we can formulate the process of the desired constant organization (i.e. keeping the total number of individuals constant in $\mathbf{P}$): the so-called *replication-deletion process*. We chose an ordered pair among elements (individuals) of $\mathbf{P}$ as follows; the first coordinate P_k of this pair is chosen with probability $\mathcal{F}_s/\mathcal{F}_\phi$, the second coordinate P_l is selected with uniform probability on $(P_i \neq P_k \,|\, i \in N)$ i.e. $1/(N-1)$. Copying P_k by an error-prone replication with single-digit error-rate p and deleting P_l completes our process.

3.2 Neutral Evolution

Neutral evolution on model landscapes has been intensively studied by analytical approaches (Derrida and Peliti, 1991) derived from the random-energy model (Derrida, 1981)both by computer simulations (Higgs and Derrida, 1991 and 1992) . More recently these studies where extended to neutral evolution on RNA folding landscapes by simulations (Huynen *et al.*, 1996) . In case of selective neutrality populations drift randomly on the corresponding neutral networks by a diffusion-like mechanism. The error-rate p of the underlying replication-deletion process acts as a temperature. By means of stochastic processes one can derive analytical expressions of pair-distance distribution in the neutrally evolving population and of the diffusion constant (Forst *et al.*, 1995; Reidys, 1995) . Using quasispecies-dynamics (see next section), transitions between two different neutral nets with equal fitness-values assigned have been studied (Weber, 1997). Here essential dependencies between chosen pair of secondary structure (which induce the two neutral networks), the topology of the corresponding neutral network – especially for the intersection region – and of the dynamics can be reported. It has been shown (ibidem) that these transitions take place at intersection regions or in close vicinity. Three distinct scenarios for the dynamic behavior of the population have been observed:

- long time fixation on one (or the other) neutral network,
- sharp transitions between the neutral networks,
- long term fuzzy transitions.

Essential parameter for these different classes of transitions are the coupling rate between the networks (dependent on the pair of chosen secondary structures) and the distribution of the intersection. Fig. 5a shows the time behavior of an evolving population in a two network scenario. In Fig. 5b the corresponding probability distribution of the population on two networks is shown (c.f. Weber, 1997).

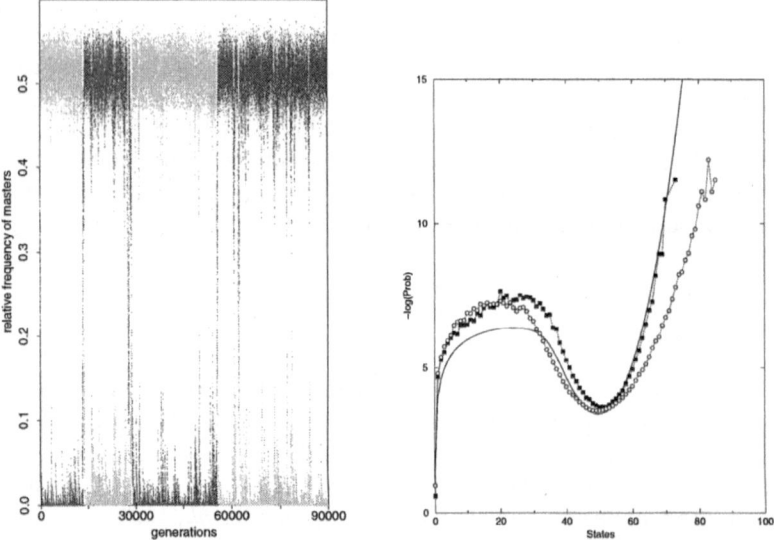

Fig. 5. Transition in a two-network scenarios are shown : a) Relative frequencies of masters on network 1 (dark) and on network 2 (light) are shown. Two secondary structures with chain-length 30 are chosen. The alphabet of the underlying sequences is binary (**G**, **C**), neutral network are formed under minimum free energy conditions (RNA-folding). The initial population size is set to $N = 1000$. b) Stationary probability distributions are shown.

3.3 Adaption and Error Thresholds

A Darwinian scenario of optimizing species of an population by evolutionary processes is easily realized as hill-climbing in a high-dimensional fitness landscape. The underlying dynamics of this process is linear (in most cases) and approaches a stationary state. One quite important class of dynamics is the so called *quasispecies dynamics* which describes a population of replicating individuals under mutational forces in an constant environment. In terms of biochemical reactions the corresponding reaction-equation are as follows:

$$I_i \xrightarrow{\ f[s(I_i)]\cdot W_{ij}\ } I_i + I_j \tag{2}$$

$$I_i \xrightarrow{\quad \Phi \quad} \emptyset$$

We denote I_i, $i = 1, \ldots, n$ as reacting species with a fixed phenotype $s(I_i)$, W_{ij} as stochastic matrix indicating the probability of reproducing I_j by replicating I_i. $s(I_i)$ corresponds to the structure (or phenotype) of the individual, and $f(s)$ is the fitness of phenotype s. Thus the top reaction-equation of Equ. 2 describes an autocatalytic, error-prone replication of I_i. the bottom equation refers to an unspecific dilution flux maintaining the total numbers of individuals constant in the system. Applying chemical reaction-kinetics yields following selection-mutation equation originally formulated by Eigen in a simpler representation:

$$\dot{x}_i = x_i \left[f(s(I_i))W_{ii} - \Phi(\mathbf{x}) \right] + \sum_{j \neq i} k_j W_{ij} x_i \qquad i = 1, \ldots, n. \tag{3}$$

Eigen in his original paper described evolution in molecular-biological systems without explicit usage of a genotype-phenotype mapping. Thus he assigned constant reaction rates $k_i \equiv f(s(I_i))$ for each genotype. He focused especially on dynamics on the so called *single-peak landscape* as mean field approach, where a single genotype has superior fitness (high reaction rate k_i implying fast replication) upon all other genotypes with equal but lower fitness. In the past both deterministic and stochastic models of quasispecies upon single-peak landscapes have been studied (Eigen *et al.*, 1988; McCaskill, 1984; Nowak and Schuster, 1989; Schuster, 1989; Schuster and Swetina, 1988; Swetina and Schuster 1982) . An essential result is the report of a so called *error-threshold*. This threshold is the maximal error-rate for the system where the organization of the population in a cloud around a master-sequence (the quasispecies) is replaced by a random distribution of individuals all over sequence space. It can be understood as *phase-transition* from an ordered and organized phase to a disordered, random phase.

Introducing genotype - phenotype mapping motivated by sequence - structure relations of RNA molecules yields an expansion of the quasispecies dynamics with totally new properties. First studies of quasispecies-dynamics on RNA secondary-structure folding-landscape have been followed by Fontana

and Schuster (1987), and by Fontana *et al.*(1989). Forst *et al.*(1995) performed evolutionary dynamics on *single-shape landscapes* where shape refers
to a fixed secondary structure. Analogue to the approach of a single-peak
landscape Forst *et al.*(1995) classified the sequence-space in fit sequences (sequences which are mapped in the distinct structure) and non-fit sequences
(all other sequences). This implies a classification of an evolving population
in masters (fit individuals) and non-masters (non-fit individuals). Starting
with low error rates the population is localized around a non-moving master as the quasi-species in the single-peak landscape. At a distinct error-rate
(the error-threshold for a single-peak landscape) the population starts moving and drifts on a neutral network analogue to a diffusion-process. for even
higher error rate the population breaks into small clusters with lifetime obeying a power law (large cluster have long lifetime, small cluster have short
lifetime) . After a sufficiently high error-rate a so called *phenotypic error-
threshold* can observed. Here the population is no longer able to conserve
the information of the phenotype but diffuses randomly distributed all over
sequence-space. Analytical expressions of stationary distributions of a population in this landscape and of diffusion constant has been reported by Forst
et al.(1995). Derivations and proofs of these formulas can be found at Reidys
(1995).

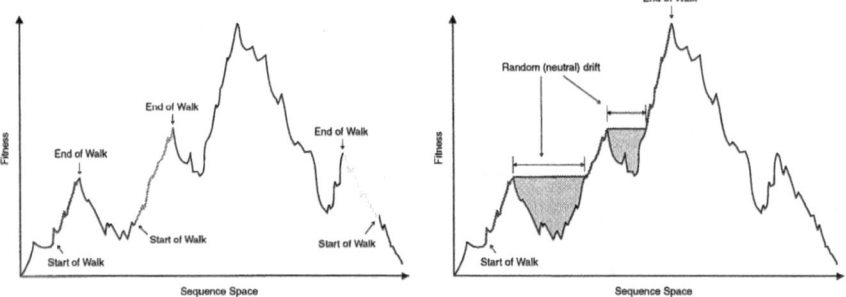

Fig. 6. Optimization on Fitness Landscapes: The difference of optimization in landscapes without and with selective neutrality is shown. On landscapes without partly
neutral properties optimization reaches a local optimum and stays there. Having
selective neutrality optimization don't have to stop at local optima but is able to
move neutrally in sequence space finding new opportunities to optimize.

As a consequence of neutral networks, a population seeking the global
optimum is likely to find its goal (c.f. Schuster, 1996). Fig. 6 shows different
dynamical behavior of an optimizing population for landscapes without and
with selective neutrality. Landscapes without partly neutral properties lead
an evolving population to the nearest local optimum. In landscapes with
selective neutrality a population reaches a local optimum but is not doomed

to stay there forever. Instead the population drifts neutrally in genotype space, changes neutral networks by transitions occasionally and looks for a better place. Only in the eyes of the impatient experimentalist this local, so called *virtual optimum* is real. After finding better blessed grounds the population heads towards new optimal phenotypes.

4 Dynamics of Catalytic Reaction Networks

To model interaction between species a canonical next step is the extension of Equ. 2 to higher order reactions. Complex reaction networks can easily performed by introducing additional second order kinetics yielding following reaction system:

$$I_i \xrightarrow{\quad f[s(I_i)] \cdot W_{ij} \quad} I_i + I_j$$
$$I_i + I_k \xrightarrow{\quad g[s(I_i), s(I_k)] \cdot W_{ij} \quad} I_i + I_j + I_k \qquad (4)$$
$$I_i \xrightarrow{\quad \Phi \quad} \emptyset$$

The new function $g[s(I_i), s(I_k)]$'s denote the rate of the catalyzed reaction. By decoupling both levels from each other yields following schema (Fig. 7): First studies of such dynamical systems (Equ. 4 with $f \equiv 0$) as homogeneous ordinary differential equations (referred to as *Replicator Equations* without mutation and distinct genotype – phenotype mapping) have already performed by Hofbauer *et al.* in the early eighties:

$$\dot{x}_i = x_i \left[\sum_k A_{ik} x_k - \sum_{jk} A_{jk} x_j x_k \right], \qquad i = 1, \ldots, n \qquad (5)$$

Similar to Equ. 3 the reaction rates $g[s(I_i), s(I_k)] \equiv A_{ik}$ have been kept constant in these studies. Due the nonlinear property even for small system size (four individual are sufficient) chaotic behavior can be observed (Forst, 1996; Schnabl *et al.*, 1991). In the last 17 years studying Replicator Equations and derivates (which can be described as perturbations of Replicator Equations by perturbation theory (Stadler and Schuster, 1992)) have become a huge research field of its own. As a recent example may deal the work upon *Autocatalytic networks with intermediates* by Hecht *et al.*(1997).

In this paper we restrict ourselves to an arbitrary assignment of values to functions f and g. It is straightforward to introduce reaction rates derived from specific properties of the involved phenotypes (e.g. structure, structural motives, combination between structure and sequence composition, structure of transition-states, ...).

Important for the dynamical characteristics of the system are the reaction-graphs of phenotypes and the error-prone replications on the level of genotypes. The topologies of the underlying neutral nets assure that there are

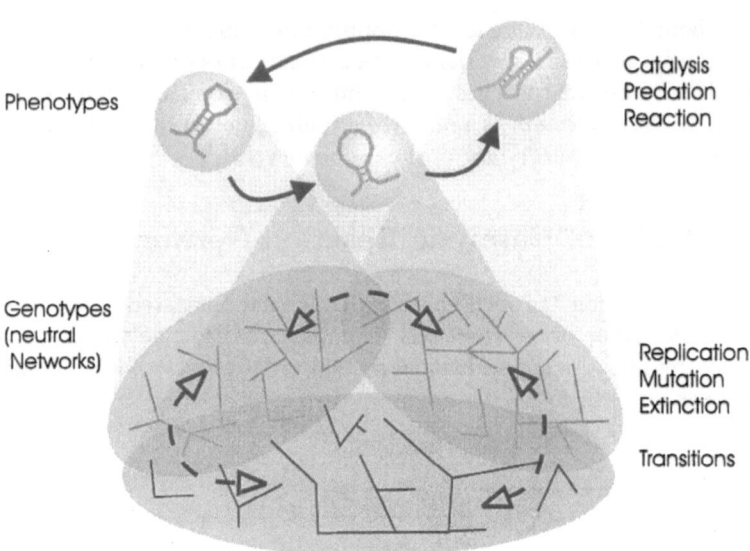

Phenotypes

Catalysis
Predation
Reaction

Genotypes
(neutral
Networks)

Replication
Mutation
Extinction

Transitions

Fig. 7. Driving forces of Evolutionary Dynamics: three molecular species are connected in a catalytic cycle. Directed catalysis of replication takes place on the level of phenotypes. On the contrary on the level of genotypes pairs of neutral networks are almost always in close vicinity to each other. By stochastic flow transitions between two neutral networks happen.

couplings between each two of them. Thus parts of the population can switch from one net to the other and thereby cause a stabilizing effect for the hypercycle. This stabilizing effect is not restricted to hypercyclic organizations and can be extended to (all) reaction networks with cooperative behavior.

4.1 Hypercycles and Parasites

As a paradigm of natural self-organization *hypercycles* have been studied intensively since the well-known work of Eigen and Schuster (1977 and 1978). A hypercycle is a special kind of catalytic network, where each individual is capable of inherent autocatalysis of its replication. In addition individual I_i catalyzes the replication of individual I_{i+1}, $i = 1, \ldots, n-1$ (I_n catalyzes I_1). Special interest have been laid in the study of structural stability of hypercyclic organization in competition with other hypercycles and parasites (Schuster *et al.*, 1978 and 1979). A sketch of a hypercycle with a parasite is shown in Fig. 8. Stability against a superior parasite in a two-dimensional spatial model of a cellular automata (CA) has been performed by Boerlijst and Hogeweg (1991). They observed a stabilization of the hypercycles in spirals against the parasite. Similar studies using partially differential equations have been done by Streissler (1992) . Cronhjort and Blomberg (1994) ob-

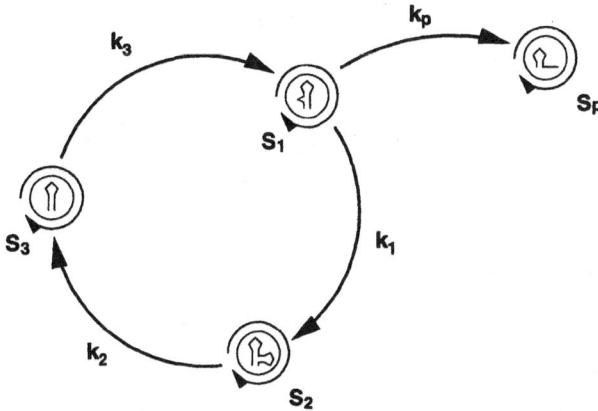

Fig. 8. Hypercycle with three members and one parasite. S_i are the corresponding structures and k_i refer to the reaction rates of catalyzed replication $i = 1, 2, 3$, and p for the parasite.

served that their results are quite different from CA-results of Boerlijst and Hogeweg. Regardless of spirals the hypercycle is vulnerable to parasites in their system. Therefore stabilizing hypercycles against parasites by spirals is highly model-dependent.

In the stochastic formulation with explicit genotype – phenotype mapping we observe the following:

In contrast to homogeneous models with infinite population size the competing behavior between parasite and hypercycle is not a drastic *live and let die* scenario. Both competing systems can coexist for a long time. Due to the (transient) periodic behavior of hypercycles with four and more elements the parasitic concentration exhibits periodic variations which cannot be observed in deterministic, homogeneous systems (Fig. 9). Do keep in mind that two competing effects influence the dynamical behavior. On the one hand undirected transitions are forcing population in a periodic change between two neutral networks. On the other hand directed catalysis drives dynamics in a distinct direction maintaining a higher organization against mixing by transitions. By omitting the driving force of catalysis for much smaller N the population would end at an average value. This phenomenon is again determined by transitions between the networks. Dependent on the population size directed catalysis dominate undirected transitions and vice versa. Depending on the local topology of the corresponding neutral networks and of the given support by catalyzing hosts the parasite is able to out-compete the hypercycle. This scenario is symmetric, thus by same effects the hypercycle can over-populate the parasite (Fig. 9). Similar effect have been discussed by Rasmussen *et al.*. Here coupled hypercycles can generate chaos and clear dominance shifts between different co-existing hypercycles can be observed.

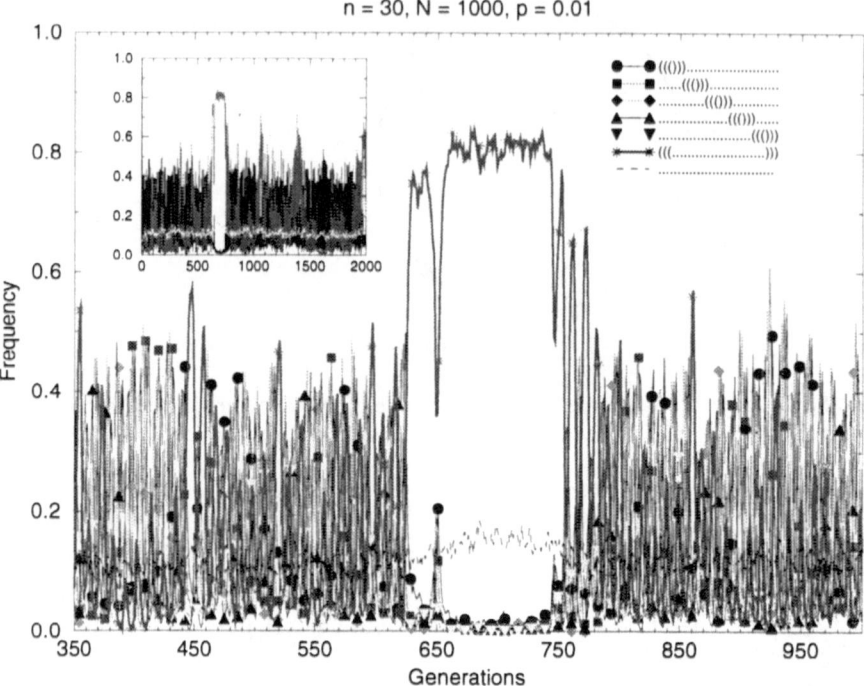

Fig. 9. Five-member hypercycle with parasite. Starting with one member of the hypercycle the parasite becomes dominant at generation 600. The hypercycle is still able to coexist with the parasite and out-competes the latter at a later generation. Parameters are: population size = 1000, chain length = 30, catalyzed reactions are the same for hypercyclic members as for the parasite. Both reactions are 10-times faster than the autocatalytic background reaction.

4.2 Voyaging Large Catalytic Nets

Biological systems (especially biomolecules) are alway object of change by mutations and reorganization; so are dynamical organizations where biomolecules are involved. Early studies about autocatalytic sets and emergence of metabolisms have been performed by various people (Bagley *et al.*, 1992; Fontana and Buss, 1994; Kauffman 1986; Rasmussen *et al.*, 1991) . An essential feature which all these systems are lacking, is an adequate genotype – phenotype mapping. Again we use our model of evolutionary dynamics and setup a large catalytic network which has to be explored by the evolving population. This *exploration* of reaction networks can be interpreted as reorganization of (bio)chemical reaction-units under selection-pressure. How such voyages happen is sketched in Fig. 10. Part of the network has already

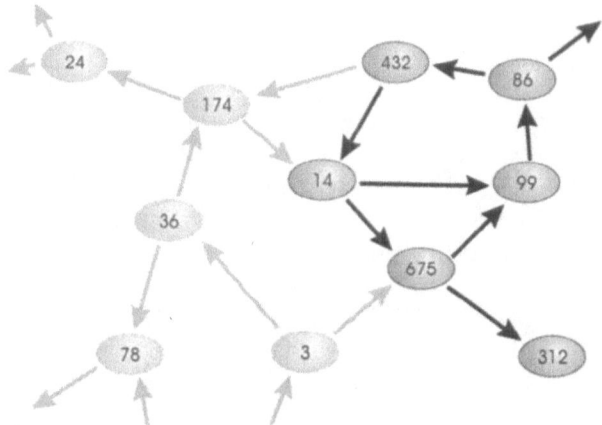

Fig. 10. Given a large catalytic network with phenotypes (structures) as vertices and randomly chosen edges to an maximal out-degree of 2. A population will establish a catalytic unit (light region). Depending on alternative catalytic pathways on the level of phenotypes or stochastic effects on the level of genotypes (neutral drift of the support or transitions between different neutral nets) new phenotypes are "built" and new dynamic organizations can emerge.

been explored by the population. Due to undirected and directed motion new regions, thus new dynamical organizations, are discovered.

Using an ordered set of 1000 secondary structures as phenotypes, a catalytic network of about 420 random hypercycles of size 3 to 7, and about 18 additional random reaction paths is constructed. This yields a total of about 2000 possible reaction-paths. By an initial population of $N = 1000$ individual with phenotype of rank 1 the catalytic network is explored. The reaction-rates g for the catalyzed reaction are randomly distributed with distinct mean and variance. The population explores the catalytic network in search for better support (Fig. 11). The following scenario can be observed: although long hypercycles are predefined in the given reaction-networks the population prefers short hypercycles of length three or four resp. This is due to small population-size relative to the capacity of the catalytic units. Only for short hypercycles the catalytic organization can overwhelm stochastic effect due to transitions.

This still quite simple model gives deep insight how new catalytically active units can be formed, how they can reorganize by neutral drift and how optimization to more efficient units take place. Important for this purpose is the error-rate on genotype level which is contained in the stochastic mutation matrices W_{ij}. This leads us to the next section.

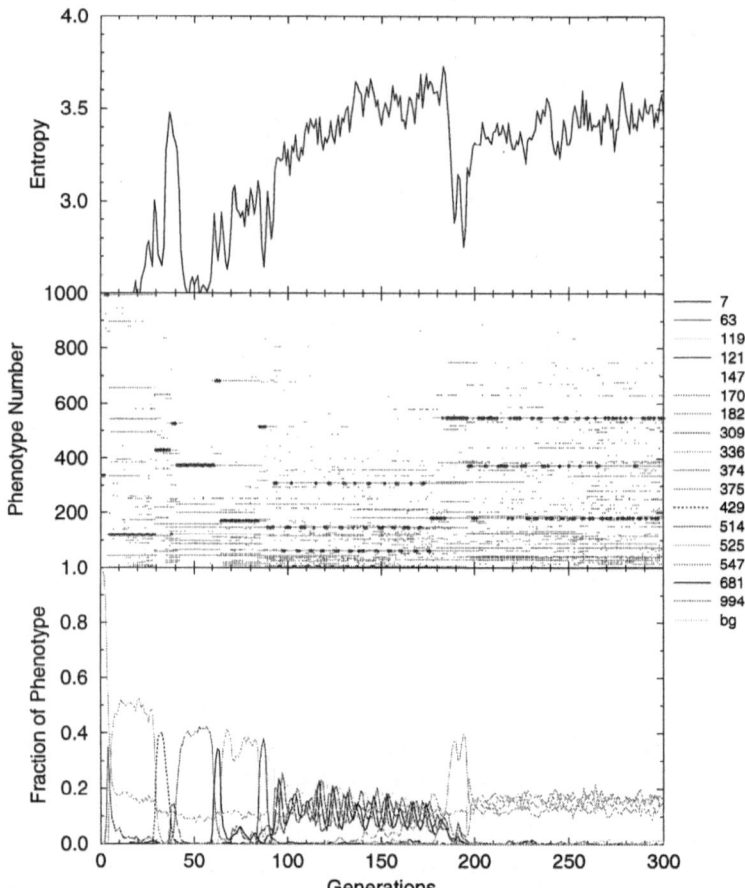

Fig. 11. Voyaging large catalytic networks: embedded in an artificial secondary-structure landscape of 1000 phenotypes 420 randomly chosen hypercycles of length three to seven are assigned. In addition 18 reaction paths between randomly chosen phenotypes are defined. In this figure an optimization from one catalytic organization (four-member hypercycle) to a three-member hypercycle is shown. In the bottom plot concentrations of only these phenotypes are shown whichever existed with maximal concentration at least once in the simulation. The middle plot shows all phenotypes in time. Stars (large dots) denote phenotypes with maximal concentration. In the upper graph the Shannon entropy of the population is plotted. Reaction rates for the catalyzed replication are randomly distributed between $2f$ and $65f$ ($f = const.$).

223

5 Conclusion

Nonlinear dynamics are an inherent property of biological systems. Not only due to the complex matter of biology itself but also due to complex feedback loops and nonlinear relationship between different subunits. By todays biotechnolgical tools these molecular-biological systems are quite easily designed and evolved (Gebinoga, 1995; Gebinoga and Oehlenschläger, 1996). Theories of molecular evolution can be proven and give novel insights for new experiments. An essential property of the dynamics of such systems are partly-neutral mappings between genotype and phenotype. Neutral networks provide a powerful medium through which evolution can become really efficient. Adaptive walks of populations, usually ending in one of the nearby minor peaks of the fitness landscape, are supplemented by random drift on neutral networks. Periods of neutral diffusion end when the population reaches areas of higher fitness values. Series of adaptive walks interrupted by phases of neutral random walk allow to approach the global fitness-maximum provided the neutral networks are sufficiently large. For more complex organized dynamical units, such as hypercycles, similar behavior can be observed. Hypercycles are able to out-compete parasites by avoiding parasitic regions of sequence-space due to neutral drift off such infected areas. By same mechanism new dynamical organizations emerges accordingly to a given selection pressure.

Summarizing, this article is a "perspective blink" to a complete and comprehensive description of molecular evolutionary dynamics by a theoretical approach. *"Molecular biology and the enormous wealth of molecular data in this field which is fastly growing at present may already contain the 'Rosetta stone' of the life sciences that allows to translate the language of physics and chemistry into the language of biology"* – Peter Schuster (1997).

Acknowledgments

I thank Prof. Peter Schuster for fruitful discussions and for the support with historic quotes. Part of this work presented here is joint research with Drs. Michael Gebinoga, Ulrike Göbel, Janos Palinkas, Christian Reidys, Peter Stadler and Jacqueline Weber which is published elsewhere. I also would like to thank the community of the Santa Fe Institute where this publication has been finished during a visit in July 1997.

References

R. J. Bagley and J. D. Farmer. (1992): Spontaneous emergence of a metabolism. In C. G. Langton, C. Taylor, J. D. Farmer, and S. Rasmussen, editors, *Artificial Life II*, volume X of *Santa Fe Institute Studies in the Science of Complexity*, chapter Origin/Self-Organization, pages 93–141. Addison Wesley, Redwood City.

R. J. Bagley, J. D. Farmer, and W. Fontana. (1992): Evolution of a metabolism. In C. G. Langton, C. Taylor, J. D. Farmer, and S. Rasmussen, editors, *Artificial Life II*, volume X of *Santa Fe Institute Studies in the Science of Complexity*, chapter Origin/Self-Organization, pages 141–158. Addison Wesley, Redwood City.

M. C. Boerlijst and P. Hogeweg. (1991): Spiral wave structure in pre-biotic evolution: Hypercycles stable against parasites. *Physica D*, **48**, 17 – 28.

M. B. Cronhjort and C. Blomberg. (1994): Hypercycles versus parasites in a two dimensional partial differential equations model. *J. theor. Biol.*, **169**, 31–49.

B. Derrida. (1981): Random-energy model: An exactly solvable model of disorderes systems. *Phys. Rev. B*, **24**(5), 2613–2626.

B. Derrida and L. Peliti. (1991): Evolution in a flat fitness landscape. *Bull. Math. Biol.*, **53**, 355–382.

T. Dobzhansky. (1973): Nothing in biology makes sense expect in the light of evolution. *Am. Bio. Teacher*, **35**, 125–129.

M. Eigen. (1971): Selforganization of matter and the evolution of biological macromolecules. *Die Naturwissenschaften*, **10**, 465–523.

M. Eigen, J. McCaskill, and P. Schuster. (1988): Molecular Quasi-Species. *Journal of Physical Chemistry*, **92**, 6881–6891.

M. Eigen and P. Schuster. (1977): The hypercycle A: A principle of natural self-organization: Emergence of the hypercycle. *Naturwissenschaften*, **64**, 541–565.

M. Eigen and P. Schuster. (1978): The Hpercycle B: The abstract hypercycle. *Naturwissenschaften*, **65**, 7–41.

M. Feinberg. (1977): Mathematical aspects of mass action kinetics. In L. Lapidus and N. Amundson, editors, *Chemical Reactor Theory, A Review*, pages 1–78, Englewood Cliffs, NJ. Prentice-Hall Inc.

W. Fontana and L. W. Buss. (1994): The arrival of the fittest: toward a theory of biological organizazion. *Bull. Math. Biol*, **56**, 1 – 64.

W. Fontana, W. Schnabl, and P. Schuster. (1989): Physical aspects of evolutionary optimization and adaption. *Physical Review A*, **40**(6), 3301–3321.

W. Fontana and P. Schuster. (1987): A computer model of evolutionary optimization. *Biophysical Chemistry*, **26**, 123–147.

C. V. Forst. (1996): Chaotic interactions of selfreplicating RNA. *SFI Preprint 95-10-093, Computers Chem.*, **20**(1), 69–84.

C. V. Forst, C. Reidys, and J. Weber. (1995): Evolutionary dynamics and optimization: Neutral Networks as model-landscape for RNA secondary-structure folding-landscapes. In F. Morán, A. Moreno, J. Merelo, and P. Chacón, editors, *Advances in Artificial Life*, volume 929 of *Lecture Notes in Artificial Intelligence*, pages 128–147, Berlin, Heidelberg, New York. ECAL '95, Springer. Santa Fe Preprint 95-10-094.

C. V. Forst, J. Weber, C. Reidys, and P. Schuster. Transitions and evolutive optimization in Multi Shape landscapes. in prep., (1995):

G. Galilei. (1968): *Opere*, page 232. Babera, Firenze, Italy, A. Favaro edition. The original quotation reads: "... It (the great book of the universe) is written in the language of mathematics and its symbols are triangles, circles, ... ".

M. Gebinoga. (1995): Hypercycles in biological systems. *J. Endocyt.* submitted.

M. Gebinoga and F. Oehlenschläger. (1996): Comparison of self-sustained sequence replication reaction systems. *Eur. J. Biochem.*, **235**, 256–261.

D. Gillespie. (1977): Exact stochastic simulation of coupled chemical reactions. *J. Chem. Phys.*, **81**, 2340–2361.

R. W. Hamming. (1986): *Coding and Information Theory*. Prentice Hall, Englewood Cliffs.

R. Hecht, R. Happel, P. Schuster, and P. F. Stadler. (1997): Autocatalytic networks with intermediates I: Irreversible reactions. *Math. Biosc.*, **140**, 33–74. Santa Fe Institute preprint 96-05-024.

P. G. Higgs and B. Derrida. (1991): Stochastic models for species formation in evolving populations. *J. Physics A*, **24**, L985–L991.

P. G. Higgs and B. Derrida. (1992): Genetic distance and species formation in evolving populations. *J. Mol. Evol.*, **35**, 454–465.

J. Hofbauer, P. Schuster, K.Sigmund, and R. wolff. (1980): Dynamical systems under constant organisation II: Homogeneous growth functions of degree $p=2$. *Siam. J. Appl. Math.*, **38**(2), 282–304.

M. A. Huynen, P. F. Stadler, and W. Fontana. (1996): Smoothness withing ruggedness: The role of neutrality in adaption. *Proc. Natl. Acad. Sci.*, **93**, 397–401.

S. A. Kauffman. (1986): Autocatalytic sets of proteins. *J. Theor. Biol.*, **119**, 1–24.

S. Kopp, C. M. Reidys, and P. Schuster. (1996): Exploration of artificial landscapes based on random graphs. In F. Schweitzer, editor, *Self-Organization of Complex Structures: From Individual to Collective Behavior*, London, UK. Gordon and Breach.

J. S. McCaskill. (1984): A localization threshold for macromolecular quasispecies from continuously distributed replication rates. *J. Chem. Phys.*, **80**, 5194–5202.

M. Nowak and P. Schuster. (1989): Error tresholds of replication in finite populations, mutation frequencies and the onset of Muller's ratchet. *Journal of theoretical Biology*, **137**, 375–395.

D. Park. (1988): *The How and Why. An essay on the Origins and Development of Physical Theory*. Princeton University Press, Princeton, NJ.

S. Rasmussen, C. Knudsen, and R. Feldberg. (1991): Dynamics of programmable matter. In C. G. Langton, C. Taylor, J. D. Farmer, and S. Rasmussen, editors, *Artificial Life II*, volume X of *Santa Fe Institute Studies in the Sciences of Complexity*, pages 211–254, Redwood City. Addison Wesley.

S. Rasmussen, E. Mosekilde, and J. Engelbrecht. (1989): Time of emergence and dynamics of cooperative gene networks. In P. Christiansen and R. Parmentier, editors, *Structure, Coherence and Chaos in Dynamical Systems*, Proceedings in Nonlinear Science, pages 315–331. Manchester University Press.

C. Reidys. (1995): *Neutral Networks of RNA Secondary Structures*. PhD thesis, Friedrich Schiller Universität, Jena.

C. Reidys, P. F. Stadler, and P. Schuster. (1997): Generic properties of combinatory maps and neutral networks of RNA secondary structures. *Bull. Math. Biol.*, **59**(2), 339–397.

W. Schnabl, P. F. Stadler, C. Forst, and P. Schuster. (1991): Full characterization of a strange attractor: Chaotic dynamics in low dimensional replicator systems. *Physica D*, **48**, 65 – 90.

P. Schuster. (1989): Optimization and complexity in molecular biology and physics. In P. J. Plath, editor, *Optimal Structures in Heterogenous Reaction Systems*, Synergetics. Springer Verlag.

P. Schuster. (1995): Artificial life and molecular evolutionary biology. In F. Morán, A. Moreno, J. Merelo, and P. Chacón, editors, *Advances in Artificial Life*, Lecture Notes in Artificial Intelligence, Berlin, Heidelberg, New York. ECAL '95, Springer.

P. Schuster. (1997): Genotypes with phenotypes: Adventures in an RNA toy world. *Biophysical Chemistry*. submitted, SFI-Preprint 97-04-036.

P. Schuster and K. Sigmund. (1983): Replicator dynamics. *J. theor. Biol.*, **100**, 533–538.

P. Schuster, K. Sigmund, and R. Wolff. (1978): Dynamical systems under constant organisation I.topologigal analysis of a family of non-linear differential equations - a model for catalytic hypercycles. *Bull. Math. Biophys.*, **40**, 743–769.

P. Schuster, K. Sigmund, and R. Wolff. (1979): Dynamical systems under constant organisation III.cooperative ond competitive behaviour of hypercycles. *J. Diff. Equ.*, **32**, 357–368.

P. Schuster and J. Swetina. (1988): Stationary mutant distributions and evolutionary optimization. *Bull. Math. Biol.*, **50**, 635.

S. Spiegelman. (1971): An approach to the experimental analysis of precellular evolution. *Quart. Rev. Biophys.*, **17**, 213.

P. F. Stadler and P. Schuster. (1992): Mutation in autocatalytic networks - an analysis based on perturbation theory. *J. Math. Biol.*, **30**, 597–631.

C. Streissler. (1992): *Autocatalytic Networks under Diffusion*. PhD thesis, Universität Wien.

J. Swetina and P. Schuster. (1982): Self-replication with errors – a model for polynucleotide replication. *Biophys. Chem.*, **16**, 329–345.

M. S. Waterman. (1978): Secondary structure of single - stranded nucleic acids. *Studies on foundations and combinatorics, Advances in mathematics supplementary studies, Academic Press N.Y.*, 1, 167 – 212.

J. D. Watson and F. H. C. Crick. (1953): Molecular structure of nucleic acid. A structure for Desoxyribo Nucleic Acid. *Nature*, **171**, 964–969.

J. Weber. (1997): *Dynamics of Neutral Evolution – A case study on RNA secondary structures*. PhD thesis, Friedrich Schiller Universität Jena.

The Retinal Spreading Depression:
A Model for Nonlinear Behavior
of the Brain

Wolfgang Hanke[1], Markus Goldermann[1], Sabine Brand[1] and Vera Maura
Fernandes de Lima[2]

[1] Institut für Zoophysiologe, Universität Hohenheim, D-70593 Stuttgart, Germany
[2] Institute of Biomedical Engineering, UNICAMP, Campinas, SP, Brazil

Abstract. The human brain may behave as an excitable medium, at least up
to a certain degree. Although increasing evidence supports this statement,
the experimental basis for investigations in the field is small. However, the
spreading depression, a propagating wave of depression of the electrical
activity in neural tissue, is a tempting example of self-organisation in the
brain and also occurs in the retina as a true part of CNS. There it can be
easily investigated and can be used as a tool for studying self-organisation in
the brain. We will shortly introduce some of the experimental advances of
the retinal model of the CNS and the retinal spreading depression in this
article, together with some conclusion from experiments which have been
done with it.

Content:

1 Introduction

Excitable media are non-linear systems far from thermodynamical equilibrium conditions. Theories for such systems can be used to describe all those findings not fitting the framework of linear or quasi-linear physics. The CNS of higher vertebrates and especially of man, including the human brain, may behave as an excitable medium, as for example the generation of ordered structures, among others, propagating excitation-depression waves, have been observed in it (i.e. Bures et al, 1974). The spontaneous onset of such waves and their propagation in the CNS, treated as a non-linear thermodynamical system, is coupled to the following conditions:

    1 the system must be thermodynamically open
    2 the system must be in more than critical distance from equilibrium
    3 the dynamical equations necessary to describe the system must be non-
      linear
    4 feedback must exist in the system
    5 the microprocesses must be co-operative.

All these condition are fulfilled in the vertebrate CNS. And as a first experimental verification for the statement that the brain may behave as an excitable medium, let us stay here with the fact that propagating waves of excitation-depression have been observed in it.

Non-linearities, which are among other requested from the above table exist in the CNS on different levels:

    1 in the cell membranes
    2 in single cells, especially neurones
    3 in assemblies of cells
    4 in parts of the brain and in the entire brain.

From point 4 of this table, taking the human brain as an assembly of about some billions of excitable cells, each of which can be coupled to thousands of other cells, the brain is the ideal excitable medium and a variety of experimental predictions from suitable theories for such systems (see chapter 3) should be fulfilled, as there are:

    1 oscillations of system parameters
    2 propagating excitation-depression waves
    3 self-sustained activity, i.e. spirals of propagating waves
    4 non-equilibrium phase transitions with critical slowing down
    5 enhancement of fluctuations near critical points.

In the experimental section of this article we will shortly show that at least some of the listed predictions in experiments already has been verified.

Of course it would be tempting to do experiments concerning question of self-organisation directly in the brain. Some studies in fact have been performed towards this direction and the experimental results are in line with our statements. However, due to technical, moral and ethical reasons as well as on limitations by laws, there is a serious need of experimental systems to investigate self-organisation in the brain.

The retina is a true part of the brain and can easily be prepared from chicken, for example. Experiments concerning the question of the retina behaving as an excitable medium can be performed in the retina much easier than in other parts of the brain. The spreading depression (SD) can be observed in the retina easily, even with the naked eye, due to the big intrinsic optical signal (IOS) accompanying it. In figure 1 two snapshots showing the temporal development of a concentric SD wave following a mechanical stimulus in the retina are shown.

30 sec   50 sec

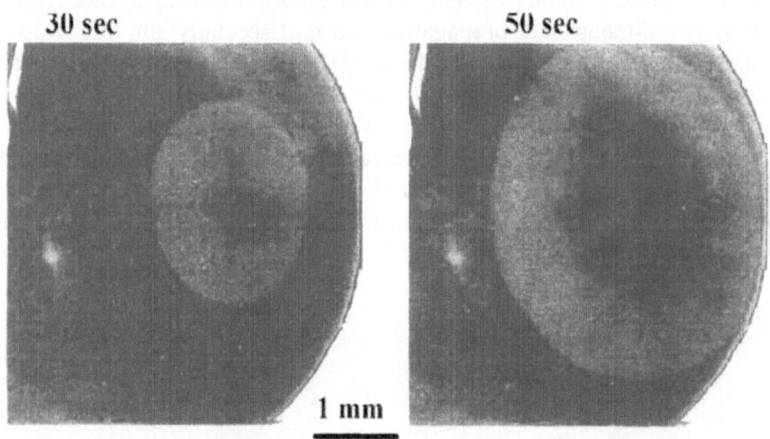

1 mm

Figure 1 *Two photos of a concentric SD wave in a chicken retina after a mechanical stimulus with a fine glass needle. The SD can be seen as an increasing concentric area of higher brightness. Time after stimulus is given above the photos.*

2 The retinal spreading depression

2.1 History

The spreading depression was first described by Leao in the 1940th (Leao, 1944) in the rat cortex as a wave of suppression of the electrical activity in the neuronal tissue. He measured the wave propagation velocity to be about 3-5 mm/sec. In the following, a variety of concomitants of the wave, as there are ion movements across the membranes, geometrical changes of the tissue, the IOS, and transmitter release from the nerve terminals, have been described (see Bures et al., 1974).

Even earlier Lashly (Lashly, 1940) described the visual hallucinations of the migraine aura to be the consequence of a propagating excitation-depression wave in the visual cortex propagating with about 3-5 mm/sec, too. Several years later Milner (Milner, 1958) was the first to point out some striking similarities between both observations and to postulate SD waves to be the mechanistical basis of the migraine aura. The relation of SD to migraine is highly interesting, but is still discussed controversial today.

SD waves were found up to today in practically all parts of the CNS including the retina. The retinal SD was in detail characterised by Martins-Ferreira, mainly using the chicken retina (Martins-Ferreira, 1983). It is accompanied by an unusual big IOS making its observation very easy and allowing the application of for example standard video imaging techniques (Hanke et al., 1996). The retina is furthermore a thin, close to two-dimensional piece of intact tissue (300 μm thick), allowing to observe the SD as a two-dimensional propagating wave. Especially on this reason, striking similarities between the retinal SD and the Belousov-Zhabotinski reaction (Zaitkin and Zhabotinski, 1970) became obvious. Bures and co-workers (Goroleva and Bures, 1983) were the first to draw the obvious conclusions and to discuss the SD, especially the retinal SD, in terms of non-linear thermodynamics. Among other things he predicted the existence of spirals of spreading depression waves in the retina. An example of a spiral SD wave in a retina is shown in figure 2.

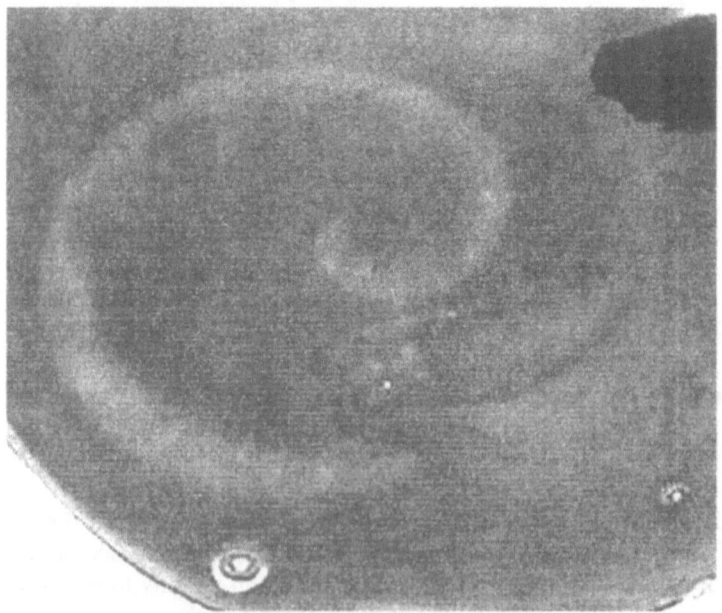

Figure 2 *Photo of a spiral of spreading depression in a retina.*

Another form of autowaves of SD in the retina was developed and described in detail by Martins-Ferreira (Martins-Ferreira et al., 1974), the so called circulating wave. The principles of experiments with such a permanently circulating wave are pointed out in sketch in figure 3.

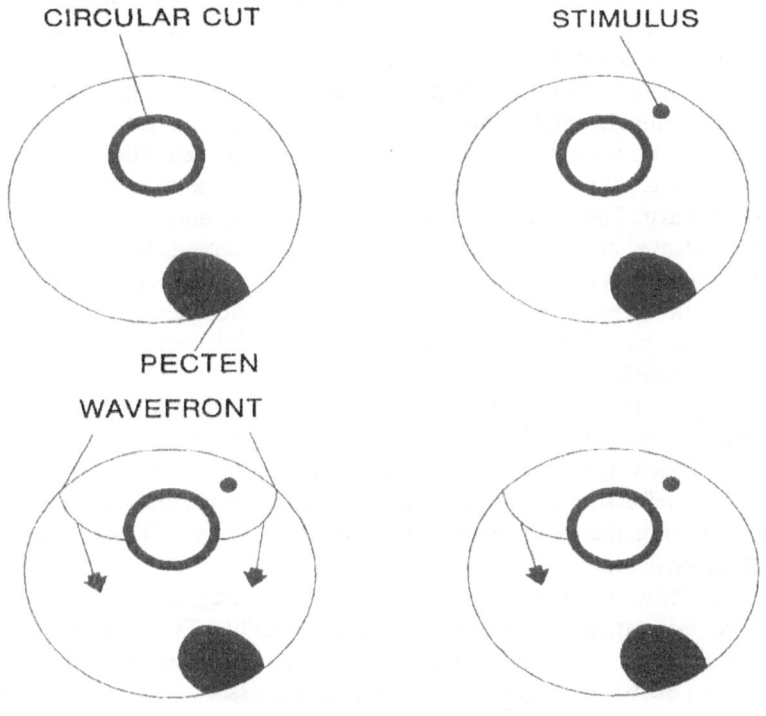

Figure 3 *Cartoon of how to elicit and obtain a circulating wave of SD in a retina. A circular cut is made in the retina and in the outer ring of tissue a wave is elicited by a proper stimulus. One of the wave fronts is wiped out (chemically or by injecting current) and the other than can circulate for hours.*

A first mechanistical model, describing the SD as a propagating wave of increased extracellular potassium concentration was given by Grafstein (1956). Later more detailed models, taking for example the transmitter release from nerve terminal into account (van Harrefeld, 1959) were launched. A first theoretical description was given by Hodgkin (1959) in an internal manuscript, and some detailed studies treating the SD as a reaction-diffusion wave (Tuckwell and Miura, 1978) have been published. Interestingly the same approach has been used successfully to describe the behaviour of the visual aura usually preceding a migraine attack (Reggia and Montgomery, 1996).

2. 2 Mechanistical basis of the spreading depression

The main events defining the behaviour of SD waves are briefly listed in the following. As a consequence of a appropriate trigger or due to fluctuations of activity in the neuronal tissue, a local increase in extracellular potassium concentration occurs. This leads to a depolarisation of cell membranes and thus to an activation of voltage dependent potassium channels, and thus to a release of potassium from the cells. The potassium diffuses to cells close by and there induces depolarisation and activation of voltage sensitive potassium channels. Following the electrochemical gradient, potassium leaves neurones and thus the extracellular potassium concentration increases further and this circle of amplification finally gives rise to a propagating potassium wave. The depolarisation of the cell membranes also triggers a massive release of neurotransmitters from the nerve-terminals which increases the effects even more (van Harrefeld, 1959). A variety of electrical and metabolic events is the consequence of these basic events and after the wavefront has passed, the ion gradients across the cell membranes can be completely abolished.

Only due to the existence of metabolic energy in the cells, the system can be re-established (mainly by ATP consuming ion-pumps). During this recovery period the system is totally refractive to further waves for about 2 min and then relative refractive for up to 20 min, however, after restoration of the ion gradients during the absolute refractory period, the next SD wave can be elicited in the tissue.

When the tissue is running down in metabolic energy, the ion-gradients cannot be re-installed and no further waves are possible. Typically in such a situation the cells die after breakdown of the ion-gradients following the passage of a SD wavefront and lesions occur in the tissue.

Although mainly neurones are discussed in this feedback circuit, glia cells are important for example for potassium clearance and influence the properties of the propagating wave. Since the importance of glia cells in SD events today is unquestionable, details are presently under discussion.

2.3 Comparison of the retina to other parts of the CNS

According to its origin and development the retina is a true part of the CNS (Dowling, 1987). It is a somewhat simplified structure with a few defined layers of neurones, forming a piece of tissue with a close to two-dimensional structure. Thus the retina is the ideal slice of neuronal tissue, especially as it can be prepared easily without significant damage. In addition, the retina of some birds is avascular, first allowing the study of neuronal tissue without interaction of vessels and second making its maintenance in experiments quite easy. The primary sensory receptors of the retina are not significantly involved in the retinal SD, they can be removed without significantly effecting the SD.

In the retina, all the neurotransmitters being present in other parts of the brain have been found, too.

Different to the other parts of the brain, the glia cells of the retina, the Müller cells are of special structure, as they are spanning the complete retina vertically. Thus they are able to fulfil the task of potassium clearance without the need of electrical coupling, as is given in other parts of the brain. This, however, does not effect the principles of SD wave propagation in the retina. The presence of gap junctions between Müller cells in the retina of some species still has not been verified unequivocally, in the chicken retina we have demonstrated that no gap junctions between Müller cells are present (Hanke et al., 1997).

The behaviour of SD waves in the retina has been shown to be very similar, if not identical compared to that in other parts of the brain. All the theoretical and experimental considerations in the following done for the retina thus are valid for other parts of the brain, too.

3. Theoretical aspects of spreading depression waves

3.1 Theories describing the retinal spreading depression

The SD in its general appearance has some aspects in common with propagated action potentials, except the velocity, which is orders of magnitude different. Starting from these observations Hodgkin (1959) launched a first mathematical model of the SD, based on the set of differential equations describing the action potential.

Later, when the retinal SD had been described and when its similarities with the Belousov-Zhabotinski reaction (BZ) became obvious, models on the same basis as used for the BZ (i.e. Dockery et al., 1988) were used for the SD, too. A detailed discussion of such a model can be found at Tuckwell (Tuckwell and Miura, 1978).

3.2 Predictions from the theories

A suitable theory should be able to explain all the experimental findings being known so far. However, besides doing that, sometimes predictions about new experiments can be made based on a theory. The verification of such predictions then is a verification of the theory itself. According to this, a variety of prediction can be made about the behaviour of SD waves based on the diffusion-reaction theories.

From the comparison of SD waves to propagating action potentials and especially due to the fact that both exhibit an explicit refractory phase, it can be concluded that annihilation should occur when two wave fronts collide. In figure 4 the verification of this statement is shown.

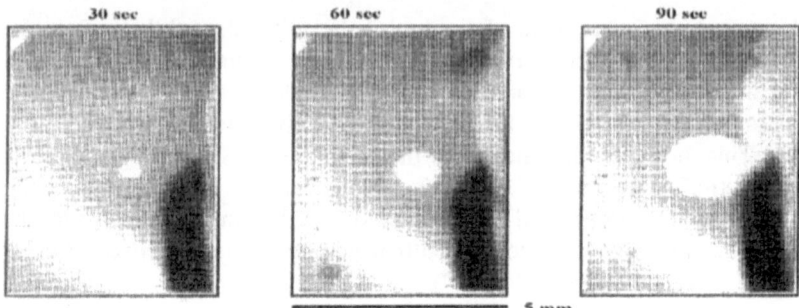

Figure 4 *Series of photos showing the collision of two SD waves in a retina, with annihilation at the point of collision.*

Also it can be concluded that the amplitude of a SD wave in the partially refractory state of the tissue after a preceding wave should be smaller. This is demonstrated in figure 5 using the IOS of a retinal SD wave as parameter.

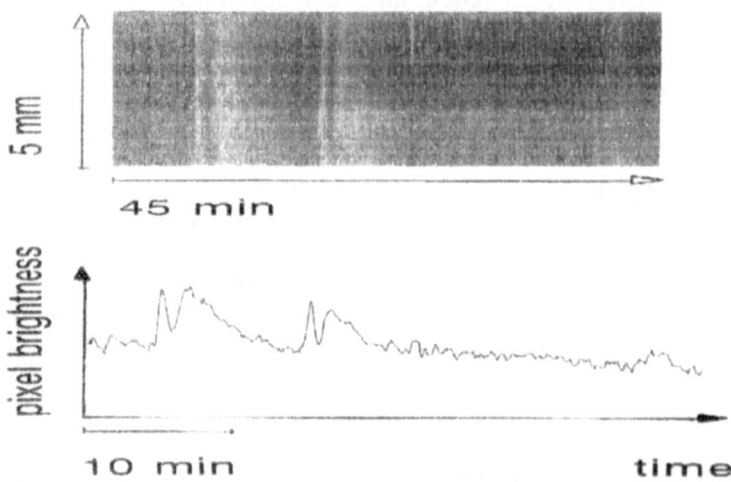

Figure 5 *IOS of two retinal SD waves. The second wave is travelling in the partially refractory period of the first, as a result the amplitude of the IOS is smaller. In the upper part of the figure a two-dimensional representation (stack of lines from consecutive video-frames) is shown, in the lower part the brightness of a small area of the retina is measure in time.*

From the comparison of the retinal SD to the behaviour of the BZ reaction and the cited theories, the possibility of SD spirals propagating in the retina already has been postulated and verified (see above).

The dispersion relation of the SD should be non-linear increasing with decreasing repetition rate of waves. It should be very similar to that of the BZ reaction according to the application of an identical theoretical approach (see figure 7).

From a more detailed look at the theories (and of course also from the mechanistic basis of SD) we predicted some more experimental results. Fluctuations of system parameters should occur at the front of the propagating wave and between stimulus and onset of wave propagation. An increased spike frequency in the ganglion cell layer of the retina should precede the SD wavefront as for example predicted from the model of Reggae and Montgomery (1996) for the visual aura of migraine. Sub-threshold stimuli should induce damped parameter fluctuations in the retina without eliciting waves. Furthermore the question was raised, whether SD waves speed up before collision.

Experiments concerning some of these predictions are described in the following section, the existence of spirals and the annihilation upon collision has already been demonstrated above.

4 Experiments

4.1 Technical notes

The simplest system to do retinal spreading experiments in, is the so called eye-cup preparation. This is the back half of an eye, with the retina attached to it, after removal of the humour vitreous (for details see: Fernandes de Lima et al., 1997). Frequently chicken eye-cups are used, as the retina of chicken is avascular, thus allowing experiments in plain neuronal tissue without disturbances from vessels. After preparation, the eye-cup (typical diameter about 5 to 10 mm) can easily be mounted in a perfusion chamber, or even simpler in a Petri-dish, under a binocular with moderate magnification. To the binocular usually video equipment is mounted. Additionally electrophysiological experiments can be performed under optical control. In figure 5 a cartoon of a set-up to perform retinal spreading depression experiments is shown. It allows parallel electrophysiological and optical recording from the retina. All the data taken can be stored on computer (electrophysiological) and video tape (optical). Data evaluation, including image processing usually is done off-line after the experiments.

Spreading depression experiments can be done in one eye-cup for several hours.

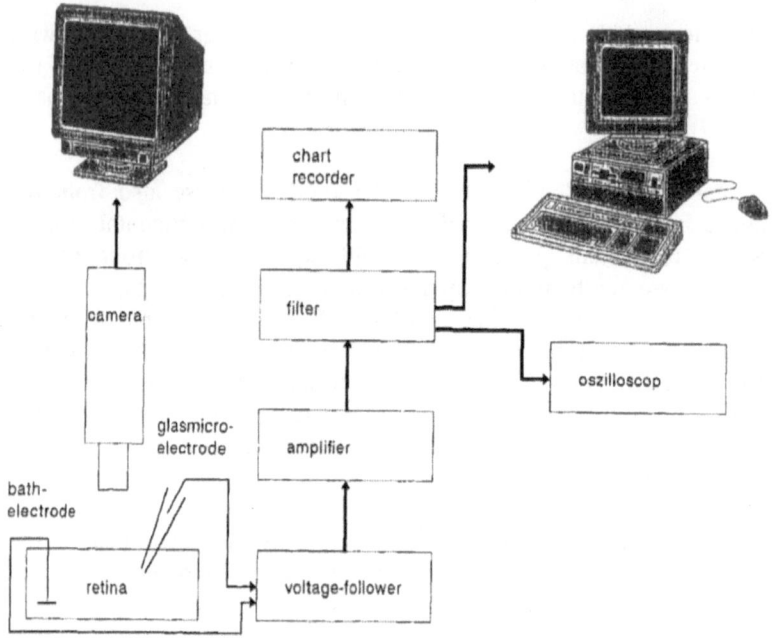

Figure 6 *Cartoon of a set-up to perform retinal spreading depression experiments. It includes a binocular, the perfused chamber and a variety of equipment for optical and electrophysiological recording. Starting from such a basic set-up nearly any degree of experimental sophistication can be added.*

The experiments described in the following section were all done with equipment similar or identical to that shown in figure 6. Although in this article we focus on retinal SD experiments, let us shortly mention that experiments in slices of any neuronal tissue can be performed using the same equipment and procedures, however, the preparation of neuronal slices often is much more difficult than that of an eye-cup, as is the maintenance of such preparations.

4. Results

Before describing specified experiments let us summarise a few results from the history of retinal SD. It is very similar in its appearance to the behaviour of the BZ-reaction including a propagation velocity of about 3-5 mm/sec for both, independent of the substantially different mechanisms. The existence of spirals and other autoactivity as well has been verified. The annihilation upon collision was already mentioned above (figure 4).

The next to show is that the dispersion relation of the retinal SD follows diffusion-reaction theories (Brand et al., 1997). In figure 7 the dispersion relation of retinal SD waves is shown, measured in double wave experiments. The first wave always is taken as reference, the second is elicited after a variable time. Then the retina is allowed to rest for 30 min and the next double wave is elicited. The wave propagation velocity of the second is plotted as a function of the temporal distance of the two waves. The dispersion relation measured this way is about that what could be expected even from a quite simple theoretical approach, the propagation velocity is monotonously non-linear increasing with decreasing repetition rate of the waves, approaching a constant velocity of about 3-5 mm/sec.
In more detailed experiments, however, a few interesting points become obvious. The retinal SD exhibits an absolute refractory period of about 2 min, which is independent on temperature changes. The relative refractory period is quite long, 10-20 min, and variable.

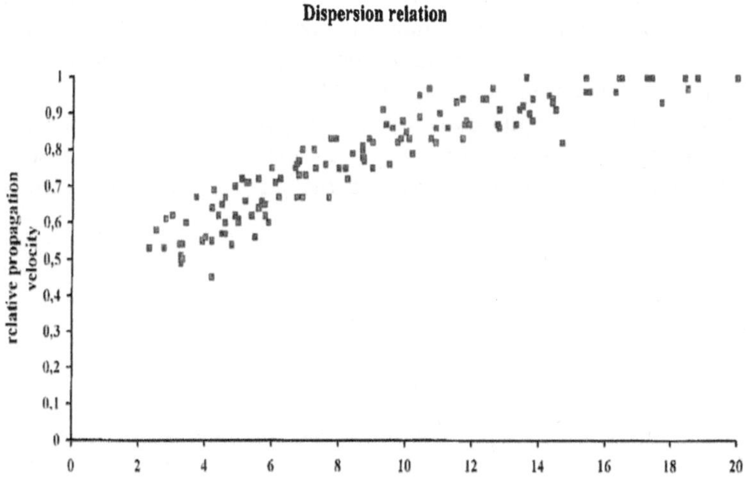

Dispersion relation

Figure 7 *Dispersion relation of the retinal SD measured in double wave experiments at 29 °C (see text for explanation). Longer times on the x-axis thus depict lower frequencies.*

A few more predictions from the theories can also be shown. At the front of the wave an increase of the spike frequency in the ganglion cell layer of the retina has been found (Egert, personal communication). Fluctuations of system parameters at the wave front or between stimulus and the onset of the wave are more difficult to verify, however, at least in the red IOS of the retina such fluctuations could be found (Fernandes de Lima and Hanke, 1997). According to the behaviour of waves at collisions we could furthermore show that the wavefronts accelerate before they collide when they are closer than about 500 µm (250 µm from each wavefront), however, only when the retinal tissue is in the partially refractory state. The geometric scale of about 250 µm seems to be somehow typical for other properties of retinal SD waves, too, i.e. fluctuations in the red IOS at the wavefront are limited to about this distance. A mechanistic basis could be given by the connectivity of the retinal cells which is spanning distances of about the same order.

An interesting approach to the SD is to ask how it can be elicited or how wave propagation sets up. Local fluctuations in system parameters should, in case occasionally crossing a certain threshold, induce an increasing instability which then can be the basis of a „spontaneous" propagating wave. Starting on the other hand from the typical experimental protocol of eliciting a wave by a mechanic stimulus, between stimulus and the onset of the wave a certain temporal delay should be present which is necessary for the amplification of the induced local fluctuations. Upon a super-threshold stimulus, increasing fluctuations should be observed between stimulus and wave onset, upon a sub-threshold stimulus the oscillations should be damped out. Although principally verified, the experimental evidence for fluctuations of system parameters is quite weak.

However, it has been unequivocally verified that there is a temporal delay between stimulus and wave onset and even more that there is a strict functional relation between this delay and the wave propagation velocity. This is shown in figure 8.

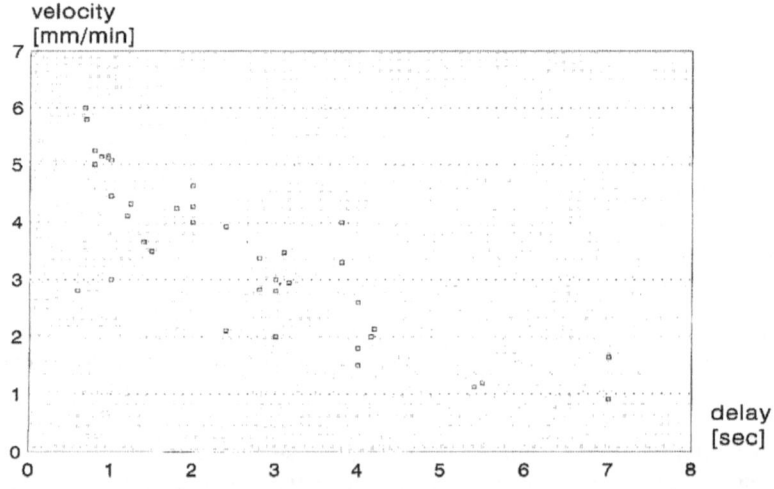

Figure 8 *Functional dependency of wave propagation velocity on the delay between stimulus and wave onset.*

Furthermore, waves do not always start at the point of stimulation, but the point of origin of wave propagation can be geometrically displaced as is shown in figure 9. Displacements up to about 300 μm were found, which is in the range of the geometrical constant already cited above.

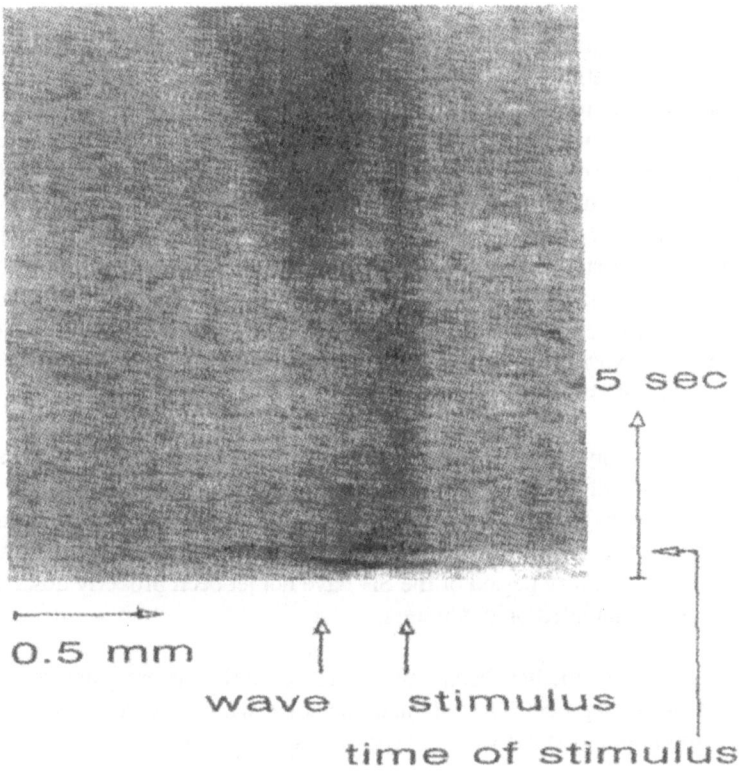

Figure 9 *Temporal and geometrical displacement of the onset of a retinal SD wave relative to the place of the mechanical stimulus, as shown in a stacked series of lines from consecutive video frames passing the place of the stimulus. A developing lesion at the place of the stimulus can also be seen.*

Some additional observations, independent on SD experiments, can also be used to support the ideas of this paper. Thus, the investigation of EEG recordings under the view of non-linear thermodynamic theories has contributed a lot to the understanding of processes in the brain, this behaving as an excitable medium, even when these studies mainly were done under pathophysiological conditions (Haken, 1985; Lopes da Silva et al., 1994; Renshaw, 1994). Also the behaviour of assemblies of excitable cells has been discussed under the aspect of chaotic behaviour (Elbert et al., 1994).

5. Computer simulation of retinal spreading depression

Before starting a more speculative view at the brain as an excitable medium, let us finally look at what has been done in using computer simulations of the retinal SD and related events. As already done in other excitable media, starting with a typical set of differential equations in numerical simulations most of the above described facts can be simulated quite well. The same is true for using cellular automata. The differences between simulating the BZ-reaction and the retinal SD are not that significant, except that the SD has a more pronounced refractory period. The things change when it is tried to include the mechanistic basis in the computer models. According to the BZ-reaction, a scheme of reaction equations is commonly excepted, describing it mechanistically, and this can be the basis for a more realistic simulation. Looking at the SD the situation is somewhat more complicated. Of course there is a principal consensus about the structure of neuronal tissue and especially the retina, however, the details about the connectivity of the cells and the structures involved in coupling are still under discussion. Nevertheless, a variety of promising compartimented models to simulate SD wave are presently under investigation.

Although obviously the principals of SD waves can be simulated quite easily with classical diffusion-reaction theories, at least to our knowledge, no models are presently available or discussed, giving for example acceleration of waves before collisions or describing the onset of waves in detail. Also details of the refractory period of the SD have not jet been properly described by theories or simulated on computers.

Without going to a further discussion, in figure 10 the general simulation of a spiral wave using a cellular automata is shown which should be compared to the original photo of a retinal SD spiral as shown above.

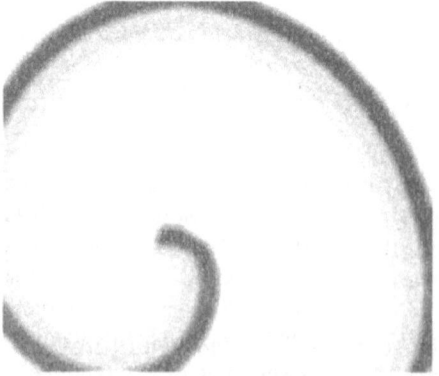

Figure 10 *Simulation of a spiral wave similar to the behaviour of retinal SD spiral waves, using a cellular automata.*

6 Conclusions and outlook

6.1 Future experiments

Due to the complicated structure of the CNS a complete mechanistic description of SD events presently is not possible. However, as discussed in this article above, a variety of experiments have been done and the principles of SD waves seem to be obvious. What is not that clear is for example the role of the glia cells. One of the aims of future experiments thus should be to clarify up to which extent glia cells are involved in the onset and propagation of SD waves. Although an extended pharmacology has been measured concerning SD waves, more experiments are necessary here. Finally more electrophysiology on the single channel level (patch-clamp experiments) during SD waves is required to better understand the mechanistic basis of SD waves.

According to the presence of SD waves in the human brain non-invasive experiments presently are rare due to the immense technical needs, however, with new machinery for visualising the activity of the human brain such experiments became realistic. Due to their medical relevance such experiments will become more important in the future anyhow.

6.2 Medical relevance of spreading depression and related phenomena

In the field of pathoneurophysiology a variety of so called functional syndromes are know, which are medically relevant disturbances without a known histological basis. Migraine as has been discussed previously already together with SD belongs to the field as well as some epileptic events and certain forms of amnesia. Although still under discussion, at least the onset of migraine and the visual aura frequently preceding it, is widely accepted to be based on a SD event. It has also been speculated that migraine and certain forms of epilepsy have the same common basis, however, evidence is weak here.

May be the understanding of some of the functional syndromes which cannot be explained presently on a mechanistical basis can be better understood when looking at the brain as an excitable medium. Thus physical theories of the brain in the future may contribute to advances in the medical treatment of some diseases, which is of special interest when having in mind that in western countries for example up to 10% of the population is suffering in migraine.

6.3 General aspects of the human brain being treated as an excitable medium

At the end of this article let us shortly ask a few more general, even philosophical questions about the matter discussed. The CNS and thus the human brain at least under certain conditions behaves as an excitable medium as has been shown above. Some of its functions obviously can be described by theories for such non-linear thermodynamical systems. Although up to now mainly pathophysiological events are seen under this aspect, it should be allowed to ask, whether other functions of the human brain which are not understood presently, follow the same physical rules. Necessarily we end at the point to look at higher mental capabilities of the brain to be the consequence of it to be a physical system, more clearly a system obeying the rules of non-linear thermodynamics.

In a more popular approach it is for example discussed, whether the human memory at least partially may be based on associative structures as are given in chaotic systems.

Although due to its complicated structure a complete theoretical description of the (human) brain presently seems to be not possible, in principle it may be reasonable, when not using mechanistic theories, but describing it as an excitable medium. Using such an approach, on the price of loosing the mechanistic (molecular) basis, a better understanding of how the brain works could be obtained.

7 References

Brand, S., Dahlem, M., Fernandes de Lima, V.M., Müller, S.M. and Hanke, W., Dispersion relation of spreading depression waves in the chicken retina.. Int. J. Bifurcation and Chaos, in press, 1997

Bures, J., Buresova, O. and Krivanek, J., The mechanism and application of Leao's spreading depression of encephalographic activity. Academic press, New York, 1974

Dockery, J.D., Keener, J.P. and Tyson, J., Dispersion of travelling waves in the Belousov-Zhabotinsky reaction. Physica D, **30**, 177-191, 1988

Dowling, J., The retina: An approachable part of the brain. Harvard University Press, Cambridge, USA, 1987

Elbert, T., Ray, W.J., Kowalik, Z.J., Skinner, J.E., Graf, K.E. and Birbaumer, N., Chaos and physiology: Deterministic chaos in excitable cell assemblies. Physiol. Rev., **74**, 1-47, 1994

Fernandes de Lima, V.M., Goldermann, M. and Hanke, W., The retinal spreading depression: A tool for interdisciplinary research. Academic Press, in press, 1997

Fernades de Lima, V.M. and Hanke, W., Excitation waves in central grey matter: The retinal spreading depression. Prog. In Retinal and Eye Res., in press, 1997

Goroleva, N.A. and Bures, J., Spiral waves of spreading depression in the isolated chicken retina. J. Neurobiol., **14**, 353-363, 1983

Grafstein, B., Mechanism of spreading depression. J. Neurophysiol., **19**, 154-171, 1956

Haken, H., Complex systems-Operational approaches. Springer Verlag, Berlin, 1985

Hanke, W., Mezler, M., Ladewig, T., Goldermann, M. and Fernandes de Lima, V.M., Patch-clamp experiments at isolated cells and in intact tissue of the chicken retina correlated to potassium channels and the retinal spreading depression. Electrochem. Acta, in press, 1997

Hodgkin, A.L., Spreading depression: Grafsteins hypothesis. Unpublished manuscript, Cambridge, 1959

Lashley, K.S., Patterns of cerebral integration indicated by the scotomas of migraine. Arch. Neurol. Psych., **46**, 331-339, 1940

Leao. A.A.P., Spreading depression of activity in the cerebral cortex. J. Neurophysiol., **7**, 359-390, 1944

Lopes da Silva, F.H., Pijin, J.P., Wadman, W.J., Dynamics of local neuronal networks: Control parameters and state bifurcations in epileptogenesis. Prog. Brain. Res., **102**, 359-370, 1994

Martins-Ferreira, H., de Olivera Castro, G., Struchinger, C.J. and Rodrigues, P.S., Circling spreading depression in isolated chick retina. J. Neurophysiol., **37**, 773-783, 1974

Milner, P.M., Note on a possible correspondence between the scotomas of migraine and spreading depression of Leao. EEG Clin. Neurophysiol., **10**, 705, 1958

Reggia, J.A: and Montgomery, D., A computational model of visual hallucinations in migraine. Comp. Biol. Med., **2**, 133-141, 1996

Tuckwell, H.C. and Miura, R.M., A mathematical model for cortical spreading depression. Biophys. J., **23**, 257-276, 1978

Renshaw, E., Chaos in biometry. J. Math. Appl. Med. Biol., **11**, 17-44, 1994

Zaitkin, A.N. and Zhabotinsky, A,M., Concentration wave propagation in two-dimensional liquid-phase self-oscillating system. Nature, **225**, 535-537, 1970

van Harreveld, A., Components in brain extracts causing spreading depression of cerebral cortical activity and contraction of crustacean muscle. J. Neurochem., **3**, 300-315, 1959

An Analytically Solvable Model of Collective Excitation Patterns in Cortical Tissue

Werner M. Kistler and J. Leo van Hemmen

Physik Department der TU-München, D-85747 Garching bei München, Germany

1 Introduction

Spontaneous pattern formation belongs to the most amazing phenomena in nature, but not until the mid of this century an appreciable insight into the underlying mechanisms has been gained. Starting with the famous works of A. Turing there is a long tradition of investigating pattern formation in the context of reaction-diffusion (RD) systems. RD systems, which are made up of two or more freely diffusible reactants, coupled by a nonlinear reaction kinetics, are the basis of as different phenomena as chemical waves in the laboratory assistant's dish or the stripes of a zebra.

Mathematically, RD systems are described in terms of nonlinear partial differential equations. The spatial derivatives account for the diffusive spread of the reactants and the temporal derivatives for the ongoing chemical reactions. Due to the essential nonlinearities, a mathematical analysis is rather demanding and in most cases no (nontrivial) analytic solutions are known. Nevertheless, there is an elaborated toolkit of mathematical methods particularly suited for the investigation of RD systems, among them bifurcation analysis and the theory of complex Ginzburg-Landau equations, to mention only the most prominent ones.

Here we consider a two-dimensional lattice of spiking neurons with local interactions. The strength and the sign of the interaction of two neurons depends on their distance. 'Spiking' means that we are not interested in the mean firing rate of the neurons, but in the single firing events when an action potential is released. The pattern, we are interested in, is the subset of all 'active' neurons, i.e., all neurons that are currently firing an action potential. The mathematical structure of this system is quite different from the RD systems, but it exhibits some phenomenological features that are surprisingly similar to those of a RD system. We find, for example, patterns of collective excitation in the form of traveling pulses and waves, rotating spirals, and expanding circles. Even the dispersion relation is qualitatively the same as in RD systems.

There are at least two reasons why this system is interesting to us. First, as for plane fronts and waves, we can solve the model analytically and obtain the dispersion relation, i.e., the relation between propagation velocity and wavelength. Second, self-organization of neuronal activity may have some biological relevance. In the twenties H. Klüver (1966) noticed that people who have ingested mescaline or other hallucinogens report that they 'see' geometric patterns like grids, waves, rotating spirals or expanding concentric circles during their intoxication. These patterns are not generated by the retina, but probably in the very next stage of visual processing, the primary visual cortex V1. It is hypothesized that these patterns result from a self-organization process in V1 under the influence of the hallucinogen which reduces the amount of overall neuronal inhibition.

2 Spike Response Neurons

Biological neurons communicate through action potentials. An action potential is a sharp voltage pulse ($\sim 100\,\mathrm{mV}$) which can travel down the axon of the neuron, resembling the propagation of a soliton. After some branchings the axon finally ends in a synapse, the place where one neuron contacts an other one. Each incoming action potential triggers a complex cascade of biochemical reactions in the synapse and in the adjacent membrane of the neighboring (postsynaptic) neuron. This results in a transient increase of the membrane voltage of the postsynaptic neuron, the so-called *postsynaptic potential* (PSP). A single neuron receives synaptic input from about 10^4 presynaptic neurons and the postsynaptic potentials from simultaneously arriving action potentials superpose accordingly. The neuron will fire an action potential itself, if its membrane voltage exceeds a certain threshold value. Immediately after a neuron has fired an action potential, the triggering of a second action potential is suppressed – this effect is called refractory behavior. We will not go into the details of neurobiology but touch just one important point: Biological neurons are *either* excitatory *or* inhibitory. Excitatory neurons raise the membrane potential of their postsynaptic targets (excitatory PSP, or EPSP) while inhibitory neurons show the opposite effect (inhibitory PSP, or IPSP).

We do not intend to build a detailed model of the electrophysiology involved in the transmission and triggering of action potentials on the level of single ion channels as the Hodgkin-Huxley-type neuron models do. Instead, we try to summarize the effect which is exerted on a postsynaptic neuron by the firing of a presynaptic neuron. We make three basic assumptions. First, all EPSPs look alike and several simultaneously arriving EPSPs sum up linearly. The synaptic contribution to the membrane voltage h_i of neuron i is

$$h_i^{\mathrm{syn}}(t) = \sum_j J_{ij} \int_0^\infty \mathrm{d}t'\, \epsilon(t - t')\, S_j(t' - \Delta_{ij}^{\mathrm{ax}}) \ . \tag{1}$$

The response kernel ϵ describes the form of the postsynaptic potential. For ϵ we could use any function that is zero for negative arguments, has a single maximum and decreases to zero reasonably fast as $t \to \infty$, but usually we take $\epsilon(t) = t \exp(-t/\tau)\, \Theta(t)/\tau^2$; cf. Fig. 1. Here, τ is the time constant of the PSP and Θ is the Heaviside function that yields zero for negative and unity for positive arguments. Furthermore, J_{ij} is the strength of the synapse connecting the presynaptic neuron j with the postsynaptic neuron i. The spike train of neuron j, $S_j(t) = \sum_f \delta(t - t_j^f)$, is a sum of δ pulses and every δ pulse corresponds to a single action potential emitted by neuron j. The upper index f labels the firing times t_j^f of neuron j. Finally, $\Delta_{ij}^{\mathrm{ax}}$ is the axonal time delay from neuron j to neuron i.

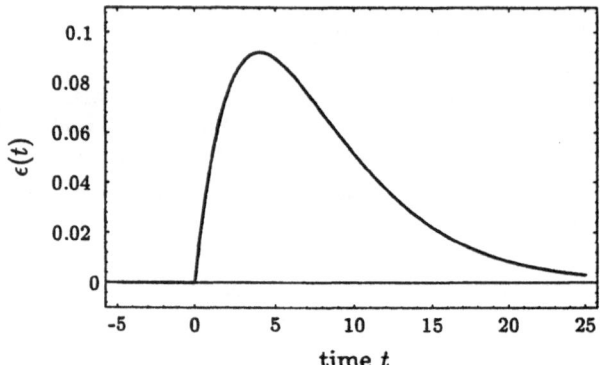

Fig. 1. The response kernel $\epsilon(t)$ describes the form of an elementary postsynaptic potential. Here we use a so-called α function $\epsilon(t) = (t/\tau^2)\exp(-t/\tau)\Theta(t)$ with time constant $\tau = 4$ [ms]. The response function vanishes identically for negative arguments, reaches its maximum at $t = \tau$, and then decays to zero monotonically.

In order to get some feeling for what this setup does, let us assume that there is only one neuron, say neuron j, that is firing a single action potential, say at time $t = 0$. We can easily rewrite Eq. (1) and obtain $h_i(t) = J_{ij}\,\epsilon(t - \Delta_{ij}^{\mathrm{ax}})$. The firing of neuron j at time $t = 0$ elicits a postsynaptic potential at neuron i. The PSP starts raising after the axonal delay $\Delta_{ij}^{\mathrm{ax}}$ and reaches its maximum at time $t = \Delta_{ij}^{\mathrm{ax}} + \tau$, τ being the time constant of the PSP. If J_{ij} is positive, the influence of neuron j on neuron i is excitatory, otherwise it is inhibitory.

The second assumption (or simplification, if you wish) concerning neurobiology is that we replace the spike triggering process by a simple threshold criterion. Our model neurons fire whenever the membrane voltage crosses a certain threshold ϑ from below. A firing event of neuron i at time t_i^f is defined by

$$t_i^f : \quad h_i(t_i^f) = \vartheta, \text{ and } \dot{h}_i(t_i^f) > 0 . \tag{2}$$

Here $\dot{h}_i$ denotes the derivative of h_i with respect to time.

We also have to take refractoriness into account. This is achieved by adding a negative afterpotential η to the membrane voltage whenever the neuron has fired. The total membrane voltage is the sum of the synaptic and the refractory contribution,

$$h_i(t) = h_i^{\mathrm{syn}} + h_i^{\mathrm{ref}}$$
$$= \sum_j J_{ij} \int \mathrm{d}t'\, \epsilon(t - t')\, S_j(t' - \Delta_{ij}^{\mathrm{ax}}) + \int \mathrm{d}t'\, \eta(t - t')\, S_i(t') \ . \qquad (3)$$

For η we can choose a function that is zero for negative arguments and yields $-\infty$ for arguments between 0 and τ^{abs}, the duration of the absolute refractory period when no spike can be triggered at all. After τ^{abs}, η approaches zero from below. In a way we can think of η as the response of the membrane voltage to a postsynaptic spike, whereas ϵ describes the response to a presynaptic spike.

The neuron model we have defined so far has several neat features. First of all, it is simple. Large networks of heavily interconnected neurons can be simulated easily without the need of solving high-dimensional differential equations. Furthermore, it is amenable to an analytic treatment. The following sections of this article present examples of such an analysis.

In this section, we have introduced the spike response model (Gerstner and van Hemmen 1992) in a rather axiomatic way. There is, however, a systematic procedure (Kistler et al. 1997a) to faithfully reduce a detailed multi-compartment and multi-ion channel model to a single-variable threshold model. In this context, the spike response model with appropriately chosen response functions ϵ and η is the first-order approximation of neuronal dynamics. This approximation can be a surprisingly good one as has been shown in the case of the Hodgkin-Huxley equations with a fluctuating input current where a first-order approximation correctly reproduces 90 percent of the action potentials with a precision of 2 ms.

3 Network Architecture

After having specified the properties of a single neuron, we now have to describe the network which interconnects these neurons. Since we are interested in the spike dynamics of biological tissue, the model network should reflect at least the most salient features of its biological paragon.

One of the best investigated areas is the primary visual cortex V1, but even in V1 surprisingly little detail is known about the connectivity of the neurons within this area. On the other hand, there is a lot of data about the response properties of V1 neurons to visual stimulation. First of all, the optic nerve connects neighboring receptor cells in the retina with neighboring neurons in V1. That is to say, the retino-cortical map is *retinotop* in that it preserves neighborhood. The region in the visual field from which a given neuron receives its input is called its *receptive field*. However, the activity pattern in V1 does not simply reflect the distribution of light and dark in the visual field, but single V1 neurons respond in a rather specific way to complex features of the visual stimulus at the position of their receptive fields. There are, for example, neurons which are triggered preferentially by stripes with a certain orientation (orientation selectivity), others discriminate moving stimuli with respect to speed and direction (direction selectivity). In a way similar to the smoothness of the retino-cortical map, the response of the neurons to features like orientation and motion is a smooth function of space, i.e., neighboring neurons have similar response characteristics. The observed response characteristics of V1 neurons arise from the specific way in which the neurons are connected to each other and to the retina, but it is still controversial whether the main effect is due to the intrinsic or to the afferent wiring of the visual cortex.

We intend to study the intrinsic dynamics of the cortex in the absence of an external stimulus ('closed eyes'). We may therefore neglect the afferent connections and concentrate on the internal wiring of the cortex. In want of better knowledge we postulate an isotropic and homogeneous connectivity. This is, of course, a very crude approximation because it is known that the connectivity of two neurons depends, for example, on the relation of their orientation selectivity. Isotropy and homogeneity are, however, essential for the solvability of our model.

The synaptic couplings are assumed to be distributed randomly such that the probability of two neurons being connected is a function of their distance. Since the connectivity between cortical neurons is very high – in fact, one neuron receives synaptic input from about 10^4 other neurons – we may define an average coupling strength $J(r_{ij}) = \langle J_{ij} \rangle$ of neurons i located in a small neighborhood of r_i and neurons j located near r_j with $|r_i - r_j| = r_{ij}$. In the following, the average coupling strength of two neurons will be assumed as a Gaussian with a certain standard deviation which describes the average range of the interaction.

We are interested in the activity of pyramidal cells because these are the neurons that carry information to 'higher' areas. While pyramidal cells are purely excitatory they can exert some inhibition to their surroundings via inhibitory interneurons. Inhibition among pyramidal cells is thus an indirect effect and the range of the inhibition is the result of a convolution of the function which describes the arborization of the pyramidal cell axon and the corresponding function for the interneuron. For Gaussian arborization functions, the net inhibitory effect also has a Gaussian distribution but with a wider range than the excitatory one so that inhibition may outrun excitation at larger distances.

For the sake of simplicity we include only pyramidal cells in our model. The influence of the inhibitory interneurons is taken care of by an effective interaction for the pyramidal cells which is *excitatory* for short distances and *inhibitory* for larger distances, i.e., we take a 'canonical' Mexican hat function for the average coupling strength $J(r)$,

$$J(r) = \sum_{i \in \{1,2\}} a_i \exp\left(-\frac{r^2}{\lambda_i^2}\right),\qquad(4)$$

with $a_1 = 0.12$, $a_2 = -0.02$, $\lambda_1^2 = 15$, and $\lambda_2^2 = 100$.

We are going to investigate a large set of spiking neurons which are arranged on a two-dimensional square lattice. The synaptic coupling from neuron j to neuron i is a function of the distance r_{ij} between these neurons, i.e., $J_{ij} = J(r_{ij})$. Analogously, we take the delay Δ_{ij} to depend on the distance r_{ij} and write $\Delta_{ij} = \Delta(r_{ij})$. However, the lateral dimensions of the visual cortex are rather small so that the delay due to the propagation of the spikes along the axons (≈ 0.1 ms) is negligible as compared to the synaptic delay (≈ 1 ms) which does not depend on the distance of the neurons at all. So we simply absorb the constant synaptic delay in the response function ϵ and neglect the axonal delay.

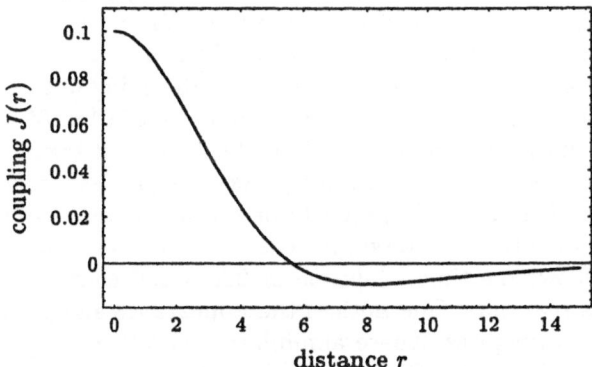

Fig. 2. The average coupling strength $J(r)$ summarizes the effect of the firing of a presynaptic neuron on postsynaptic neurons separated by a distance r. It has a so-called Mexican-hat shape; cf. Eq. (4). Positive values of $J(r)$ correspond to excitation, negative ones to inhibition. Dale's principle (see, e.g., Shepherd 1988) stating that a neuron cannot be both excitatory and inhibitory does not apply since J incorporates inhibitory interneurons as well.

4   Simulations

We have simulated this system in the absence of an external input but with some noise added to the membrane potential of each neuron. Depending on the noise amplitude, the spiking threshold ϑ, and the mix of excitation and inhibition (fraction $\alpha = a_1/a_2$, 'drug level') we observe the formation of geometrical patterns of single-spike activity such as traveling waves, rotating spirals, and expanding circles; cf. Figs. 3 to 7.

Starting with a 'dressed vacuum' consisting of incoherently, accidentally firing neurons and increasing α, Fohlmeister et al. (1995) have found a sequence of four scenarios of coherent excitation: I. Short single stripes which are moving (not shown here). II. Rotating x-armed spirals with $1 \leq x \leq 4$, shown in the two top and middle left pictures of Fig. 3. For the same parameter values one can also find moving parallel stripes or simple *plane waves* (middle right in Fig. 3) whose analytic treatment in Sects. 6 and 7 is a key issue of the present paper. III. Expanding concentric rings which form a target pattern or 'tunnel'; see the lower left of Fig. 3. IV. Complex pulsating patterns are shown in the lower right-hand corner of Fig. 3.

The inhibitory interaction between the neurons deserves special attention. Fohlmeister et al. (1995) had taken a *local* shunting inhibition into account through a self-inhibitory delay loop. The very same construction has been used for Figs. 3–5 and 7. In Fig. 6, however, the loops have been replaced by inhibitory interneurons, which do reproduce the very same scenario III as in Fig. 3. In fact, neither scenario I nor scenarios III and IV are modified but scenario II is to some extent. Since this specific issue is of no direct relevance to the present paper we refer the reader to forthcoming work.

Figures 4 and 5 exhibit the temporal evolution of $10^3 \times 10^3 = 10^6$ neurons in scenario II during 3 seconds *real* time. The snapshot taken at $t = 100\,\mathrm{ms}$ still resembles scenario I rather strongly. As time proceeds only three-armed spirals survive. Identifying large parts of the 3 s picture (lower right-hand corner of Fig. 5) with plane waves is nearly a matter of taste.

Figure 6 shows the time evolution of 250×250 neurons in scenario III during 165 ms real time. One notices the appearance and growth of several expanding-ring structures. These annihilate each other as they hit because of the refractory behavior after neurons have fired.

Figure 7 shows a novel structure, a 'paternoster'. Its name stems from an old-fashioned elevator system consisting of two parallel vertical rows of open cabins that move upwards and downwards and form an endless string. The stripes on the right above the main diagonal are ascending, those below the main diagonal are descending – hence the name. The paternoster of Fig. 7 appeared in scenario II after 0.5 s and remained stable for about 2 s real time. Note that it is *not* the boundary of two spirals with equal chirality, as a glance at the upper left- and lower right-hand corner readily reveals.

5 Continuum Approximation

The present model can be solved analytically if we replace the discrete lattice by a continuous sheet of neurons. This can be justified, if the neurons are densely packed, i.e., if the distance between two neurons is small as compared

Fig. 3. Increasing excitation as compared to inhibition in spatially homogeneous cortical tissue, one expects four scenarios (Fohlmeister et al. 1995) of which three are shown here: spirals with one to three arms (upper row and middle left) and plane waves (middle right) belonging to scenario II. Expanding rings (lower left) constitute scenario III whereas for very high levels of excitation one ends up with complex pulsating patterns (lower right) that run for scenario IV.

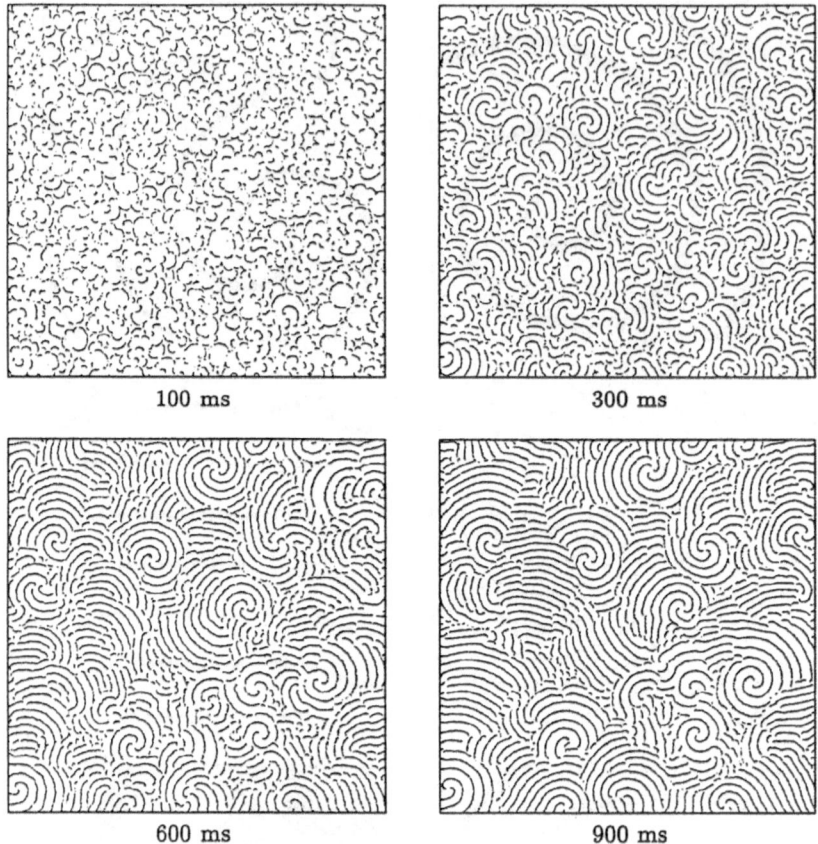

Fig. 4. Figures 4 and 5 belong together in that they show a series of snapshots of scenario II for real time increasing from 0.1 to 3 seconds. The system size $10^3 \times 10^3$ is so large (altogether a million neurons) that several spirals appear out of the many short, moving stripes at $t = 0.1$ s. The initial state ($t = 0$) is a 'dressed vacuum' with incoherently firing neurons. As time proceeds, only some of the 3-armed spirals survive. Closer inspection reveals that 2-armed spirals show up only transiently (lower right-hand corner of the 600, 900 and 1000 ms snapshot) and are overwhelmed by the 3-armed bandits.

to the characteristic length scale of the coupling function $J(r)$. Furthermore, we neglect all boundary effects in that we assume the sheet to extend to infinity in all directions. Under these premises, we may replace the sum over all neurons in Eq. (1) by an integral over space. We choose the length scale to be such that the neuron density equals unity. In the continuum limit, Eq. (1) reads

$$h^{\text{syn}}(r, t) = \int d^2 r' \, J(|r - r'|) \int_0^\infty dt' \, \epsilon(t') \, S\Big[r', t - t' - \Delta(|r - r'|)\Big]. \quad (5)$$

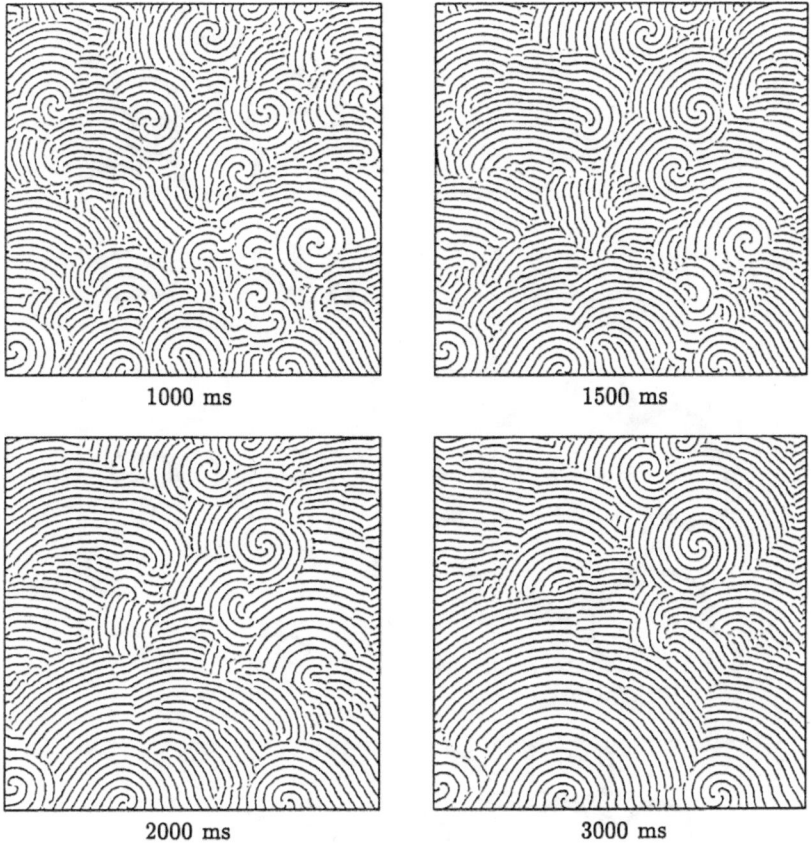

1000 ms   1500 ms

2000 ms   3000 ms

Fig. 5. Continuation of Fig. 4.

This is the synaptic contribution to the local field of a neuron at position r and time t given the spike activity $S(r, t')$ of the network for all times $t' < t$.

A spike is triggered if the local field crosses the threshold ϑ from below. We therefore put a δ function everywhere in the net where the local field h equals ϑ and has a positive time derivative,

$$S(r, t) = \delta[h(r, t) - \vartheta] \ |\nabla_r h(r, t)| \ \Theta[\dot{h}(r, t)] . \tag{6}$$

We have included the norm of the gradient of the local field in order to obtain a *normalized* δ function when integrating S over space. That is to say, integrating along a normal to the curve we get one and integrating over a two-dimensional domain D we obtain the length of the curve contained in D.

For the moment we neglect the time delay $\Delta(|r - r'|)$ and the refractory contribution h^{ref} to the local field. The former approximation is justified to some extent by the speed of spike propagation in the cortex which is fast as compared to the rise time of the postsynaptic potentials; cf. Milton (1996).

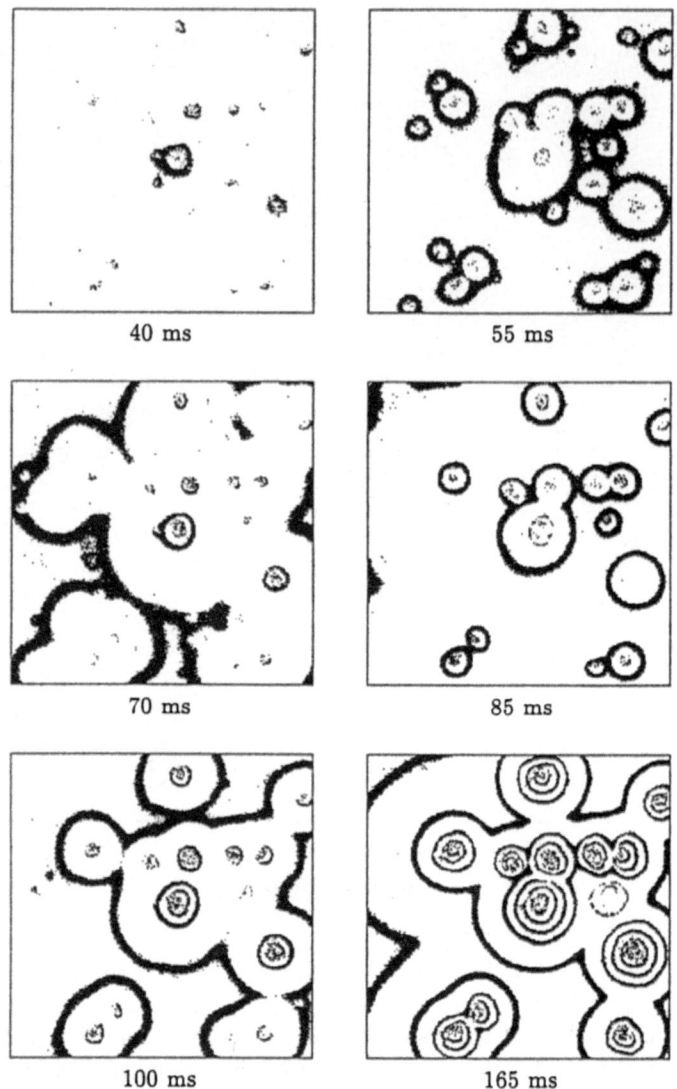

Fig. 6. A series of snapshots (40 ms $\leq t \leq$ 165 ms) leading to scenario III in a system with inhibitory interneurons. The initial state at time $t = 0$ was a dressed vacuum (cf. Fig. 4) and the system size 250 × 250 was so large that several sets of expanding rings could appear. As two rings merge, they annihilate each other's activity because of refractory neuronal behavior that directly follows firing.

 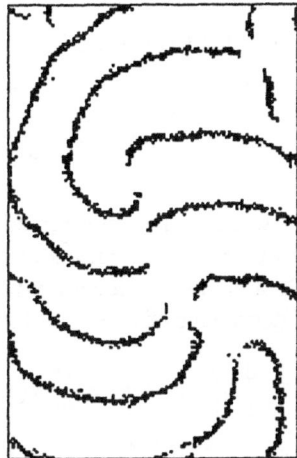

Fig. 7. Paternoster. The highlighted area on the left is shown enlarged on the right. With the main diagonal as a line of reference, the stripes on the right steadily rise, round the corner, and then descend until they round the corner once again and restart their trip. This snapshot is taken from a simulation of a system with 500×500 neurons. The paternoster remained stable for about 2 s real time.

The latter approximation is fair as long as we consider low firing rates. In situations where every neuron fires only once refractoriness does not show up at all and we may omit it from the beginning. The dynamics of the network is then determined by the integral equation

$$h(\boldsymbol{r}, t) = \int d^2 r' \, J\big(|\boldsymbol{r} - \boldsymbol{r}'|\big) \int\limits_{0}^{\infty} dt' \, \epsilon(t') \, S(\boldsymbol{r}', t - t'), \tag{7}$$

with

$$S(\boldsymbol{r}, t) = \delta\big[h(\boldsymbol{r}, t) - \vartheta\big] \, |\nabla_{\boldsymbol{r}} h(\boldsymbol{r}, t)| \, \Theta\big[\dot{h}(\boldsymbol{r}, t)\big]. \tag{8}$$

The first equation is linear in S whereas the nonlinearity required for pattern formation is introduced through the second equation.

Field equations describing neuronal activity have been introduced by Wilson and Cowan (1973), Amari (1977), and Ermentrout and Cowan (1979) . In these works, however, a mean-field theory is developed for *firing rates* rather than for single-spike activity as we do here.

Unfortunately, Eq. (7) gives a rather implicit description of the *temporal evolution* of spike activity. In order to derive equations for the evolution of a pattern, we start by assuming that an activity pattern can be described by single time-dependent curve

$$\Gamma(t) : \mathbb{R} \supset I \to \mathbb{R}^2, \quad s \mapsto \gamma(s, t). \tag{9}$$

Here t denotes time, $I \subset \mathbb{R}$ is an interval, and $s \in I$ parameterizes the curve. The local field can then be written as a line integral over Γ,

$$h(r,t) = \int\limits_{0}^{\infty} dt' \; \epsilon(t') \int\limits_{\Gamma(t-t')} dr' \; J\big(|r - r'|\big) \,. \tag{10}$$

This can be easily generalized to a set of curves. We simply have to add the contributions of all curves in order to obtain the local field.

On a curve, the local field is constant because it equals ϑ. Hence the total time derivative of this quantity vanishes,

$$0 \equiv \frac{d}{dt} h[\gamma(s,t),t] = \frac{\partial}{\partial t}\gamma(s,t) \cdot \nabla_r h(r,t)|_{r=\gamma(s,t)} + \frac{\partial}{\partial t} h[\gamma(s,t),t] \,. \tag{11}$$

The vector $\nabla_r h[\gamma(s,t),t]$ is perpendicular to the curve because $\Gamma(t)$ is the contour line of $h(r,t)$. The dot product in the righthand-side of Eq. (11) is therefore proportional to the normal velocity $v_\perp$ of the activity curve, i.e.,

$$v_\perp(s,t) = -\frac{\frac{\partial}{\partial t} h(r,t) \, \nabla_r h(r,t)}{|\nabla_r h(r,t)|^2}\bigg|_{r=\gamma(s,t)} \tag{12}$$

Later on we will rely on this formula to numerically calculate the solution for expanding circular rings.

6 Plane Fronts and Plane Waves

We start by looking for the simplest collective phenomenon, that is, we study a traveling excitation front. To this end, we make an ansatz for the spike activity in the form of a single plane front with propagation velocity v,

$$S(x,y,t) = \delta(x - v\,t)\,. \tag{13}$$

With this ansatz we can calculate the corresponding local field from Eq. (7) analytically. Since the spike activity is a function of $x - v\,t$ only, so is the local field and it suffices to state the time dependence of the local field at the origin. We obtain

$$\begin{aligned}
h^{\text{front}}(r = 0, t) = \frac{\pi}{2\,\tau\,v} \sum_{i \in \{1,2\}} a_i \lambda_i^2 \Bigg\{ & \frac{\lambda_i}{\tau\,v\,\sqrt{\pi}} \exp\left[-\left(\frac{t\,v}{\lambda_i}\right)^2\right] \\
& + \left(\frac{t}{\tau} - \frac{\lambda_i^2}{2\,\tau^2\,v^2}\right) \exp\left[-\frac{t}{\tau} + \left(\frac{\lambda_i}{2\,\tau\,v}\right)^2\right] \\
& \times \operatorname{erfc}\left(\frac{\lambda_i}{2\,\tau\,v} - \frac{t\,v}{\lambda_i}\right) \Bigg\}\,.
\end{aligned} \tag{14}$$

Here erfc is the complementary error function.

We still have to determine the allowed values for the propagation speed v that shows up as a free parameter in the ansatz (13). Since the neurons will fire only if the local field reaches the threshold ϑ, we have a self-consistency condition which states that the local field *at the front* must equal ϑ,

$$h(x = vt, y, t) = \vartheta \quad \text{and} \quad \left.\frac{\partial h}{\partial t}\right|_{x=vt} > 0. \tag{15}$$

This establishes a relation between the threshold ϑ and the propagation speed v,

$$\vartheta \overset{!}{=} h^{\text{front}}(0,0)$$

$$= \sum_{i \in \{1,2\}} \frac{a_i \lambda_i^3}{2\,\tau^2\,v^2} \left\{ \sqrt{\pi} - \frac{\lambda_i\,\pi}{2\,\tau\,v} \exp\left[\left(\frac{\lambda_i}{2\,\tau\,v}\right)^2\right] \text{erfc}\left(\frac{\lambda_i}{2\,\tau\,v}\right) \right\} \tag{16}$$

As can be seen from Fig. 8 there is a limited range of possible velocities and a reciprocal relation between threshold and velocity: a low threshold value results in a high speed of propagation and vice versa.

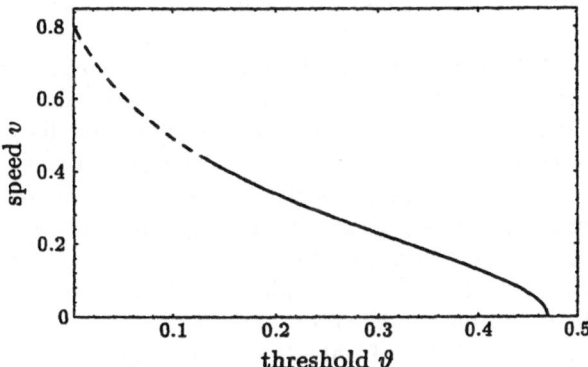

Fig. 8. Speed of propagation v plotted as a function of the threshold value ϑ for plane fronts of excitation. As will be shown in Sec. 7, only fronts with a speed below a certain critical value are stable (solid line), those with a higher speed of propagation are unstable (dashed line).

The above analysis can be extended easily so as to investigate plane waves of excitation. We consider an arrangement of equidistant plane fronts that travel in the positive x-direction,

$$S(r, t) = \sum_{n=-\infty}^{\infty} \delta(x - vt - n\lambda). \tag{17}$$

We have two parameters in this ansatz, viz., the wavelength λ and the phase velocity v. Both parameters have to be determined in a similar fashion as before. First, we have to calculate the corresponding local field. Since the expression for the local field in Eq. (7) is linear in S this is just a superposition of the local fields induced by the single wave fronts,

$$
\begin{aligned}
h^{\text{wave}}(0, t) &= \int d^2 r'\, J(|r'|) \int_0^\infty dt'\, \epsilon(t')\, S(r', t - t') = \\
&= \sum_{n=-\infty}^{\infty} \int d^2 r'\, J(|r'|)\, \epsilon\left(t + n\frac{\lambda}{v} - \frac{x'}{v}\right) = \\
&= \sum_{n=-\infty}^{\infty} h^{\text{front}}\left(0, t + n\frac{\lambda}{v}\right).
\end{aligned}
\tag{18}
$$

As before we have a self-consistency condition which states that the local field at the pattern must equal ϑ,

$$
h^{\text{wave}}(v\,t + n\,\lambda, y, t) = \vartheta \quad \text{and} \quad \left.\frac{\partial h^{\text{wave}}}{\partial t}\right|_{x = v\,t + n\,\lambda} > 0, \quad \forall n \in \mathbb{Z}. \tag{19}
$$

Through this condition we obtain a relation between the threshold ϑ, the wavelength λ, and the phase velocity v. For fixed ϑ, this relation can be interpreted as a *dispersion relation* for the frequency $\omega = 2\pi v / \lambda$ and the wave number $k = 2\pi / \lambda$. This dispersion relation relates a 'temporal' frequency ω and a 'spatial' frequency k and is plotted in Fig. 9 for several threshold values.

We can discern three different regimes in the dispersion diagram of Fig. 9. For low spatial frequencies, i.e., for large wavelengths, the wave fronts do not 'feel' each other. The phase velocity is *independent* of the wavelength and the temporal frequency ω increases linearly with the spatial frequency k because the phase velocity is just the ratio of ω and k. We have the very same situation as in the case of a single plane front. In particular, the relation between threshold and phase velocity is the same as that between threshold and propagation speed of plane fronts. If we reduce the wavelength, then the dispersion relation bends towards smaller temporal frequencies and the phase velocity drops below the propagation speed of the corresponding single plane front. Finally, the slope of the curve and therewith the group velocity become even negative. This should show up in a reversed traveling direction of small density fluctuations in the wave.

In calculating the dispersion relation, we have neglected the influence of a refractory afterpotential but, with the benefit of hindsight, this can be corrected. The dispersion diagram of Fig. 9 is basically a contour plot of the local field on one of the wave fronts as a function of ω and k. Clearly, the afterpotential is a function of the temporal frequency ω only. Now it is easy to imagine what will happen, if the afterpotential is added to the local field. If we account for total refractoriness only, i.e., use an afterpotential that yields $-\infty$ during a short time interval immediately after an action

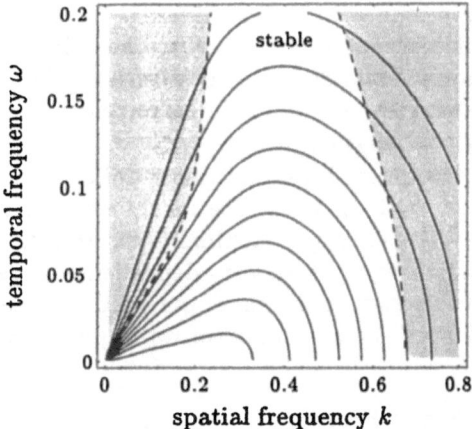

Fig. 9. Dispersion relation for plane waves of excitation propagating in a locally connected neuronal network. The plot shows the angular frequency $\omega = 2\pi v / \lambda$ as a function of the wave number $k = 2\pi / \lambda$ for several values of the threshold parameter ϑ. The threshold increases from top to bottom from $\vartheta = 0$ to $\vartheta = 0.45$ in steps of 0.05. In addition, this figure summarizes the result of the stability analysis (cf. Sect. 7): The shaded regions correspond to *unstable*, the region in between to *stable* waves.

potential and zero elsewhere, the dispersion relation will be clipped at the frequency corresponding to the length of the refractory period. In addition to the total refractory period, we can include relative refractoriness too. This would be achieved by an afterpotential that smoothly increases to zero after the total refractory period, and results in an inhomogeneous stretching of the dispersion relation in the ω direction.

7 Stability Analysis

Up to now we have verified the existence of some simple solutions to the integral equation (7). The existence of such a solution, however, does not imply that this solution can be observed in a physical system. A necessary condition for a solution to be observable is its stability towards small perturbations. In this section we will present a stability analysis for plane fronts and waves demonstrating that indeed only some, but not all, of the analytic solutions are stable.

The canonical method of investigating stability is to consider a small perturbation, linearize the equations of motion, and to see whether the amplitude of the perturbation increases or decreases. In linear perturbation theory it suffices to study sinusoidal perturbations of, say, a plane front because an arbitrarily shaped perturbation can be constructed by linear superposition of several sinusoidal perturbations with different wave numbers. Due to the linearization, the different modes of the perturbation do not interact and each mode grows or decays independently of the others. A solution is said to be stable, if it is stable with respect to *all* sinusoidal perturbations whatever the wave number of the perturbation.

In our case, however, there is an additional difficulty that stems from the history dependence of the local field. The local field at time $t = 0$ cannot be calculated from the spike activity S at time $t = 0$. In fact, the whole history of S for all times $t < 0$ is required. History dependence holds for the evolution of any perturbation to a given solution of our model. In general, the growth rate of a perturbation will be a function of its amplitude *in the past*.

7.1 Stability of Plane Fronts

We start with a perturbed front, e.g.,

$$\Gamma(t) : \mathbb{R} \to \mathbb{R}^2, \quad s \mapsto \gamma(s,t) = \begin{pmatrix} v\,t + a_\kappa(t)\,\sin(\kappa\,s) \\ s \end{pmatrix}, \qquad (20)$$

Here $a_\kappa(t)$ is the amplitude of a sinusoidal perturbation with wave number κ; cf. Fig. 10.

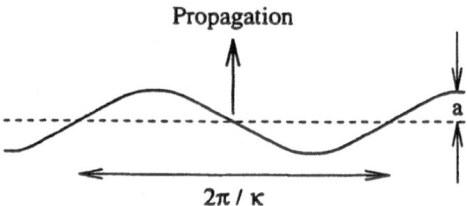

Fig. 10. Perturbation of a plane front. The dashed line presents the unperturbed plane front. The sinusoidal perturbation (solid line) has amplitude a, period $2\pi/\kappa$, and is orthogonal to the direction of propagation, i.e., transversal.

It is tempting to use the explicit expression (12) for the evolution of the perturbed front, but it turns out that this equation does not allow for a conclusive stability analysis. Nevertheless, it is interesting to see why it does not work.

Substituting the ansatz (20) into Eq. (11), we obtain

$$[v + \dot{a}_\kappa(t)\,\sin(\kappa\,s)]\,\frac{\partial}{\partial x}h(r,t)\Big|_{r=\gamma(s,t)} = -\frac{\partial}{\partial t}h[\gamma(s,t),t]. \qquad (21)$$

It is good to recall that the local field $h(r, t)$ depends on the history of the front. The above equation is therefore a non-anticipatory integro-differential equation for the amplitude $a_\kappa(t)$ of the perturbation,

$$-\dot{a}_\kappa(t) \sin(\kappa s) = v + \frac{\frac{\partial}{\partial t} h(\gamma, t)}{\frac{\partial}{\partial x} h(\gamma, t)} . \tag{22}$$

In contrast to the case of an ordinary differential equation, the right-hand side of the above equation is not just a function of $a_\kappa(t)$, but a non-linear operator operating on $a_\kappa(.)$ and being evaluated at time t. In both cases, however, we have to linearize the right-hand side around $a_\kappa = 0$. We replace the non-linear operator by the sum of its value for $a_\kappa = 0$ and its linear derivative at $a_\kappa = 0$ so that

$$-\dot{a}_\kappa(t) \sin(\kappa s) = v + \underbrace{\frac{\frac{\partial}{\partial t} h^{\text{front}}(v\,t, t)}{\frac{\partial}{\partial x} h^{\text{front}}(v\,t, t)}}_{=0} + (\mathcal{L}a_\kappa)(t) + O\left(a_\kappa^2\right) . \tag{23}$$

Here $\mathcal{L}$ is the linear operator that approximates the right-hand side of Eq. (22) for $a_\kappa \approx 0$. The operator $\mathcal{L}$ is defined by the functional derivative with respect to a_κ, evaluated at $a_\kappa \equiv 0$,

$$\mathcal{L} := \frac{\delta}{\delta a_\kappa} \left[\frac{\frac{\partial}{\partial t} h(\gamma, t)}{\frac{\partial}{\partial x} h(\gamma, t)} \right] , \tag{24}$$

where $\delta/\delta a_\kappa$ denotes the functional derivative with respect to a_κ. If we collect the first-order terms in a_κ (the zeroth-order terms drop out anyway) we find

$$\dot{a}_\kappa(t) + \int_0^\infty dt'\, \zeta_\kappa(t')\, a_\kappa(t - t') = 0, \quad t \in \mathbb{R}, \tag{25}$$

where ζ_κ is the integral kernel corresponding to the linear operator $\sin^{-1}(\kappa s)\, \mathcal{L}$.

There is no appropriate way to specify a boundary or an initial condition for (25) except regularity at $t = \pm\infty$. The unique solution is $a_\kappa(t) = 0$ for all $t \in \mathbb{R}$. This is not too surprising since, if we start with $a_\kappa(-\infty) = 0$ at $t = -\infty$, there is no reason for the perturbation to appear for $t > -\infty$. We therefore have to explicitly include an external 'force' $f(t)$ which triggers the perturbation. This leads to a Volterra integro-differential equation defined on the whole line,

$$\dot{a}_\kappa(t) + \int_0^\infty dt'\, \zeta_\kappa(t')\, a_\kappa(t - t') = f(t), \quad t \in \mathbb{R}. \tag{26}$$

We assume that the support of the external force $f(t)$ is bounded from above, say, to the left of $T < \infty$, meaning that f is to vanish for $t > T$. Stability can be judged from the asymptotic behavior of $a_\kappa(t)$ for $t > T$. If a_κ decays to zero as $t \to \infty$ the plane front is stable with respect to the mode κ.

In order to determine the integral kernel $\zeta_\kappa(s)$ we put $a_\kappa(t) = \alpha \delta(t)$ and differentiate with respect to α,

$$
\begin{aligned}
\zeta_\kappa(s) &= \frac{1}{\sin \phi} \frac{\partial}{\partial \alpha} \left[\frac{\frac{\partial}{\partial t} h(\gamma, t)}{\frac{\partial}{\partial x} h(\gamma, t)} \right]_{\alpha=0} (s) \\
&= \left[\frac{\frac{\partial}{\partial \alpha} \frac{\partial h}{\partial t} + v \frac{\partial}{\partial \alpha} \frac{\partial h}{\partial x}}{\sin \phi \frac{\partial h}{\partial x}} \right]_{\alpha=0} (s), \quad \text{with } a_\kappa(t) = \alpha \delta(t).
\end{aligned}
\tag{27}
$$

After some algebra we find

$$
\zeta_\kappa(s) = \frac{2\sqrt{\pi}\, v}{\frac{\partial}{\partial x} h(\gamma, t)} \sum_i \frac{a_i}{\lambda_i} \left\{ s\, \dot\epsilon(s) + \epsilon(s) \left[1 - 2 \left(\frac{v s}{\lambda_1} \right)^2 \right] \right\}
$$

$$
\times \exp\left[- \left(\frac{v s}{\lambda_i} \right)^2 - \left(\frac{\kappa \lambda_i}{2} \right)^2 \right]
\tag{28}
$$

with

$$
\frac{\partial}{\partial x} h(\gamma, t) = 2\sqrt{\pi} \sum_i a_i \left(\frac{\lambda_i}{2\tau v} \right)^4 \left\{ \frac{4\tau v}{\lambda_i} - \sqrt{\pi} \left[2 + \left(\frac{2\tau v}{\lambda_i} \right)^2 \right] \right.
$$

$$
\left. \times \exp\left[\left(\frac{\lambda_i}{2\tau v} \right)^2 \right] \operatorname{erfc}\left(\frac{\lambda_i}{2\tau v} \right) \right\}.
\tag{29}
$$

The solution of the linearized integro-differential equation (26) and, therewith, the stability properties of the front are determined by the form of the integral kernel $\zeta_\kappa(s)$. If $\zeta_\kappa(s) < 0$ for all $s > 0$, then it is easy to see that the solution of (26) blows up, if the external force is, for example, positive during a short interval and zero elsewhere. On the other hand, in the case of a positive kernel $[\zeta_\kappa(s) > 0$ for all $s > 0]$, it is known that there is a unique solution of Eq. (26) if and only if $[i\omega + \hat\zeta_\kappa(\omega)]$ has no real roots. Here $\hat\zeta_\kappa$ is the Fourier transform of the kernel ζ_κ; cf. Gripenberg et al. (1990, Theorem 3.7). The solution is given in terms of a differential resolvent $r \in L^1(\mathbb{R}; \mathbb{C})$,

$$
\hat{r}(\omega) = \left[i\omega + \hat\zeta_\kappa(\omega) \right]^{-1},
\tag{30}
$$

so that

$$
a_\kappa(t) = (r * f)(t), \quad t \in \mathbb{R}.
\tag{31}
$$

The solution a_κ has some nice properties, e.g., if $f \in L^p(\mathbb{R}; \mathbb{C})$ with $p \in [1, \infty]$, then $a_\kappa \in L^p(\mathbb{R}; \mathbb{C})$; cf. Gripenberg et al. (1990, Theorem 3.10). In this case, the amplitude of the perturbation approaches zero asymptotically as $t \to \infty$, whatever the external force f.

In our case, however, the situation is not as simple as in the two cases described above, because $\zeta_\kappa(s)$ in Eq. (28) is neither strictly positive nor strictly negative; see Fig. 11. Furthermore, the integral over the kernel ζ_κ vanishes identically for every κ and v. If the external force has the form of a short square pulse, the amplitude a_κ of the perturbation will start growing with positive slope as soon as the external force is positive. The convolution of a function with positive slope with a kernel as it is depicted in Fig. 11 yields a *positive* result. From Eq. (26) we conclude that the slope of $a_\kappa(t)$ remains positive and the amplitude will grow forever. Note that if the sign of ζ would be the other way round, things would be quite different: Convolution of a monotonous function with a kernel that is positive for small arguments and negative for larger ones so that its integral vanishes identically, yields a result with the inverted sign as compared to the slope of the function and leads to oscillatory solutions of Eq. (26).

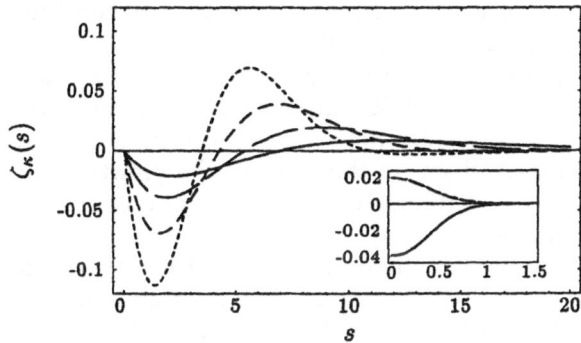

Fig. 11. The integral kernel $\zeta_\kappa(s)$ with $\kappa = 0$ plotted for the front velocities $v = 0.2$ (solid line), $v = 0.4$ (long dashed), $v = 0.6$ (dashed line), and $v = 0.8$ (dotted line). An increase of κ basically decreases the amplitude of $\zeta(s)$ without substantially changing its shape. The inset shows the maximum and the minimum of $\zeta(s)$ as a function of κ for $v = 0.4$.

The salient features of the integral kernel ζ_κ, namely vanishing integral and negative initial slope, are independent of the front velocity v and the mode κ of the perturbation. This indicates that *none* of the propagating front solutions of the explicit equation (12) is stable. This is, however, in contrast to our observation in the simulations. The reason for this discrepancy is that we have not made use of the explicit equation (12) in the simulation but rather employed the implicit ones; viz., (7) and (8). In deriving the explicit equation we have exploited the fact that the local field on the front is constant and equals ϑ. In calculating the total time derivative of the local field, the threshold parameter ϑ disappeared from the expression. The resulting equation thus describes the velocity with which the front has to move in order to keep its local field constant rather than to keep it constantly equal to

ϑ. If, now, there is some kind of external force which deflects the formerly unperturbed front, then the local field on the front will be changed too. After the external perturbation is turned off again, the explicit equation will propagate the front in a such way that its local field will remain equal to this *new* value. That is to say, it is marginally stable.

Summarizing, the explicit equation (12) describes the propagation of fronts correctly, but the stability properties of its solutions are not the same as those of the implicit equations (7) and (8).

7.2 Stability of Plane Fronts, Alternative Method

We can do better, if we start from the implicit equation (7) and (8). As explained before, the dynamics of this model is *not* determined by the actual activity pattern but by the *past* spike activity. If we want to investigate the evolution of a perturbation to a known solution, we therefore have to specify the past of the perturbation as well.

We are interested in the hypersurface in parameter space that separates stable and unstable solutions. We aim at a linear stability analysis and expect the growth rate of the perturbation to be continuous across this hypersurface. Hence this surface is given by the zeros of the growth rate. It is therefore natural to use a perturbation with *constant* amplitude as a starting point for the stability analysis.

The idea of the stability analysis is the following. We consider a perturbed front as shown in Fig. 10. We assume that the amplitude $a(t)$ of the perturbation is kept constant for times $t \in (-\infty, 0]$ and calculate the resulting local field at time $t = 0$ to linear order in a. Stability can be judged from the sign of the linear correction to the local field *at the front*. We discern three cases.

- If the correction is zero, the local field on the perturbed curve equals ϑ and the perturbed front with *constant* amplitude a is a solution of Eq. (7), at least to first order of the amplitude of the perturbation. This is the indifferent case that separates stable and unstable solutions.
- If the correction is positive for those neurons that hurry ahead of the unperturbed front, the perturbation has *increased* their local field and the front will be pushed forward, away from the unperturbed solution. This results in an increasing amplitude of the perturbation — the solution is unstable.
- Otherwise, if the perturbation has decreased the local field of those neurons that have fired too early, they will tend to fire later and the amplitude of the perturbation decreases. If this holds true for all wave numbers κ of the perturbation, then the solution is bound to be stable.

In order to calculate the local field we use the parameterization

$$\Gamma(t) : \quad s \mapsto \gamma(s, t) = \begin{pmatrix} v\, t + a\sin(\kappa\, s + \phi) \\ s \end{pmatrix} . \tag{32}$$

If we define spike activity by means of a time-dependent curve, the local field is given by

$$
h(r,t) = \int_0^\infty dt'\, \epsilon(t') \int_{\Gamma(t-t')} d\gamma\, J\left(\|r-\gamma\|\right)
$$
$$
= \int_0^\infty dt'\, \epsilon(t') \int ds\, J\left(\|r - \gamma(s,t-t')\|\right)\, \|\partial_s \gamma(s,t-t')\| \, . \quad (33)
$$

We need an expression for the local field at time $t = 0$ and position $x = a\sin\phi$, $y = 0$, corresponding to $s = 0$. We linearize with respect to a,

$$
h_{a,\kappa}^{\mathrm{front}}(a\sin\phi,0,0) = h^{\mathrm{front}}(0,0,0) + a\,\delta h_\kappa^{\mathrm{front}}(\phi) + O(a^2) \, , \quad (34)
$$

and obtain

$$
\delta h_\kappa^{\mathrm{front}}(\phi) = \int_0^\infty dt'\, \epsilon(t') \sum_{i\in\{1,2\}} a_i \frac{2\,v\,t'}{\lambda_i^2} \exp\left[-\left(\frac{v\,t'}{\lambda_i}\right)^2\right]
$$
$$
\times \int_{-\infty}^\infty ds\, [\sin(\kappa s + \phi) - \sin\phi]\exp\left[-\left(\frac{s}{\lambda_i}\right)^2\right] \, . \quad (35)
$$

The integral over s vanishes identically for $\phi = 0$ and $\phi = \pi$. Hence the local field does not change at the points where perturbed and original solution intersect. On the other hand, for $\phi = \pm\pi/2$ we find

$$
\delta h_\kappa^{\mathrm{front}}\left(\frac{\pm\pi}{2}\right) = \pm \sum_{i\in\{1,2\}} \frac{\sqrt{\pi}\, a_i\,\lambda_i^2}{2\,\tau^2\,v^2}\left\{1 - \exp\left[-\left(\frac{\kappa\,\lambda_i}{2}\right)^2\right]\right\}
$$
$$
\times \left\{\frac{\lambda_i}{\tau\,v} - \left(1 + \frac{\lambda_i^2}{2\,\tau^2\,v^2}\right)\Gamma\left[\frac{1}{2},\left(\frac{\lambda_i}{2\,\tau\,v}\right)^2\right]\exp\left[\left(\frac{\lambda_i}{2\,\tau\,v}\right)^2\right]\right\} \, . \quad (36)
$$

Here $\Gamma(a,z)$ is the incomplete gamma function $\Gamma(a,z) = \int_z^\infty t^{a-1}\,e^{-t}\,dt$. The quantity $\delta h_\kappa^{\mathrm{front}}(\pi/2)$ is the correction to the local field for points with maximal convex curvature, i.e., the foremost point of the perturbed solution. Analogously, $\delta h_\kappa^{\mathrm{front}}(-\pi/2)$ is the correction to the local field for points with maximal concave curvature. As explained before, the solution is unstable if $\delta h_\kappa^{\mathrm{front}}(\pi/2) > 0$ or, equivalently, $\delta h_\kappa^{\mathrm{front}}(-\pi/2) < 0$.

As can be seen from Fig. 12, only fronts that travel with a velocity *below* a certain critical velocity v_c are stable. This is equivalent to saying that we have to choose the threshold parameter ϑ *above* a certain critical value ϑ_c in order to obtain stable plane fronts.

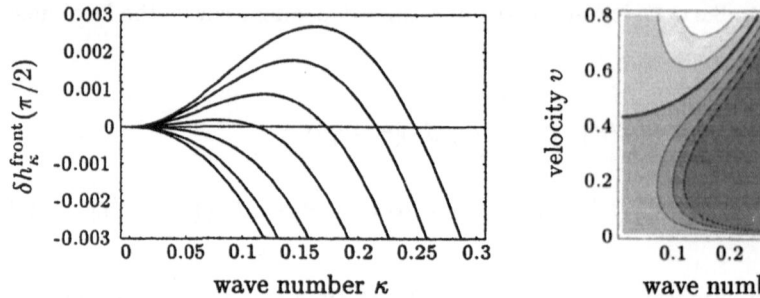

Fig. 12. The linear correction $\delta h_\kappa^{\text{front}}(\pi/2)$ for the local field at those points of the perturbed front that have maximum convex curvature. A plane front is stable if $\delta h_\kappa^{\text{front}}(\pi/2)$ is negative for all wave numbers κ of the perturbation. The left figure shows a plot of $\delta h_\kappa^{\text{front}}(\pi/2)$ as a function of κ for different velocities v, increasing from bottom to top from $v = 0.2$ to $v = 0.8$ in steps of 0.1. The right figure exhibits a contour plot of $\delta h_\kappa^{\text{front}}(\pi/2)$ as a function of κ and v. The boldface contour line that intersects the vertical v-axis at about $v_c = 0.43$ separates regions with negative (darker) and positive (brighter) values of $\delta h_\kappa^{\text{front}}(\pi/2)$.

7.3   Stability of Plane Waves

We now turn to a plane wave with wavelength λ and phase velocity v and perturb it in such a way that only one of its wave fronts is affected. The wave number of the perturbation is denoted by κ and the amplitude by a; cf. Fig. 13. This perturbation is generic in a sense that it includes transversal perturbations ($\kappa \neq 0$) as well as longitudinal perturbations [$\kappa = 0$, $\phi = \pi/2$ in (32)]. The local field at the perturbed front is made up of the contribution of the perturbed front itself and of the contribution of the neighboring (unperturbed) fronts, i.e.,

$$\hat{h}_{a,\kappa}^{\text{wave}}(a\sin\phi, 0, 0) = h_{a,\kappa}^{\text{front}}(a\sin\phi, 0, 0) + \sum_{\substack{n = -\infty \\ n \neq 0}}^{+\infty} h^{\text{front}}\left[0, (n\lambda - a\sin\phi)/v\right]$$

$$= h^{\text{wave}}(0, 0, 0) + a\,\delta h_\kappa^{\text{wave}}(\phi) + o(a^2)\,. \tag{37}$$

The linear correction to the local field of the unperturbed wave is given by

$$\delta h_\kappa^{\text{wave}}(\phi) = \delta h_\kappa^{\text{front}}(\phi) + \sum_{\substack{n = -\infty \\ n \neq 0}}^{+\infty} \delta h_\kappa^{\text{env}}(0, 0, n\lambda/v)\,. \tag{38}$$

$\delta h_\kappa^{\text{front}}(\phi)$ is the correction due to the perturbed front itself; cf. Eq. (35). The correction induced by a neighboring front is

$$\delta h_\kappa^{\text{env}}(0, 0, t) = \frac{\partial}{\partial a} h^{\text{front}}[0, 0, t - (a\sin\phi)/v]\bigg|_{a=0} =$$

$$= \sin \phi \sum_{i \in \{1,2\}} \frac{\pi \, a_i \, \lambda_i^2}{2 \, \tau^2 \, v^2} \left\{ \frac{\lambda_i}{\sqrt{\pi} \, \tau \, v} \, \exp\left[-\left(\frac{v \, t}{\lambda_i}\right)^2 \right] + \right.$$

$$+ \left[-2\left(\frac{\lambda_i}{2 \, \tau \, v}\right)^2 + \frac{t}{\tau} - 1 \right] \exp\left[\left(\frac{\lambda_i}{2 \, \tau \, v}\right)^2 - \frac{t}{\tau} \right]$$

$$\left. \times \mathrm{erfc}\left[\frac{\lambda_i}{2 \, \tau \, v} - \frac{v \, t}{\lambda_i} \right] \right\}. \tag{39}$$

According to the argument of the previous section, only those waves are stable that fulfill the relation $\delta h_\kappa^{\mathrm{wave}}(\phi = \pi/2) < 0$ for all κ.

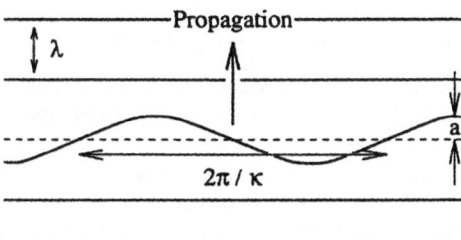

Fig. 13. Transversal perturbation of a plane wave. The wavelength of the wave is denoted by λ and the perturbation is periodic along the wave front with period $2\pi/\kappa$. The amplitude of the perturbation is a. The difference between Fig. 10 and the present one is that here the perturbation is embedded in a plane wave that repeats itself indefinitely with wavelength λ. In the context of linear perturbation theory, this kind of perturbation suffices.

Figure 14a shows the dispersion relation for plane waves again. As compared to Fig. 9, the shaded regions now correspond to unstable solutions. For low spatial frequencies or, equivalently, for large wavelengths, we find the same result as we have obtained before in the case of a single plane front: Only solutions with a (phase) velocity below the critical velocity v_c are stable. For intermediate wavelengths the wave fronts tend to stabilize each other. In this region there exist stable waves with a phase velocity even above the stability limit of single fronts. If the wavelength becomes too small, the wave fronts attract each other and the wave dissolves. This can be seen from Fig. 14b which plots the linear correction to the local field for purely longitudinal perturbations ($\kappa = 0$) as a function of ω and k. In the domain of small wavelength the correction $\delta h_0^{\mathrm{wave}}$ is positive, indicating that the wave is unstable with respect to *longitudinal* perturbations.

a

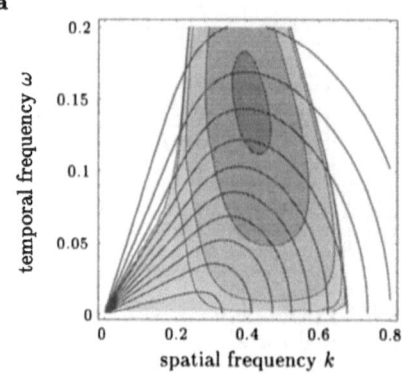
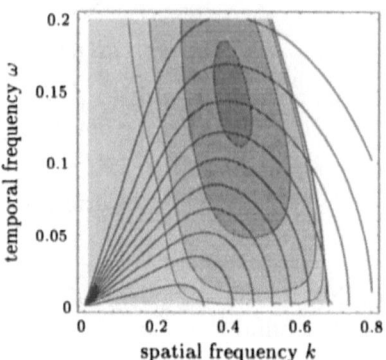

Fig. 14. Dispersion relation for plane waves of excitation in a locally connected neuronal network. **a.** In addition to Fig. 9, this diagram includes a contour plot of the maximum of the linear correction, $\max\limits_{0\leq\kappa<\infty} [\delta h_\kappa^{\text{wave}}(\phi = \pi/2)]$, summarizing the result of the stability analysis. The shaded region corresponds to negative values of $\delta h_\kappa^{\text{wave}}(\phi = \pi/2)$ and thus to stable waves, whereas the unshaded regions to the left and to the right correspond to unstable waves. **b.** A contour plot of the linear correction $\delta h_\kappa^{\text{wave}}(\phi = \pi/2)$ with $\kappa = 0$ that describes the stability of a wave with respect to *longitudinal* perturbations. The shading indicates regions with $\delta h_0^{\text{wave}} < 0$. The unshaded domain in the upper-right corner corresponds to positive values of δh_0^{wave} and, thus, to instability. Waves with parameters ω and k from this domain are susceptible to longitudinal perturbations, i.e., there is no longer a repulsive force between two wave fronts which would stabilize their spacing. The border of this domain coincides with the right border of the stability region in Fig. 14a.

8 Discussion

We have presented a simple model of the visual cortex that exhibits an amazing richness of dynamical behavior, even in the absence of external input. The design of the network and the type of the neuron model allow on the one hand efficient simulations of very large networks of up to 10^6 neurons and on the other hand an *analytic* treatment of some of the most prominent features observed in the simulations. Full details of the present work and of our results on target patterns can be found in a paper submitted to Physica D (Kistler et al. 1997b).

Even though the patterns seen in the single-spike activity of the network resemble concentration patterns in common reaction-diffusion (RD) systems, the structure of the underlying equations is different. Ordinary RD systems are described by partial differential equations with the spatial variables appearing in the term accounting for diffusion. The diffusion term contains the Laplacian of some concentration variables and is strictly local. Apart from that, RD systems may contain spatial averages of their dynamic variables. These non-local terms are used to include effects such as limitation of resources and result in a competition between several domains. In our equations, however, the whole spatial interaction is described by a convolution kernel $J(r)$. This accounts for a *non-local* interaction among the neurons. Consequently, fronts of excitation in our model do not show a unique dependence of the propagation speed upon the (local) curvature of the front, as is usually the case in systems with diffusion. Instead, we have a pronounced dependence upon the neural activity *in the past*.

We were able to calculate the dispersion relation for our model analytically. It describes not only the phase velocity of plane waves as a function of their wavelength. In addition the rotation frequency of spirals can be computed as a function of their asymptotic wavelength. Despite the structural differences, there is qualitative agreement between the dispersion relation for our model and the dispersion relation obtained experimentally for the Belousov-Zhabotinskii (BZ) reaction (Belmonte and Flesselles 1996) and that obtained analytically (but with some approximations) for a model of the BZ reaction (Keener and Tyson 1986). In all cases there is a domain in the phase diagram where the phase velocity increases with increasing period until it saturates finally for large periods. The analysis of Keener and Tyson (1986) also shows that there are waves which have the same period but different phase velocities. The very same phenomenon is present in the dispersion relation of our model but the stability properties are slightly different. Whereas in the case of the BZ system only one out of the two solutions is stable, our model exhibits a parameter regime where in fact two waves with the same period but different phase velocity may coexist.

References

S. I. Amari. Dynamics of pattern formation in lateral-inhibition type neural fields. *Biol. Cybern.*, 27:77–87, 1977.

A. Belmonte and J.-M. Flesselles. Experimental determination of the dispersion relation for spiral waves. *Phys. Rev. Lett.*, 77(6):1174–1177, 1996.

G. B. Ermentrout and J. D. Cowan. A mathematical theory of visual hallucination patterns. *Biol. Cybern.*, 34:137–150, 1979.

C. Fohlmeister, W. Gerstner, R. Ritz, and J. L. van Hemmen. Spontaneous excitations in the visual cortex: stripes, spirals, rings, and collective bursts. *Neural Comput.*, 7:1046–1055, 1995.

W. Gerstner and J. L. van Hemmen. Associative memory in a network of 'spiking' neurons. *Network*, 3:139–164, 1992.

G. Gripenberg, S. O. Londen, and O. Staffans. *Volterra integral and functional equations*, volume 34 of *Encyclopedia of Mathematics and Its Applications*. Cambridge University Press, Cambridge MA, 1990.

J. P. Keener and J. J. Tyson. Spiral waves in the Belousov-Zhabotinskii reaction. *Physica*, 21:307–324, 1986.

W. M. Kistler, W. Gerstner, and J. L. van Hemmen. Reduction of the Hodgkin-Huxley equations to a single-variable threshold model. *Neural Comput.*, 9(5):1015–1045, 1997a.

W. M. Kistler, R. Seitz, and J. L. van Hemmen. Modelling collective excitations in cortical tissue. Submitted to *Physica D*, 1997b.

H. Klüver. *Mescal and the mechanism of hallucinations.* The University of Chicago Press, Chicago, IL, 1966. This is a reprint of two papers which appeared in 1928 and 1942.

J. Milton. *Dynamics of small neural populations.* Amer. Math. Soc., Providence, RI, 1996.

G. M. Shepherd. *Neurobiology.* Oxford University Press, New York, 2nd edition, 1988.

H. R. Wilson and J. D. Cowan. A mathematical theory of the functional dynamics of cortical and thalamic nervous tissue. *Kybernetik*, 13:55–80, 1973.

Interaction of Meandering Spiral Waves in Active Media

H. Brandtstädter, M. Braune, and H. Engel

Institut für Theoretische Physik, Technische Universität Berlin,
Hardenbergstraße 36, 10623 Berlin, Germany

Abstract: Within a light-sensitive Belousov-Zhabotinskii medium we study the interaction of meandering spiral waves. Using an open gel-reactor we find experimental evidence that a small pair of spiral waves undergoes a symmetry-breaking instability, where one member of the spiral pair overwhelms its neighbor and pushes it to the periphery of the medium. To avoid this instability we consider the interaction of a spiral wave with its virtual mirror image close to a plane boundary impermeable to diffusion. The drift of this pseudo-bound state parallel to the boundary occurs on a time scale up to two orders of magnitude larger than the rotation period. The experimental data agree qualitatively with results of numerical simulations obtained within a modified kinematic approach.

1. Introduction:

Spiral waves are generic excitations in active media. They have been observed in systems of quite different nature: as rotating waves of chemical activity in the Belousov-Zhabotinskii (BZ) reaction [1]; in coverage patterns of adsorbed species on platinum single crystal surfaces during CO oxidation under ultrahigh vacuum conditions [2], as cAMP waves in aggregating social amoeba colonies like the slime mould Dictyostelium discoideum, where they control the chemotactic cell motion [3]; as circulating waves of neuromuscular activity in cardiac muscle tissue, where they are associated with cardiac arrhythmia [4]; and as spiral waves of intracellular calcium release [5].

The dynamics of isolated spiral waves has been studied intensively both experimentally [6] and theoretically [7]. Frequency selection, annihilation of colliding wave fronts, and the meander instability are well-known phenomena. It was find out that spiral wave dynamics can be complicated ranging from one-frequency (simple) rotation to quasiperiodic (compound) rotation called meandering by Winfree.

Comparatively less is known about the interaction of spiral waves. Experimentally one finds that fully developed spiral waves do not influence each other when the distance between their cores exceeds several wavelengths. The waves emitted by the spiral source provide an effective screening for the influence of other spirals. Thus, a possible interaction is expected to be short ranged.

In a series of papers by Aranson et al. [8] considered the problem of spiral wave interaction in the framework of the complex Ginzburg-Landau equation (CGLE). This equation is universal as far as it describes any oscillatory active medium close to the oscillation threshold which should be a supercritical Hopf bifurcation. It was found that the interaction between spiral waves decays exponentially at large distances of the spiral cores. At smaller separation, the interaction results in a relative motion of the spiral cores. The velocity of this motion possesses a radial component, v_r, acting along the connecting line of the core centres, and perpendicular to it a tangential component, v_t. Depending on the parameters of the medium, interaction is attractive ($v_r < 0$) or repulsive ($v_r > 0$). In a narrow parameter range at small separation bound states are possible ($v_r = 0$). If the bound state consists of two counterrotating spiral waves with different topological charges, then it is axis symmetric and the tangential velocity components point in the same direction. Axis symmetric spiral pairs drift with constant velocity along the axis of symmetry. Two spiral waves with equal topological charge eventually form a bound state that has central symmetry. Then the tangential components v_t are directed opposite, and the spiral pair as a whole rotates with constant angular velocity around the common centre of symmetry. In both types of bound states, the stationary distance between the spiral cores is about the wave length. In the CGLE model a spiral pair undergoes a symmetry-breaking instability, where one member of the spiral pair overwhelms its neighbour and pushes it to the periphery [8].

Regarding excitable media, already in 1989 Ermakova et al. [9] concluded on the basis of numerical simulations with the FitzHugh-Nagumo model that bound states of spiral waves can exist. Recently, the problem of spiral competition in excitable media was considered on the basis of numerical simulations of a three-component reaction-diffusion model [10]. Again, symmetry-breaking in a spiral pair was found, leading to one spiral suppressing and expelling another.

The above mentioned results apply to the regime of simple rotation. In the following we discuss the interaction of spiral waves performing compound rotation where two or more frequencies are involved. To be more precise, the range of parameters with so-called outward meandering is in the focus of our investigations. We present experiments on a light-sensitive BZ medium carried out with an open gel-reactor. The experimental method is outlined in chapter two. Experimental evidence for the symmetry-breaking spiral pair instability we discuss in chapter three. To avoid this instability, in chapter four we consider the interaction of a single spiral wave with a plane boundary impermeable to diffusion. In this arrangement the spiral and its virtual mirror image model two interacting spiral waves with different topological charges. The experimentally observed motion of the spiral tip is compared to results obtained within a modified kinematical approach applicable to outward meandering spiral waves.

2. Experimental method

Based on the theoretical results summarized in the introduction, effects resulting from the weak interaction between spiral waves as, for example, the drift of axis symmetric spiral pairs, we expect on a time scale that is large compared to other characteristic times as the period of rotation, the refractory period of the medium, etc. Therefore, instead of the traditional batch reactor (Petri dish), we use a new continuously fed gel-reactor developed for the light-sensitive BZ medium. Schematically the experimental setup is shown on Fig. 1. The reactor has two separate reservoirs of 70 ml volume each. Separate supply of oxidizing (bromate) and reducing (malonic acid) components of the BZ solution is possible to prevent uncontrollable formation of intermediate species outside of the active medium. However, in the experiments presented here, the same premixed reaction solution is pumped continouosly into both reservoirs. The recipe concentrations used in the following are given in the caption of Fig.2.

The active medium in the reactor is a silica-hydrogel disc of 54 mm diameter and 0.8 mm width in which the catalyst is fixed. The preparation of comparatively stable, nonfragile gel layers will be described elsewhere. On each side the gel disc is covered with a 0.8 mm thick gel layer not loaded with the catalyst. By diffusive transport through the inactive layers the active layer is in contact with the reservoirs. Diffusive coupling ensures for spatially

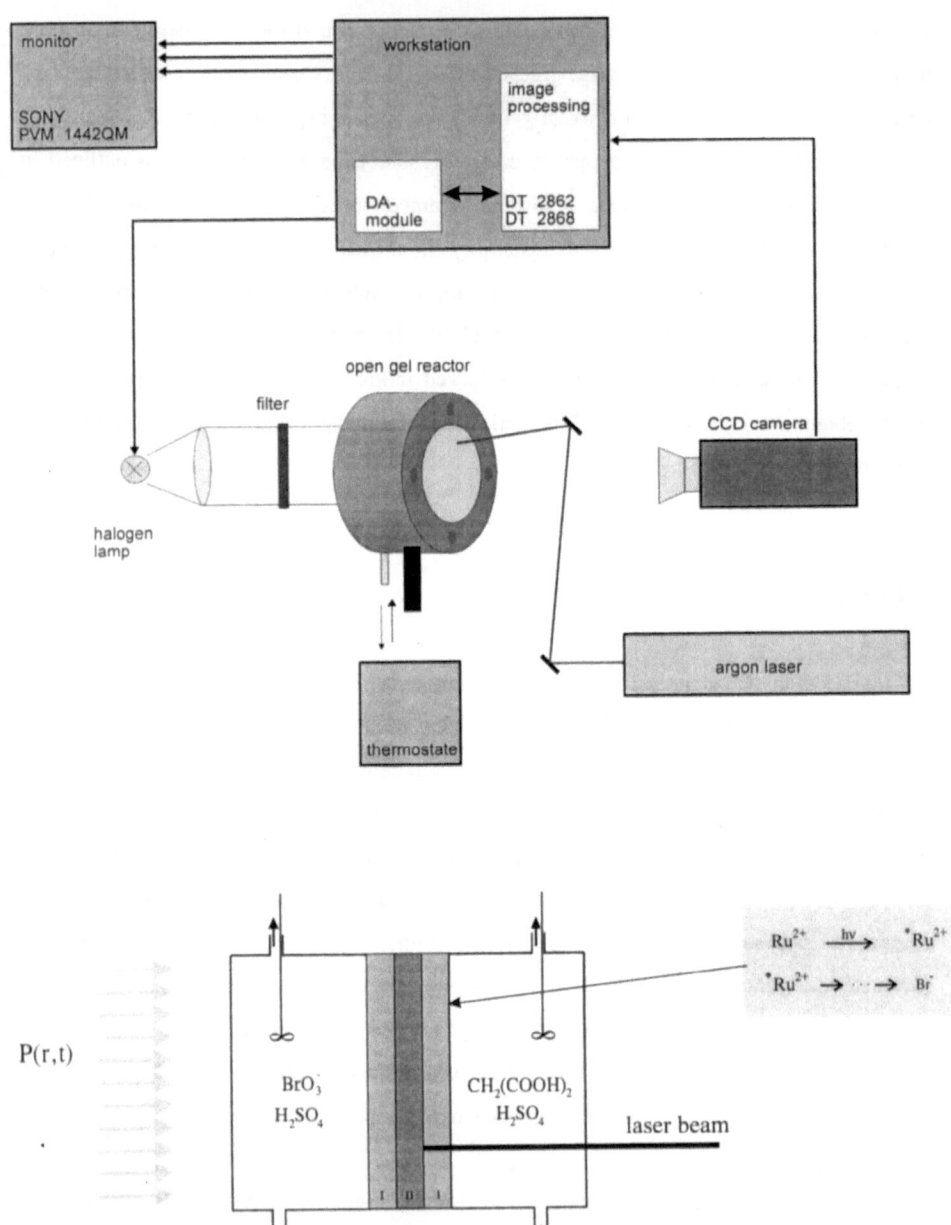

Fig. 1: Schematic representation of the experimental setup. Below, the continously fed gel-reactor for the light-sensitive BZ medium is drawn in section. The catalyst, $Ru(dmby)_3^{2+}$, is fixed in a 0.8 mm thick silicic acid hydrogel matrix. The active layer is covered on both sides by silicahydrogel layers of same thickness not loaded with catalyst to prevent influence of pumping and stirring. The active layer may be affected globally using an appropriate distribution of incident light $P(r,t)$ or locally with the spot of a laser beam. Under illumination the reduced form of the catalyst becomes photochemically excited. This excited complex is rapidly reducing bromate and its subsequent oxobromine species down to the inhibitor bromide.

uniform supply of the reactants eliminating the influence of convective motion in the reservoirs caused by the pump or by stirring. We observed that the flow through the reactor can induce systematic drift of a spiral wave. Therefore, the use of inactive protective layers is crucial in our experiments. Circulating water from a thermostat maintains the temperature at 25.0 ± 0.5 °C.

To visualize patterns formed in the active gel layer the light-sensitive catalytic complex $Ru(bpy)_3^{2+/3+}$ (where bpy$=2,2'$-bipyridyl) provides a comparatively low colour contrast during the change of the redox states from faint blue-green (oxidized state) to orange (reduced state). Better contrast is achieved using the tris(4,4´-dimethyl-2,2´-bipyridyl) ruthenium-complex, further abbreviated by $Ru(4,4'-dm-bpy)_3^{2+/3+}$ which is more sensitive and produces brighter colour contrast. An important additional advantage of $Ru(4,4'-dm-bpy)_3^{2+/3+}$ is that it sticks strongly to the gel matrix than $Ru(bpy)_3^{3+/2+}$ and is more stable against decomposition and aging.

To initiate a spiral pair we cut out a small piece from a flat travelling pulse inhibiting the remaining parts by illumination with light. The two free ends curl up and a spiral pair evolves. The distance between the cores of the counterrotating spirals depends on the size of the wave piece left after the optical initiation. If the size is sufficiently large, then the two spirals behave as individual ones, i.e. there is no interaction between them.

3. Spiral pair instability

For the recipe parameters used in our experiments the orbit of the tip of an isolated spiral wave is close to a hypocycloid. Let denote P the symmetry centre of that tip path pattern. To study the interaction between spiral waves we follow the distance between their symmetry centres P_1 and P_2, thus eliminating oscillatory variations due to spiral meander.

Fig. 2 shows the typical evolution of spiral pairs as observed in our experiments. Initially, the wave pattern is symmetric. After some time a symmetry-breaking instability destroys the bound state. Finally, one member of the spiral pair is completely suppressed. In Fig. 3 the position of the symmetry centres P_1 and P_2 during the whole process is given. The increase in the distance between P_1 and P_2 is accompanied by a slow rotation of the spiral pair as a whole: During the experiment the connectivity line P_1P_2 rotates around an angle of about

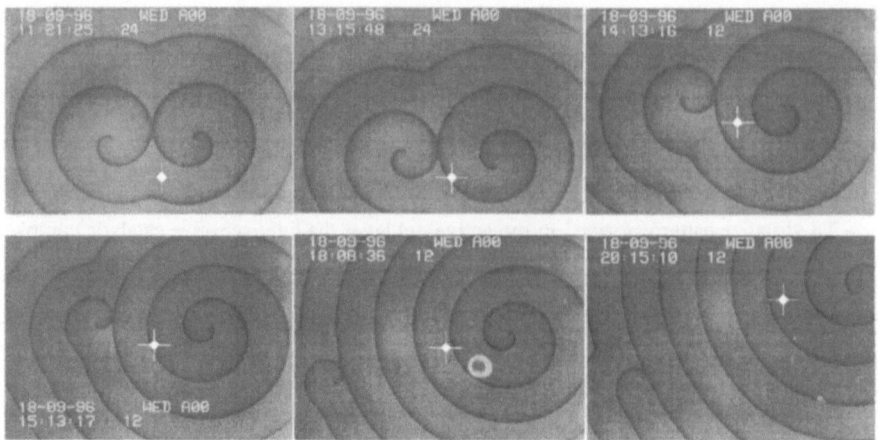

Fig. 2: Breakdown of a spiral pair. The images show the wave pattern at the indicated time moments. The spiral pair was generated 20 min before the first image was taken. Recipe parameters: $[BrO_3^-] = 7.35 \times 10^{-2}$ M, $[H_2SO_4] = 3.36 \times 10^{-1}$ M, $[MA] = 6.3 \times 10^{-2}$ M, $[Br^-] = 6.7 \times 10^{-3}$ M. The flow rate is 70 ml/h. In this medium an isolated spiral wave develops a seven lobe tip path pattern. The size of the imaged area is 15.4 x 21.7 mm^2.

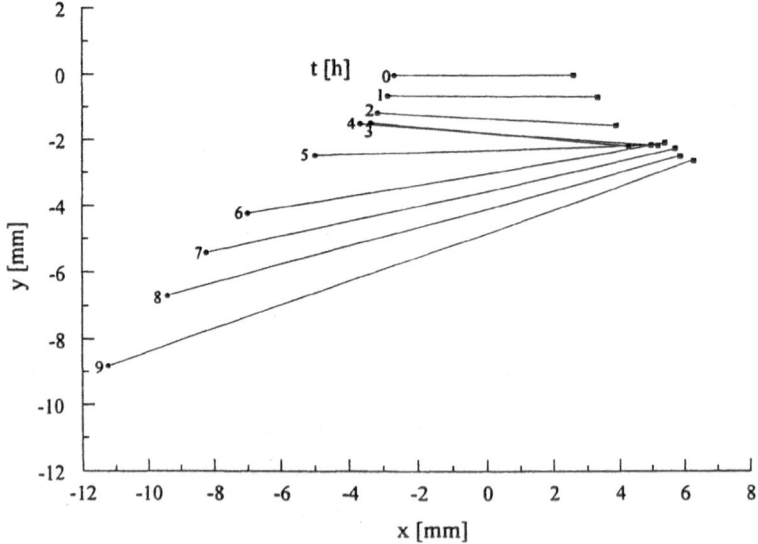

Fig. 3: Position of the two symmetry centers P_1 and P_2 of the interacting spiral waves shown in Fig. 2 at different time moments. The interval between subsequent numbers is one hour.

20 degrees. It is evident from Fig. 3, that initially both spirals move off from each other. After the symmetry of the bound state has been broken, the centre of the dominating spiral, P_2, remains almost fixed in space, whereas P continues to move away. This process is associated with morphological changes of the spiral pattern. While the dominating spiral increases its sphere of influence continuously, the inferior spiral transforms into a small remainder whose bare core is not protected from invading wave fronts.

We suggest a simple explanation for the behaviour observed after the symmetry-breaking instability. It is well-known that in the regime of compound rotation, far from the symmetry centre of the tip path pattern the medium is excited periodically with some frequency f_1. When the tip path is close to a hypocycloid, f_1 corresponds to the rotation about the primary circle. Therefore, in the later stage of the interaction process the core region of the smaller spiral is subjected to periodic forcing generated by the larger spiral. This periodic perturbation occurs near the main resonance $f_1 = f_{forc}$. Recently it was found that a meandering spiral wave under resonant forcing exhibits a drift motion of its symmetry centre [11]. Exactly at resonance, the drift trajectory is a straight line, close to the resonance the tip traces a meander trajectory along a circular orbit of large radius which approaches infinity as the resonance condition becomes valid. Obviously, this mechanism is responsible for the displacement of the left spiral in Fig. 2 after the symmetry-breaking instability has taken place. Conversely, the left spiral cannot influence the right one in the same way, because the core region of the latter is effectively screened by the wave fronts emitted from that spiral wave source. Thus, in the situation characterized by an increasing asymmetry of the spiral pair, the final outcome of its evolution is a dominating spiral wave with almost fixed tip path centre, and the other spiral wave drifting away under resonant periodic forcing. The described scenario has been observed repeatedly in our experiments. Crucial for this mechanism is that after the instability has occurred, the arm of the smaller spiral cannot grow up. In contrary, due to annihilation of the colliding wave fronts in the course of time the arm becomes shorter and shorter leaving an unscreened core that is subjected to periodic forcing caused by the dominating spiral.

4. Artificial stabilization of bound states: Spiral interaction with plane boundaries impermeable to diffusion

An elegant possibility to avoid the spiral pair instability is to place a single spiral wave close to a plane boundary impermeable to diffusion. In this case the spiral wave interacts with its virtual mirror image. Obviously, this arrangement is equivalent to the interaction between two counterrotating spiral waves. In the regime of simple rotation this phenomenon has been studied numerically by Mikhailov and Zykov [12].

Within the light-sensitive BZ medium, in a narrow range of experimental parameters we find meandering spiral waves whose tip path centre move with constant velocity parallel to the boundary. The drift of the whole pattern is superimposed on the complex motion of the tip due to spiral meandering. The tip trajectory may be approximated by a hypocycloid whose outer and inner diameter are, respectively, 1.3 mm and 0.4 mm. The periods of rotation on the primary (secondary) circle is 124 s (720 s); this corresponds to a time interval of loop formation $T_{loop} = 106$ s. Fig. 4 shows the tip path pattern 0.3, 0.5, etc. hours after generation of the spiral wave close to the boundary. The drift parallel to the boundary can be unambiguously identified only in longtime experiments running over several hours. For the chosen recipe parameters we find $v_{drift} \approx 1.2$ mm $h^{-1} = 0.02$ mm min^{-1} which must be compared to a wave velocity $v_{wave} = 1.4$ mm min^{-1}. In conclusion, the drift velocity of the

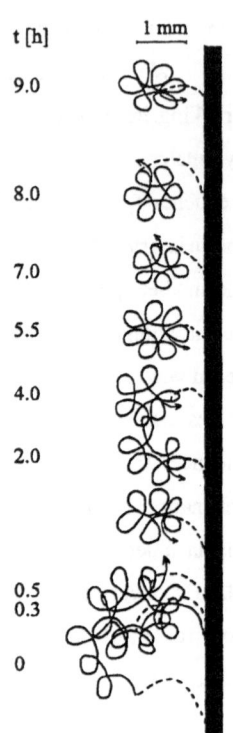

Fig. 4: Tip trajectory of a spiral wave propagating close to an unpermeable boundary. The recipe parameters are the same as in Fig. 2.

spiral pair may be up to two orders of magnitude smaller than the pulse velocity in the same medium. The stationary distance of the symmetry centre from the boundary was approximately 1.6 mm, which is nearly half the wave length $\lambda = 2.9$ mm of isolated spiral waves in the medium. The length of the wave piece left at the beginning of the experiment determines the initial distance of the evolving spiral wave from the boundary. In the same medium where we observed the parallel drift of spirals close to the boundary, spiral waves far from the boundary remained unaffected. On the other hand, starting with very small pieces, cut out of a wave front propagating perpendicular to the boundary, does not lead to spiral formation but to the disappearance of the wave piece.

To model the experimental results qualitatively we use a modified kinematic description developed recently [13]. This modified approach includes curvature effects near the tip to describe tip path patterns with curvature maxima outside. The starting point is a closed equation for the curvature $K(l,t)$ of the wave front as a function of the arc length, l, measured from the tip, and the time, t ,

$$\frac{\partial K}{\partial t} + \left(\int_0^l K V d\xi + G \right) \frac{\partial K}{\partial l} = - K^2 V - \frac{\partial^2 V}{\partial l^2} , \qquad (1)$$

where

$$V(K) = V_0 - D K , \ G(K_0) = G_0 - \gamma K_0 ,$$
$$K_0 = K(l=0,t) = K_0(t) , \ K(l \to \infty, t) = 0 . \qquad (2)$$

Among the phenomenological parameters in eq. (2) V_0 and G_0 denote, respectively, the normal velocity of a planar wave and its tangential velocity at an open end, and D and γ are positive constants with the dimension of a diffusion coefficient. Based on purely geometric considerations eq. (1,2) was derived by Mikhailov et al. [12]. In contrast to their treatment we use a modified boundary condition to account for curvature effects near the tip that revealed to be crucial for the onset of outward meandering

$$\frac{\partial K_0}{\partial t} = -G \left(\frac{\partial K}{\partial l} \right)_{l=0} , \qquad \text{if} \quad K_0 < K_{max} ,$$
$$K_0 = \hat{K}_0 \qquad , \qquad \text{if} \quad K_0 \geq K_{max} . \qquad (3)$$

Condition (3) means that when the curvature at the tip exceeds the maximum possible value for stable pulse propagation in the medium, $K_{max} \sim V_0 / D$, then we reset this overcritical value to K_0, introducing the new parameter $\hat{K}_0 < K_{max}$. In other words, the curvature at the tip abruptly jumps back to a subcritical value $\hat{K}_0$, from which it resumes its prior increase. The new phenomenological parameter K_0 determines the shape of the tip path pattern (for details compare [12]).

We have solved eqs. (1-3) numerically. From the curvature we can reconstruct the shape of the spiral and the position of the tip as a function of time. The results are presented on Fig 5.

a)

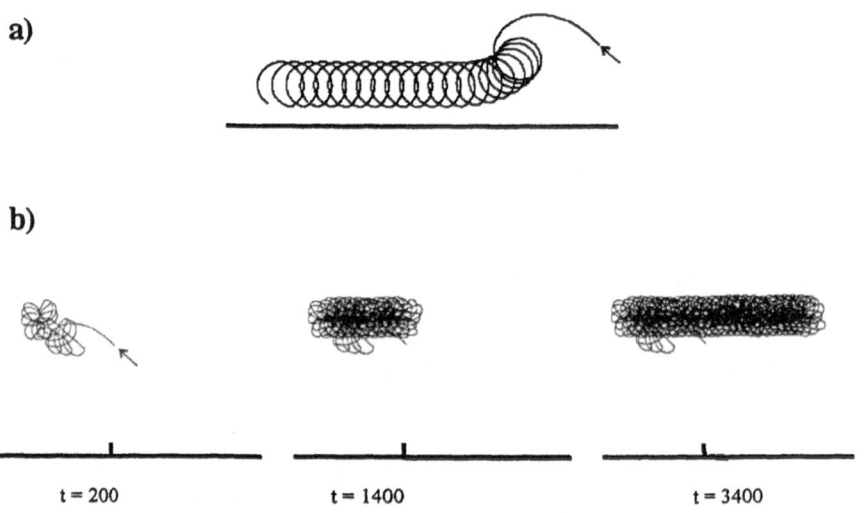

b)

$t = 200$ $t = 1400$ $t = 3400$

Fig. 5: Simulation results for spiral motion close to the boundary obtained within the modified kinematical approach. Regime of simple rotation (a) with parameters $V_0 = 1.0$, $D = 1.0$, $G_0 = 0.65$, and $\gamma = 1$. In the regime of compound rotation (b) the direction of the drift is reversed. Parameter values: $V_0 = 0.97$, $D = 1.0$, $G_0 = 0.8$, $\gamma = 1$, and $\dot{K}_0 = 0.6$.

In agreement with experimental observation, in a narrow parameter range we find that a meandering spiral wave can drift parallel to the boundary. Note, that in Fig. 5 initially the drift is directed to the left. After reaching the stationary distance of the tip path centre P from the boundary, the spiral starts to move in the opposite direction. Such reversion of the drift direction we found in numerous experimental runs.

5. Discussion

Studying dynamic processes in active media that proceed on large time scales as the interaction of spiral waves, or as near threshold phenomena with critical slowing down, strictly stationary nonequilibrium conditions should be maintained. Those experiments cannot be carried out in traditional batch reactors because of the unavoidable relaxation of the system to thermodynamic equilibrium. Recently, several types of open reactors have been developed to sustain chemical dissipative structures [14]. Here we have presented an open gel-reactor developed for the light-sensitive BZ medium. We prefer this active medium because of its methodological adventages in controlling wave dynamic. For example, varying the intensity of applied illumination it is possible to control the local excitation threshold.

We extend the results on the interaction of spiral waves obtained recently by Ruiz-Villarreal et al. [15] in several directions. These authors carried out experiments with the ferroin-catalyzed BZ system in a Petri dish covered by a feeding solution that was exchanged continuously. First, to improve the experimental conditions we have developed a genuine open gel reactor. Through protecting gel layers the active medium is in diffusive contact with the reservoirs to ensure homogeneous supply of the reagents. With that any influence of the flow through the reactor or of stirring in the reservoirs on the spiral dynamics is reliable prevented. Our results are not restricted to the regime of simple rotation. The range of parameters in our experiments belongs to meandering spiral waves. Finally, we use the equivalence between a bound state of two counterrotating spiral waves and a single spiral interacting with a boundary impermeable to diffusion to avoid the symmetry-breaking instability. This allows to follow the drift of the bound state over several hundred rotation periods.

We describe the experimental results qualitatively using a modified kinematic approach. Some results indicate a reversion of the drift direction of a bound state when simple rotation becomes unstable and is replaced by outward meandering. However, up to now this conjecture could not be confirmed experimentally. Altogether, for a quantitative comparison between experiment and theory the phenomenological parameters of the kinematical theory should be determined from experimental data or from the underlying reaction-diffusion equations. This work is under progress.

References

[1]   A.N. Zaikin, A.M. Zhabotinskii, Nature **225**, 535 (1970); A.T. Winfree, Science **175**, 634 (1972);
        Science **181**, 937 (1973).

[2]   G. Ertl, Science **254**, 1750 (1991); M. Eiswirth, and G. Ertl, Pattern Formation on Catalytic Surfaces,
        in: R. Kapral, K. Showalter, (eds) *Chemical waves and Patterns* (Kluwer Academic Publishers,
        Dordrecht (1995), p. 447; S. Jakubith, H.H. Rotermund, W. Engel, A. v. Oertzen, and G. Ertl, Phys.
        Rev. Lett. **65**, 3031 (1990).

[3]   G. Gerisch, Naturwissenschaften **58**, 420 (1983); K.J. Lee, E.C. Cox, and R.E. Goldstein, Phys. Rev.
        Lett. **76**, 1174 (1996).

[4]   J.M. Davidenko, A M. Pertsov, R. Salomonsz, W. Baxter, and J. Jalife, Nature **355**, 349 (1992).

[5]   J. Lechleiter, S. Girad, E. Peralta, and D. Clapham, Science **252**, 123 (1991).

[6]   S.C. Müller, Th. Plesser, and B. Hess, Physica D **24**, 71 (1987); G.S. Skinner, H.L. Swinney, Physica
        D **48**, 1 (1991); H. Engel, M. Braune, Physica Scripta **49**, 685 (1993); M. Braune, H. Engel, Chem.
        Phys. Lett. **204**, 257 (1993); Chem. Phys. Lett. **211**, 534 (1993); G. Li, Q. Quang, V. Petrov, and
        H.L. Swinney, Phys. Rev. Lett. **77**, 2105 (1996).

[7]   D. Barkley, M. Kness, and L.S. Tuckerman, Phys. Rev. A **42**, 2489 (1990); D. Barkley, Phys. Rev.
        Lett. **68**, 2090 (1992); Phys. Rev. Lett. **72**, 164 (1994); in *Chemical waves and Patterns* (Ref. [2]), p.
        163; A.T. Winfree, Chaos **1**, 303 (1991); W. Jahnke, W.E. Skaggs, and A.T. Winfree, J. Phys. Chem.
        93, 740 (1989).

[8]   I. Aranson, L. Kramer, and A. Weber, Physica D **53**, 376 (1991); Phys. Rev. E **47**, 3231 (1993); **48**,
        R9 (1993); Phys. Rev. Lett. **72**, 2316 (1994); I. Aranson, D. Kessler, and I. Mitkov, Phys. Rev. E **50**,
        R2395 (1994).

[9]   E.A. Ermakova, A.M. Pertsov, E.E. Shnol, Physica D **40**, 185 (1989).

[10]   I. Aranson, H. Levine, and L. Tsimring, Phys. Rev. Lett. **76**, 1170 (1996).

[11]   M. Braune, A. Schrader, H. Engel, Chem. Phys. Lett. **222**, 358 (1994); V.S. Zykov, O. Steinbock, and
        S.C. Müller, Chaos **4**, 509 (1994); A. Schrader, M. Braune, H. Engel, Phys. Rev. E **52**, 98 (1995).

[12]   A.S. Mikhailov, V.S. Zykov, in *Chemical Waves and Patterns*, (Ref. [2]), p. 119.

[13]   M. Braune, H. Engel, *"Kinematical description of meandering spiral waves in active media"*.
        In: H. Engel, F.-J. Nidernostheide, H.-G. Purwins, and E. Schöll (eds.), *"Self-Organization
        in Activator-Inhibitor Systems: Semiconductors, Gas-Discharges and Chemical Active Media"*
        (Wissenschaft & Technik-Verlag, Berlin, 1996), p. 94; Phys. Rev. E (1997), in press.

[14]   Q. Quang, H.L. Swinney, Nature **352**, 610 (1991); Chaos **1**, 411 (1991); P. De Kepper, V. Castets,
        E. Dulos, and J. Boissonade, Physica D **49**, 161 (1991).

[15]   M. Ruiz-Villarreal, M. Gomez-Gesteira, C. Souto, A.P. Munuzuri, and V. Pérez-Villar, Phys. Rev.
        E **54**, 2999 (1996).

Spatiotemporal Patterns in a Passivating Electrochemical System

R.D. Otterstedt[1], N.I. Jaeger[1], P.J. Plath[1] and J.L. Hudson[2]

[1]Institut für Angewandte und Physikalische Chemie
Universität Bremen, FB 2 - Biologie/Chemie
Postfach 330 440, D-28334 Bremen, FRG

[2]Department of Chemical Engineering, Thornton Hall,
University of Virginia, Charlottesville, Virginia 22903-2442, USA

1   Introduction

The development of spatiotemporal patterns in chemical systems far from equilibrium as well as their modeling based on chemical reaction and coupling mechanisms is well established in homogeneous reaction-diffusion systems [1] and in heterogeneous catalysis on single crystal surfaces as well as on wires, foils and supported catalysts [2-4].

Electrochemical reactions, e.g. electrocatalysis and the electrodissolution of metals, frequently exhibit electrode potential-current characteristics with a region of negative slope [5]. This class of reactions shows bistability if a large enough resistor is connected in series with the cell [6] as well as oscillations and excitability provided an appropriate feedback mechanism exists. Temporal oscillations and spatiotemporal phenomena in electrochemical reactions have been known for a long time [5, 7] However, the latter has attracted attention in the context of the study of non-linear systems only recently [8-10].

Bistability and spatiotemporal patterns in current density and potential distribution can be observed in the bulk of non-linear electronic materials like semiconductors which exhibit regions of negative impedance, i.e. Z- or S-shaped current voltage characteristics [11, 12].

The spatiotemporal behavior of a subclass of electrochemical reactions, namely the electrodissolution of metals which exhibits passivity, has received special attention, i.e., iron in sulfuric acid [9, 10], nickel in sulfuric acid [8] and cobalt in phosphoric acid [13]. Figure 1 depicts a schematic drawing of an experimental current-voltage curve showing the active-passive transition at the Flade potential, E_F. Following the transition the electrode surface is covered by an oxide film blocking the current flow.

The curve shows negative impedance. The solid line shows the experimental current vs. the applied voltage, V_a; the dashed line represents the unstable branch of the polarization curve which is not seen in the experiment. The dotted line shows the current vs. electrode potential E, which is the applied potential V_a corrected by the ohmic potential drop in the electrolyte $E = V_a - I$ (total) x R, i.e., the potential drop across the electric double layer in front of the electrode and across the passive oxide film. The transition from the active to the passive state occurs when the potential drop across the double layer exceeds the Flade potential $(E > E_F)$. In the passive state $E \sim V_a$ due to the negligible ohmic potential drop in the electrolyte. The active state can be reestablished by setting the applied voltage $V_a \sim E < E_F$. In addition to bistability the system can exist in an oscillatory or excitable state due to the pH dependence of the Flade potential $E_F = E_o - \frac{RT}{nF} \ln a_H{}^+$. The pH-value increases in front of the electrode due to migration in the active state and shifts back to the bulk value due to diffusion in the passive state.

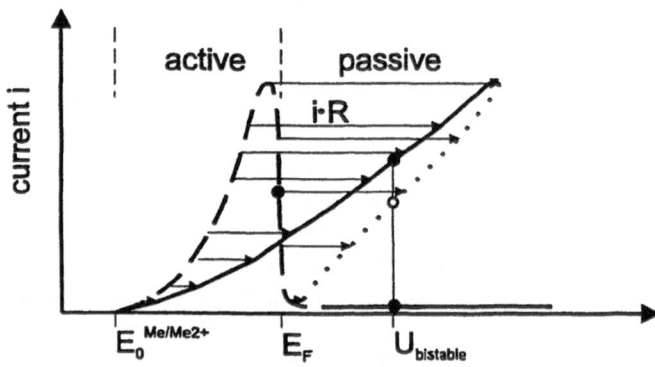

Fig. 1: Schematic of the electrode potential - current curve of a passivable metal (dashed curve, E_0: rest potential, E_F: Flade potential) and the corresponding applied voltage - current curve in presence of a resistor R in series with the working electrode. The dotted line denotes the instable branch. The system is bistable at the applied voltage $U_{bistable}$.

The oscillatory behavior of the Fe/H_2SO_4-system has been modeled by Franck and FitzHugh in 1961 [14]. Even though propagation of an active region had been experimentally observed [15], no attempt was made in the model to include spatiotemporal phenomena.

Progress in the experimental and theoretical investigation of spatiotemporal patterns in electrochemical systems will be discussed. It will be demonstrated in the case of the electrodissolution of cobalt that electrochemical systems are well suited to adjust and classify the range of coupling (local, non-local and global) in correlation with experimentally observed spatiotemporal patterns.

2 Experimental

In general, electrochemical experiments are carried out either potentiostatically or galvanostatically, i.e. the potential of the working electrode or the current flowing through the electrochemical cell is controlled by an electronic device called potentiostat or galvanostat, respectively.

In potentiostatically controlled experiments, the type of experiments reported in this work, a three-electrode arrangement in a vessel containing the electrolyte is used: The potentiostat keeps the potential between the grounded working electrode (WE) and the reference electrode (RE) constant by adjusting the potential of the counter electrode (CE). The current flowing between working- and counter electrode is measured during the experiment.

Experimental results obtained in two different electrode geometries are reported in the subsequent section. Geometry I, situated in a rectangular vessel (Fig. 2) consists of a horizontally leveled 50x0.5 mm ribbon of cobalt facing upward into the electrolyte as working electrode, surrounded by a 150x100 mm rectangular counter electrode bent from a copper wire of 2 mm diameter. Both electrodes are arranged in the same plane. The reference electrode is placed 10 mm away from the sides of a corner of the counter electrode.

Top View:

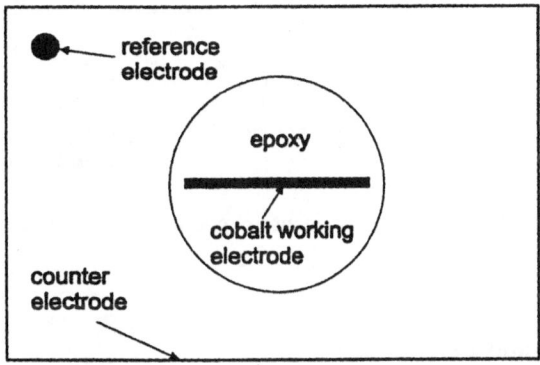

Side View:

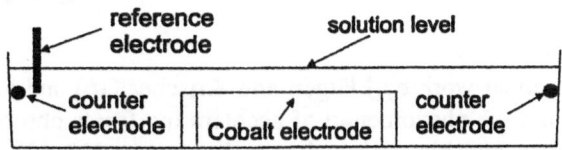

Fig. 2: Experimental setup for electrode geometry I. The quasi-1-dimensional (50x0.5 mm) working electrode (WE) is located in the center of the rectangular counter electrode (CE, 150x100 mm). The reference electrode is placed in a corner of the counter electrode. The three electrodes are located in the same plane.

The electrode geometry II, placed in a cylindrical vessel, (Fig. 3), is a ring of cobalt with a diameter of 12.5 mm and a width of 1.25 +/- 0.25 mm surrounded by ring of copper of diameter 70 mm as counter electrode; the tip of the capillary of the reference electrode is placed in the center of the working electrode, slightly above (ca. 1 mm) the plane comprising working and counter electrode.

Buffered phosphoric acid of different concentrations and pH-values serves as electrolyte.

The spatiotemporal potential distribution at the working electrode is monitored by an array of twelve potential probes adapted to the geometry of the working electrode whose signals are AD-converted and composed to two-dimensional projections of equipotential lines in space and time.

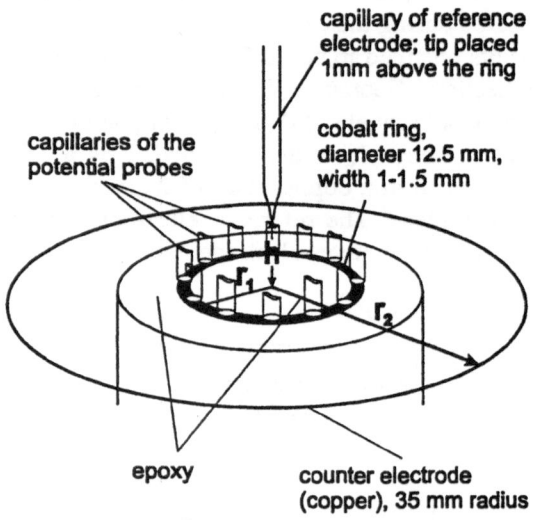

Fig. 3: Electrode arrangement in geometry II. The setup shown is placed in a cylindrical vessel.

3   Results and Discussion

3.1 Passive-Aactive Transitions Governed by Long-Range, Positive, Spatial Coupling

Recent theoretical work by Flätgen and Krischer [16] and Mazouz et al. [17] has shed light onto the phenomenon of accelerating fronts observed experimentally in quite different bistable electrochemical systems [18, 19]. Their analysis of a general model [16] was based on bistable local dynamics and Laplace's equation for the potential distribution in the electrolyte in two spatial dimensions. They showed, for the case of a reference electrode parallel to the working electrode, that the range of spatial coupling increases from local to non-local to global [17] as a geometrical system parameter (containing the ratio of the distance between working and reference electrode to the length of the working electrode) is increased. Fronts of transition have constant velocity, i.e. behave as in reaction-diffusion systems, when the range of coupling is small due to a small value of the aforementioned parameter. Increasing this parameter leads to an acceleration of the front. In case of the parameter being large, the transition from the less to the more stable state occurs almost uniformly. The spatial coupling is everywhere positive on the working electrode in the geometry of the model of Flätgen and Krischer [16].

In the bistable system, realized with geometry I and an electrolyte containing unbuffered 0.1 M phosphoric acid with pH 2.2, a visible transition from the passive to the active state can be initiated by locally destroying the protective passive oxide layer. This can be performed either mechanically by scratching the surface with a glass rod or touching it with a blank cobalt wire which leads to a local electrochemical dissolution of the protective film.

Figure 4a shows the position time plot of a front initiated by scratching taken from digitized video images. The velocity of the front increases with time which can be inferred from the upward concave shape of the curve as well as the nonlinear rise of the current (Fig. 4b).

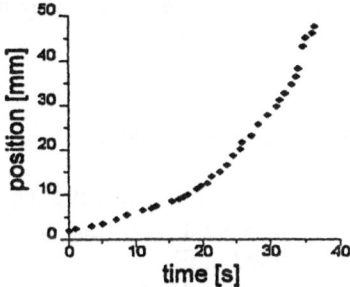

Fig. 4a: Position-time curve of an accelerating front in the bistable system.

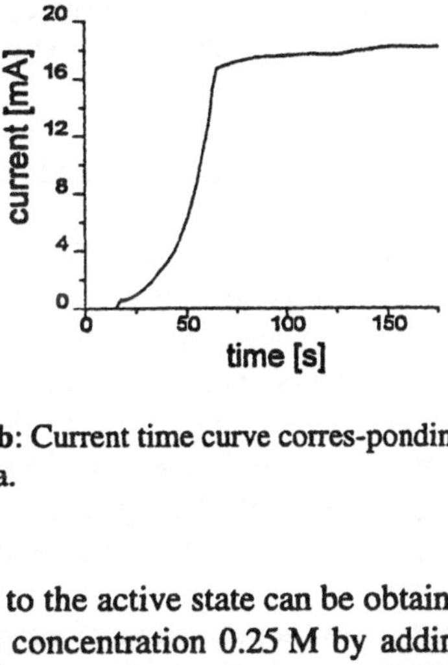

Fig. 4b: Current time curve corres-ponding to Fig. 4a.

An almost uniform transition from the passive to the active state can be obtained in geometry I with an electrolyte of pH 1.7 and concentration 0.25 M by adding an external resistor of 25 Ω in series to the working electrode. Figure 5, a projection of equipotential lines in space and time on the plane, shows that the transition from the passive to the active state occurs almost simultaneously on the whole ribbon. The resistor mediates a global spatial coupling in the system because the ohmic potential drop at the resistor due to the current flowing locally on the ribbon changes the potential of the metal of the working electrode, hence affecting the whole working electrode equally.

The effect of the large external resistor in this experiment is thus comparable to the global coupling caused by the large value of the parameter in the model of Flätgen, Mazouz and Krischer [16, 17].

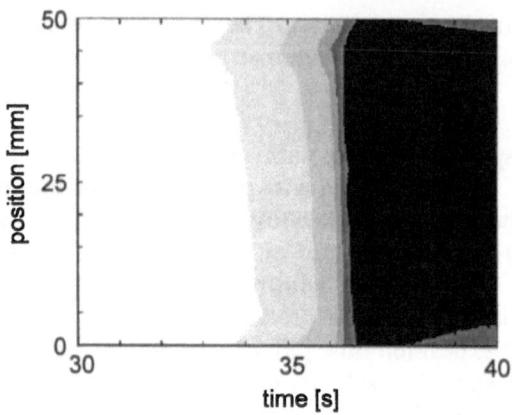

Fig. 5: Projection of equipotential lines on the ribbon on the time space plane. The area between two adjacent equipotential lines has been filled with a grey tone, which is the darker, the lower the potential.

3.2 Waves on a Ring Electrode Due to Negative Spatial Coupling

In the preceding section the influence of the range of *positive* spatial coupling was discussed. In this section experimental results are presented, revealing *negative* spatial coupling due to the electrode arrangement depicted in Fig. 3.

The experiments have been done in an electrolyte of pH 1.1 with a total phosphorous concentration of 1.0 M. The applied potential between the metal of the working electrode and the tip of the capillary of the reference electrode has been set to 1100 mV/Ag/AgCl, i.e. a value about 50 mV cathodic with respect to the Flade potential. Under these conditions, the dynamics are supposed to be oscillatory.

Figure 6 shows a projection of equipotential lines onto the time-space plane. The scale on the y-axis shows degrees. The width of the ring shows some asymmetry and the narrowest segment is located at 180 degrees. Areas between two adjacent equipotential lines are displayed in the same gray tone, the dark gray areas denoting the most active and the white areas the most passive ones. The medium gray at t=215 s denotes a potential of 1100 mV.

In the beginning of the experiment, not shown in Fig. 6, a single rotating wave develops [20]. At time t=160 s, the wave breaks up into a pair of waves at the position 270 degrees, annihilating each other at 90 degrees, i.e. 180 degrees from their point of formation. After about 20 s, where the whole electrode is passive, a new pair of waves appears at the same position where the first pair had formed. One travels clockwise the other one counterclockwise, annihilating each other again at the position 90 degrees. The shape of the single as well as of the pair of rotating waves does not remain constant in time, i.e. size and amplitude of the active area is growing and shrinking in an irregular manner. This phenomenon will be treated in the following section.

The negative spatial coupling manifests itself in the white and light gray shaded areas where the potential is more anodic than the applied potential of 1100 mV.

When there is no current flowing, for example at t=215 s, the potential everywhere on the ring is at 1100 mV and there is no spatial coupling through100 the electric field. If there is a current flow, i.e. an active area exists somewhere, then there are also areas with a potential more anodic than 1100 mV, i.e. activity on one side of the ring induces passivity or at least a stronger tendency to passivity on the opposite side of the ring.

When the tip of the reference electrode is placed more then about 3 mm away from the plane of the ring, then, homogeneous oscillations do occur instead of a rotating wave. It seems, that the spatial coupling through the electric field is no longer negative anywhere on the ring, so that no symmetry breaking occurs.

There is a relationship between the phenomenon observed in this work and symmetry breaking due to an integral negative feedback, i.e. an integral constraint. The electric field mediates an instantaneous spatial coupling, like in idealized models of a catalytic surface reaction in contact with a well mixed gas-phase [21].

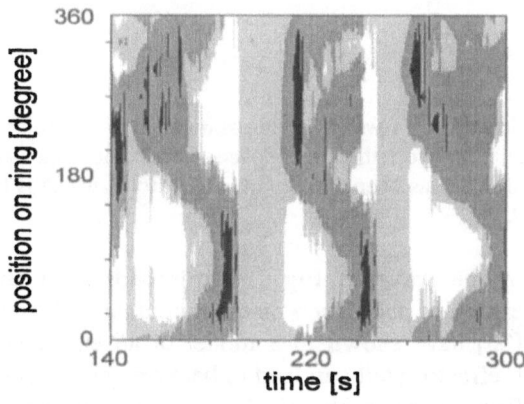

Fig. 6: Negative spatial coupling on a ring electrode. Projection of equipotential lines on the ring on the time space plane. The area between two adjacent equipotential lines has been filled with a grey tone, which is the darker, the lower the potential.

3.3 Modulated Traveling Waves

As already seen in section 2, the electrodissolution of cobalt into phosphoric acid displays another phenomenon: the modulation of waves [22]. This phenomenon occurs both in the oscillatory and in the excitable system, if the applied potential is close to the Flade potential.

Figure 7 shows an example of a modulated wave in the excitable case on the ribbon electrode in geometry I. The electrode has been passivated for 15 s at 1150 mV in an electrolyte with pH 1.67 and a phosphorous concentration of 1.25 M and then has been touched with a blank cobalt wire with 0.8 mm² active surface. The resulting active area starts propagating onto the ribbon while the amplitude and the size of the active area is shrinking and growing almost periodically in the direction reverse to the initial direction of propagation. The average amplitude of the area is rising until the size of the active area reaches system size at t=37 s. After this active event, the system falls back to the passive state almost uniformly at t=38 s and remains passive.

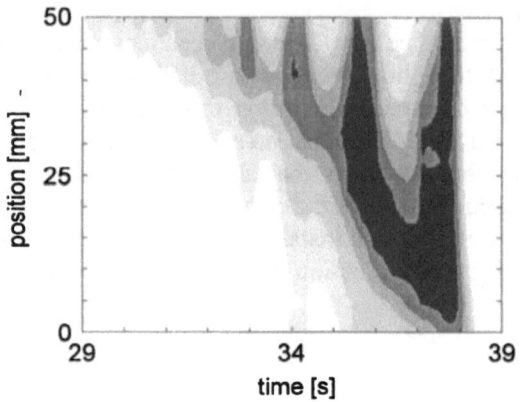

Fig. 7: Modulation of a wave in the excitable system. Projection of equipotential lines on the ribbon on the time space plane. The area between two adjacent equipotential lines has been filled with a grey tone, which is the darker, the lower the potential.

The modulation of the wave in Fig. 7 corresponds to a situation where the refractory tail appears to be short such that reexcitation in the reverse direction can occur. Bär et al. [23] have shown in a model of an excitable reaction-diffusion system that a short refractory tail can lead to backfiring of pulses in one dimension. We return to backfiring in electrochemical experiments in a subsequent paper.

4 Conclusion

In electrochemical systems, the range of the spatial coupling through the electric field can vary widely, giving rise to new spatiotemporal phenomena, i.e. the acceleration of fronts. In the axial symmetric electrode arrangement, the non-local coupling can be even negative, giving rise to phenomena typical for an integral constraint, i.e. a rotating wave in a quasi one dimensional, oscillatory system with periodic boundary conditions.

In addition to the more general aspects of spatial coupling in different electrode geometries, the electrodissolution of cobalt into phosphoric acid shows modulated waves in the excitable system.

Acknowledgement. Financial support by the Deutsche Forschungsgemeinschaft (Ja 346/16-1) and NATO (CRG.950057) is gratefully acknowledged.

References

[1]   Z. Noszticius, W. Horsthemke, W.D. McCormick, H.L. Swinney, Y.W. Tam; Nature, 329 (1987) 581.

[2]   F. Schüth, B.E. Henry, L.D. Schmidt; Adv. Catal., 39 (1993) 51.

[3]   R. Imbihl; Progr. Surf. Sci., 44 (1993) 185.

[4]   M.M. Slin'ko, N.I. Jaeger; Stud. Surf. Sci. Catal., 86, 1994.

[5]   J. Woijtowicz, in Modern Aspects of Electrochemistry, ed. J.O.M. Bockris and B.E. Conway; Plenum Press, New York, 1973, vol. 9, p. 47.

[6]   M.T.M. Koper, J.H. Sluyters; J. Electroanal. Chem. 352 (1993) 51.

[7]   J.L. Hudson, T.T. Tsotsis; Chem. Eng. Sci. 49 (1994) 1493.

[8]   O. Lev, M. Sheintuch, L.M. Pismen and Ch. Yarnitzky; Nature (London), 336 (1988) 458.

[9]   J.L. Hudson, J. Tabora, K. Krischer, I.G. Kevrekidis; Phys. Lett. A., 179 (1994) 355.

[10]   J.C. Sayer, J.L. Hudson; Ind. Eng. Chem. Res. 34 (1995) 3246.

[11]   A. Wacker, E. Schöll, J. Appl. Phys. 78 (1195) 7352

[12]   Ch. Radehaus, K. Kardell, H. Baumann, D. Jäger, H.-G. Purwins; Z. Phys. B-Condensed Matter 65 (1987) 515

[13]   R.D. Otterstedt, N.I. Jaeger, P.J. Plath; Int. J. Bifurcation Chaos 4 (1994) 1265.

[14]   U.F. Franck, R. FitzHugh; Z. Elektrochemie 65 (1961) 156.

[15]   U.F. Franck, Z. Elektrochemie,55 (1951)154.

[16]   G. Flätgen, K. Krischer; J. Chem. Phys. 103 (1995) 5428.

[17]   N. Mazouz, G. Flätgen, K. Krischer; Phys. Rev. E 55 (1997) 2260.

[18]   G. Flätgen, K. Krischer; Phys. Rev. E 51 (1995) 3997.

[19]   R.D. Otterstedt, P.J. Plath, N.I. Jaeger, J.C. Sayer, J.L. Hudson; Chem. Eng. Sci. 51 (1996) 1747.

[20]   R.D. Otterstedt, P.J. Plath, N.I. Jaeger, J.C. Sayer, J.L. Hudson; J. Chem. Soc., Farad. Trans. 92 (1996) 2933.

[21]   M.D. Graham, U. Middya, D. Luss; Phys. Rev. E 48 (1993) 2917.

[22]   R.D. Otterstedt, P.J. Plath, N.I. Jaeger, J.C. Sayer, J.L. Hudson; Phys. Rev. E 54 (1996) 3744.

[23]   M. Bär, M. Hildebrand, M. Eiswirth, M. Falcke, H. Engel, M. Neufeld; Chaos 4 (1994) 499.

Instabilities of Pollutant Concentrations in the Troposphere Due to Chemical Reactions

H. Lustfeld

Forum Modellierung
and
Institut für Festkörperforschung,
Forschungszentrum Jülich, D 52425 Jülich, Germany

December 10, 1997

Abstract. In this paper we investigate chemical reaction equations of pollutants in the troposphere and compare it with a simple model. We show that time averages are possible, in the model they represent a very good approximation. The averaged equations are of a special polynomial type, thus the formalism of Groebner bases is one possibility to detect *all* fixed points. Of particular interest are fixed points in the physical regime. Varying the source strengths - they serve as control parameters - bifurcations originating from these fixed points are expected. We have tested this in the model: The formalism of Groebner bases let us find 12 fixed points. However, in the chosen range of the control parameters there exists only one physically important fixed point. This can become unstable undergoing a Hopf bifurcation. Beyond the instability strong variations of the pollutant concentrations occur. Use of the center manifold theorem can reduce the dimension of the problem considerably.

1   Introduction

The concentrations of pollutants in the troposphere depend on the atmospheric currents (described by the Navier Stokes equations) and the chemical reactions[1] between the pollutants (described by chemical reaction equations).

There are several codes computing the pollutant concentrations on large scales[1][2] and local scales[3].

These codes allow a forecast for a given situation by integrating the system of combined partial and ordinary differential equations. However, if we ask under which cicumstances large changes of pollutant concentrations can occur we have a problem: we cannot simply use these programs because of the vast possibilities of initial values in a huge phase space. Instead we have to try to get insight into the possible solutions to these equations. In full generality this is a formidable task and therefore it is reasonable to discuss various parts separately. From a physical point of view this is allowed in particular for the chemical part: If, due to weather conditions, the mixing of the pollutants is nearly complete in a certain region the chemical reaction equations govern the properties of the solutions.

---

[1] Moreover the concentrations depend on cloud formation, humidity, ice etc.

At first sight the chemical part seems to represent an unsolvable problem in itself because there exist several hundreds of pollutants. However, in a pioneering work[4] it was shown that many pollutants being chemically very similar can be approximated by a typical representative. In this way the number of *relevant* pollutants is reduced in a consistent way to 63 with 157 reactions taking place between them. The resulting equations are solved by several codes[5] for given initial conditions[6].

Still we have the same problem as before: we cannot get an overview over the solutions of the chemical reaction equations by going through all possible initial values. But because of their simple form it is tempting to apply tools of nonlinear dynamics for getting insight into the structure and instabilities of their solutions. We will suggest these tools and show that they work in a simple model: This model is described in detail elsewhere[7][8][9]. It consists of 6 pollutants, NO_2, O_3, CO, NO, OH and HO_2, it's phase space dimension is 6. 13 typical reactions are included, two typical pollutant sources, NO and CO, are incorporated.

In Sect. 2 we will present general properties of the reaction equations. Due to the very low concentration in air the probability of an interaction between 3 pollutants is negligible and therefore the reaction equations are at most quadratic in the concentrations of the pollutants. Moreover the equations are dissipative and explicitly time dependent. In Sect. 3 we discuss the averaging over 24 hours in general and compare the exact solutions of our model with the corresponding ones obtained from the averaged equations. We find very good agreement. In Sect. 4 we discuss the time averaged, autonomous equations. Since they are polynomial equations of second order one can in principle find *all* fixed points by applying the theory of Groebner [10][11] bases. We discuss the meaning of fixed points in the *physical* part of the phase space[2] in the *negative* part of the phase space[3] and in the *complex* part of the phase space[4]. In our model the scheme of the Grobner bases works: We detect 12 fixed points. For the chosen strength of the pollutant sources we get only *one* fixed point in the physical part of the phase space. This fixed point can become unstable and undergo a Hopf bifurcation leading to periodic - and what is more important - strongly varying concentrations. Indeed the pollutant concentrations can change by more than one order of magnitude in a few days. In Sect. 5 we apply the center manifold[12] formalism. In zeroth order it is the tangent space approximation leading to simple relations that can be interpreted as conservation laws. We show that in our model calculations this approximation is a good one. We obtain 4 'conservation laws' reducing the dimension of our model from 6 to 2. This explains why we do not find chaotic solutions in our calculations. More important it suggests that the center manifold theorem will play an important role when

---

[2] all concentrations are nonnegative.
[3] At least one pollutant concentration is negative.
[4] The imaginary part of at least one pollutant concentration is nonzero.

looking for instabilities of the reaction equations. The conclusion ends the paper.

2 The Reaction Equations of Pollutants

The chemical reaction equations describing the change of the pollutant concentrations c_n can always be written in the form

$$\dot{\mathbf{c}} = \mathbf{f}$$
$$with \tag{1}$$
$$f_n = -d_n \cdot c_n + p_n, \ n = 1...N, \ N = \text{number of pollutants}$$

Here d_n and p_n are the decay rates and the production terms respectively. Both terms have the property

$$d_n = d_n(c_1...c_N) \geq 0 \text{ in the physical part of phase space} \tag{2}$$
$$p_n = p_n(c_1...c_N) \geq 0 \text{ in the physical part of phase space}$$

The ratios of the decay rates are very different, up to 10^{10}! This means that the equations contain many different timescales. Solving these equations requires special mathematical routines. Physically important is the fact that nonstationary solutions change their time scale. This is already well known from the brusselator[13].

Because of the low concentrations at most two pollutants participate in a chemical reaction. Therefore the d_n are at most linear and the p_n at most quadratic functions of the c_i. Moreover there are no autocatalytic reactions between the pollutants. Therefore

$$\frac{\partial}{\partial c_n} p_n = 0 \tag{3}$$

and thus

$$\nabla \cdot \dot{\mathbf{c}} < 0 \tag{4}$$

This means that the reaction equations are strictly dissipative.

Since pollutants decay by and by their concentrations approach zero unless there are pollutant source terms. The larger the strengths of these terms the larger the contribution of the nonlinear terms. Thus the strengths of the sources act as control parameters of the system.

In passing we mention another control parameter. It is the height in which the chemical reactions take place. In fact, with increasing height all the processes depending quadratically on the concentrations become less important: Therefore with increasing height the system becomes more linear.

The differential equations are not autonomous. They are explicitly time dependent since the photolytic reactions have a daily and a yearly period[5].

In this and the following sections we compare our results and suggestions with a simple model[7][8][9]. This is a strongly simplified version of the chemical reaction equations for pollutants in the troposphere. Taken into account are 6 typical pollutants and radicals only: NO, NO_2, O_3, CO, OH, HO_2. 13 typical reactions are included. Two typical source terms, CO and NO are included as well.

3 Time Averages

The explicit time dependence of the reaction equations consists of a short period (one day) and a long period due to the seasonal changes (one year). First we can discard of the one year period since no atmospheric current can be neglected for one year. Instead we treat the season as a parameter.

An ordinary differential equation having a *periodic* time dependence with period T can be transformed into an autonomous difference equation by constructing Poincaré's map. The resulting difference equation, however, can only be represented numerically and therefore is not easily accessible. In our case there is an experimental difficulty too, since it is awkward to take into account data measured solely at certain moments.

Instead of using the method of Poincaré we proceed here in a different manner by looking at the averaged concentrations

$$\bar{c}_n(t) = \frac{1}{T} \int_{-T/2}^{T/2} c_n(t + t')dt' \tag{5}$$

A good approximation of the $\bar{c}_n(t)$ can be computed easier than the $c_n(t)$: Assuming that the relevant time scale is much larger than 24 hours the differential equation can be written

$$\frac{d}{dt}\mathbf{c} = \mathbf{f}(\mathbf{c}, t/\epsilon) \tag{6}$$

and making a time scale transformation

$$\tau = t/\epsilon \tag{7}$$

we obtain

$$\frac{d}{d\tau}\mathbf{c} = \epsilon \mathbf{f}(\mathbf{c}, \tau) \tag{8}$$

---

[5] There are other time dependences as well: e.g. the temperature, pressure and pollutant source strengths fluctuate during the day. These fluctuations can be averaged over in the same manner as those of the photolytic reactions, cf next section.

This is a well known form that has been discussed extensively in the literature[12][14][15]. The following statements can then easily be proven:

$$c(\epsilon\tau) = y(\epsilon\tau) + \epsilon w(y(\epsilon\tau), \tau) \qquad (9)$$

Here y is the solution of the *averaged* differential equation

$$\frac{d}{d\tau}y = \epsilon\bar{f}(y)$$

$$with \qquad (10)$$

$$\bar{f}(y) = \frac{1}{T}\int_{-T/2}^{T/2} f(y, t'/\epsilon)dt'$$

$$w(y, \tau) = f(y, \tau) - \bar{f}(y)$$

From this it is clear that

$$\bar{c} = y + \mathcal{O}(\epsilon^2) \qquad (11)$$

Therefore, up to an error of $\mathcal{O}(\epsilon^2)$ we can compute $\bar{c}$ from

$$\frac{d}{dt}\bar{c} = \bar{f}(\bar{c}) + \mathcal{O}(\epsilon^2), \quad \bar{f}(y) = \frac{1}{T}\int_{-T/2}^{T/2} f(y, t'/\epsilon)dt' \qquad (12)$$

In contrast to the difference equation of the Poincaré map this equation has a simple analytic structure: $\bar{f}$ agrees with the original f, the only difference being that the time dependent reaction strengths are replaced by their mean value. For all further investigations we will use eq.(12). In our model we have compared the two $\bar{c}$, one, obtained from eq.(1) together with eq.(5), the other, computed directly from eq.(12). The agreement is very good indeed, cf Fig.1 and Fig.2.

For long times $t \gg \mathcal{O}(1)$ the solutions of eq.(12) may deviate exponentially fast from the exact results[6]. However, in our case this effect is irrelevant. Indeed, for those long times the results become spurious anyway because the dynamics of the troposphere is no longer negligible.

4   Fixed Points and Attractors

In this section we discuss the solutions of eq.(12) in the phase space of concentrations.

We are interested in trajectories of *attractors* because they are representative for a finite volume of phase space - the basin of attraction.

Fixed points are of particular interest. Many attractors either contain an unstable fixed point[7] in their closure or at least contain one in the closure of their basin of attraction. Fixed points are solutions of

$$\bar{f} = 0 \qquad (13)$$

---

[6] This is a general feature of averaging methods[12][14][15].
[7] cf the definition of a strange attractor[12].

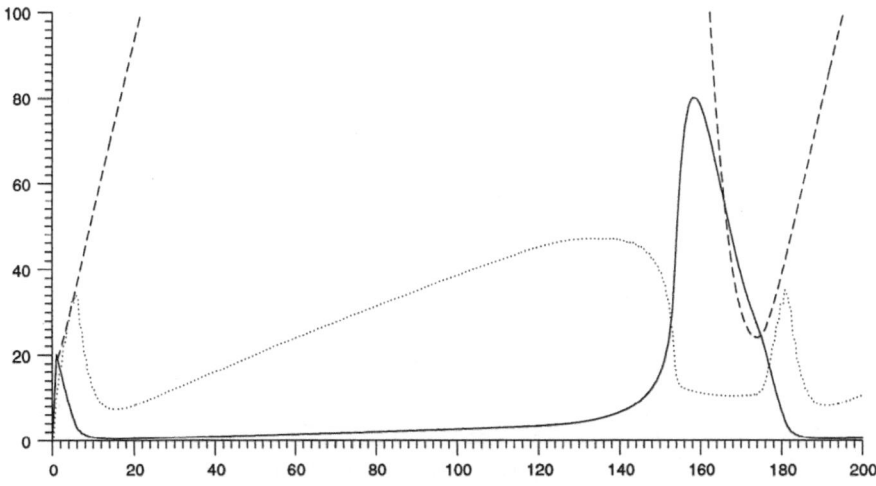

Fig. 1. averaged concentrations of pollutants computed in our model using eq.(1) and eq.(5). Horizontal axis: time in days. Vertical axis: concentrations. Solid line: $[O_3]$ in units of ppb, dashed line: $[CO]$ in units of ppb, dotted line: $[NO_2]$ in units of 10^{-1}ppb. Strength of pollutant sources: $CO_{source} = 13.0 \cdot 10^5 cm^{-3} sec^{-1}$, $NO_{source} = 13.0 \cdot 10^5 cm^{-3} sec^{-1}$.

and can lie in the *physical, negative* or *complex* part of the phase space.

An extended attractor (i.e. one that is not a fixed point) in real space cannot have a basin of attraction in the complex space. This is shown in appendix A. Consequently fixed points located in the *complex* part of the phase space can neither be in the closure of these attractors nor in the closure of their basin of attraction. Such fixed points are rather uninteresting as long as they do not penetrate into the real part of the phase space due to a change of the control parameters.

An extended attractor in the *negative* part of the phase space becomes unstable as soon as it touches the physical part of the phase space. The reason is that for both differential equations, eq.(1) and eq.(12), trajectories once being in the physical part remain there forever. Consequently fixed points in the negative part of the phase space cannot be in the closure of an attractor located in the physical part of the phase space. But unstable ones can be in the closure of it's basin of attraction. Therefore one can start trajectories in the neighborhood of these fixed points to get to a physically relevant attractor. Moreover - as in the previous case - the fixed points can migrate into the physical part of the phase space.

Obviously the most interesting fixed points are those in the physical part of the phase space. For small pollutant source strength all the pollutant concentrations are small, the influence of the nonlinear terms in eq.(12) is small

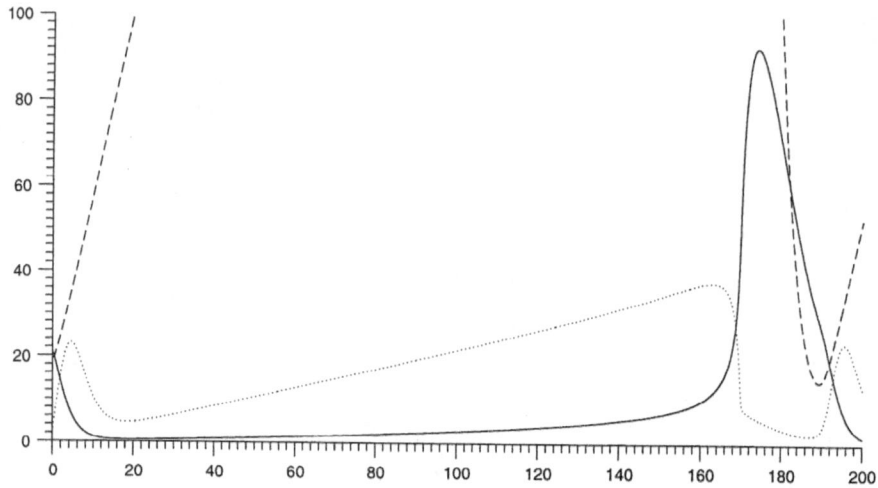

Fig. 2. averaged concentrations of pollutants computed in our model using eq.(12). Notations and source strengths as in Fig.1.

and we expect one stable fixed point. With increasing strength the fixed point may become unstable and new - simple - attractors occur[16]. These again can become unstable leading to more complicated attractors[17][18]. However we may detect all these by carefully following the bifurcation diagram. Besides, unstable fixed points located in the physical part of the phase space can be in the closure of an attractor also located in the physical part of the phase space. And they can be in the closure of it's basin of attraction.

The above considerations demonstrate clearly that the knowledge of the fixed points yields valuable information about the attractors of eq.(12). In general it is difficult to find the fixed points in high dimensional spaces. However, the set of coupled quadratic equations, eq.(13), belongs to the class of coupled polynomial equations. This class is special because *all* fixed points can be obtained *exactly* by applying the formalism of Groebner[10][11] bases[8].

We applied that formalism successfully to our model (in which the phase space dimension is 6). For simplicity we kept the CO source strength fixed at

$$CO_{source} = 13.0 \cdot 10^5 cm^{-3} sec^{-1} \tag{14}$$

and kept the NO source strength in the range

$$5.0 \cdot 10^5 cm^{-3} sec^{-1} \leq NO_{source} \leq 13.0 \cdot 10^5 cm^{-3} sec^{-1} \tag{15}$$

---

[8] To do so one has to use an algebraic manipulation program. For our computations we used the package maple[20]. Unfortunately there do not exist good estimates, how much time the determination of the fixed points will cost. This is not too surprising though since a *linear* system will have just *one* solution while N coupled quadratic equations can have up to 2^N solutions.

We got the following results: For $NO_{source} < 8.98 \cdot 10^5 cm^{-3} sec^{-1}$ there exist 12 fixed points, 6 are complex, 5 are real but contain negative components and only one is in the physical part of the phase space. For $NO \geq 8.98 \cdot 10^5 cm^{-3} sec^{-1}$ there exist 12 fixed points as well, 8 are complex, 3 are real but contain negative components and again only one is in the physical part of the phase space. The latter fixed point is stable for $NO_{source} < 7.15 \cdot 10^5 cm^{-3} sec^{-1}$, cf Fig.3 and Fig.4. At $NO_{source} = 7.15 \cdot 10^5 cm^{-3}$ it

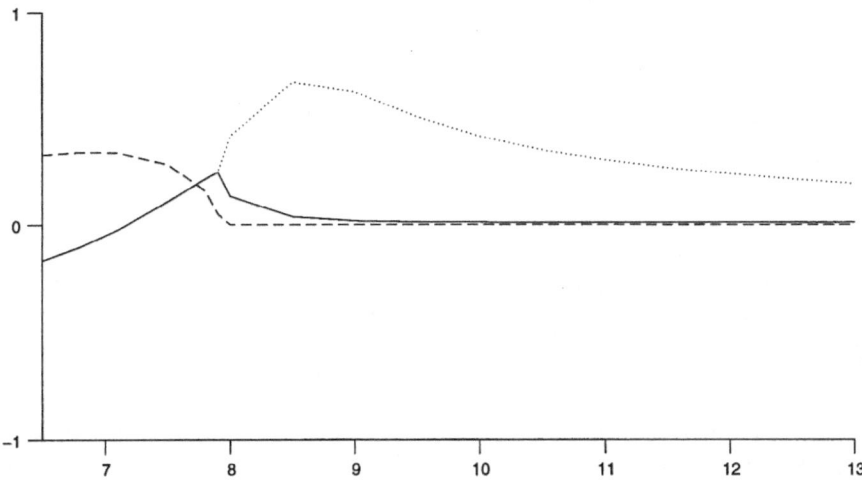

Fig. 3. The two largest (with respect to the real part) eigenvalues of the fixed point matrix. Horizontal axis: NO_{source} strength in units of $10^5 cm^{-3} sec^{-1}$. Scale of the vertical axis: $10^{-5} sec^{-1}$. Solid line: real part of the first eigenvalue. Dotted line: real part of the second eigenvalue. Dashed line: imaginary part of the first eigenvalue. CO_{source} strength as in Fig.1.

becomes unstable and undergoes a Hopf bifurcation, cf Fig.3. Above that value the concentrations $\bar{c}_i$ are periodic, cf Fig.5. A further bifurcation is not observed in the chosen parameter range.

Although the periods are rather long ($\mathcal{O}(100\,days)$) and the concentrations vary slowly over a wide range yet they also change rather abruptly (within a few days) by more than an order of magnitude, cf Fig.5 and Fig.2. Such a behavior is characteristic for solutions of stiff differential equations[19]. The relevance of the periodic solutions is the following: In reality we cannot observe a full period because of the varying weather conditions, but we can see 'windows' of it. In particular the rather rapid change within a few days is a dynamics of the chemical reaction equations that should be oservable.

In the parameter range under consideration all trajectories starting in the neighborhood of negative fixed points either moved to infinity or to the

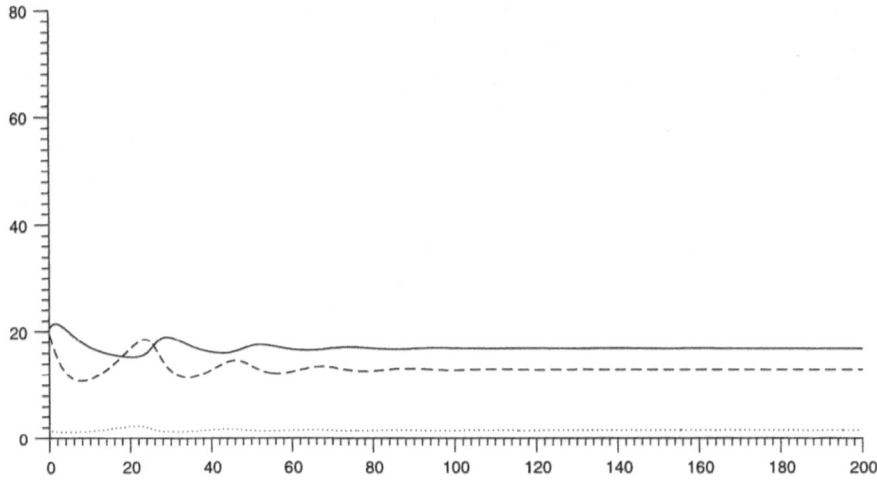

Fig. 4. averaged concentrations of pollutants computed in our model using eq.(12). Notations as in Fig.1. Source strengths: $CO_{source} = 13.0 \cdot 10^5 cm^{-3} sec^{-1}$, $NO_{source} = 7.0 \cdot 10^5 cm^{-3} sec^{-1}$.

stable attractor (fixed point or periodic orbit) in the physical part of the phase space. This is a strong indication that it is the only attractor in this regime[9].

5 Conservation Laws and Relevant Dimension

We can ask the question - important also for chemists - are there linear combinations of the concentrations that do not depend on time? It is clear that if they exist they must exist in particular at a fixed point in the physical part of the phase space. But at a fixed point the answer is simple and affirmative: Let the eigenvectors of the adjoint fixed point matrix be $\mathbf{a}^{(l)}$ (with eigenvalue $\lambda^{(l)}$) we obtain the relations[10]

$$\mathbf{a}^{(l)} \cdot \bar{\mathbf{c}} = const, \text{for } \Re\{\lambda^{(l)}\} < 0 \qquad (16)$$

These equations do not hold at once but after the time $t_{ini} \approx \mathcal{O}((\Re\{\lambda\})^{-1})$ which means that the lth equation is the less strict the larger $\Re\{\lambda_l\}$ is. In our

---

[9] We also did brute force calculations by computing trajectories for many different initial values. they always ended on the forementioned attractor.

[10] When taking a new coordinate system with its origin at the fixed point and choosing as a subspace the space spanned by the stable directions then the projection of the trajectory on this subspace approaches 0 as long as the trajectory remains in the neighborhood of the fixed point.

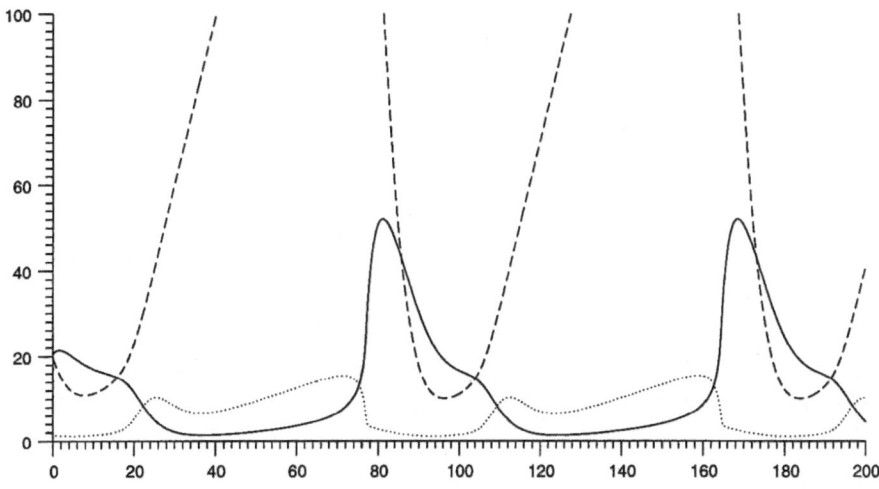

Fig. 5. averaged concentrations of pollutants computed in our model using eq.(12). Notations as in Fig.1. Source strengths: $CO_{source} = 13.0 \cdot 10^5 cm^{-3} sec^{-1}$, $NO_{source} = 7.2 \cdot 10^5 cm^{-3} sec^{-1}$.

model we find the following: There are 4 eigenvalues $\lambda^{(l)}$ having negative real parts. In the whole range considered here t_{ini} is $\mathcal{O}(1\,day)$ for the eigenvalue with largest real part.

Note that the $a^{(l)}$ depend on the control parameters and the components of the $a^{(l)}$ need not be integers at all (as the chemists would like).

The relations of eq.(16) can be regarded as conservation laws if there is an attractor in a sufficiently small neighborhood of the fixed point. We tested the validity of eq.(16) in our model by inserting eq.(16) in eq.(12). The result is plotted in Fig.6 and the agreement with Fig.2 is very good indeed.

If the conservation laws are valid and their number is L the relevant dimension of the problem is reduced by L. That means for our model that the relevant dimension is 2. Because of this it is not surprising that chaotic behavior could not be found.

What we have done here is nothing but an application of the *center manifold theorem* in zero[th] order, the so called *tangent space approximation*[12]. The fact that already this simple approximation leads to very good results in our model gives rise to the supposition that the *center manifold theorem* is an important tool in the context of reaction equations like eq.(12). This point will be discussed in a forthcoming paper.

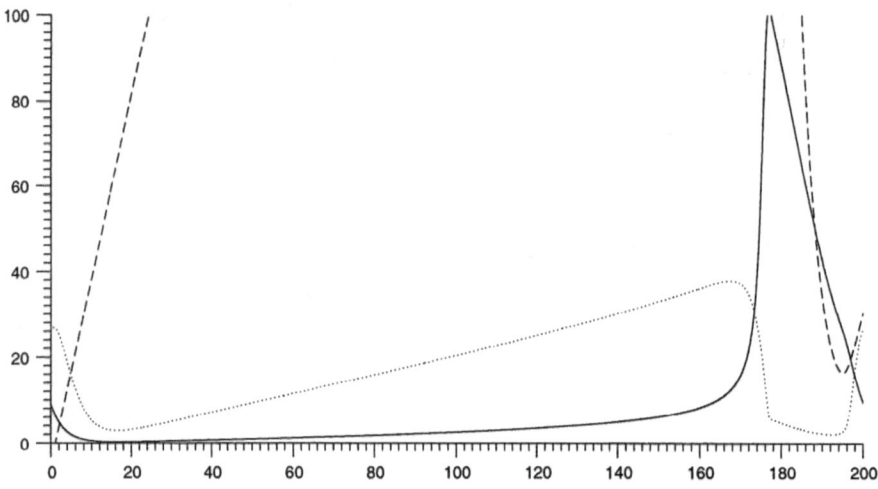

Fig. 6. averaged concentrations of pollutants computed in our model using eq.(12) together with eq.(16). Notations and source strengths as in Fig.1 and Fig.2.

6 Conclusion

The reaction equations of pollutants in the troposphere present a high dimensional nonlinear problem. In fact, the number of pollutants is large (about 60) and the reactions contain terms that are quadratic in the pollutant concentrations. Therefore it is not at all trivial to obtain the properties of the solutions. Nevertheless one needs to know these properties: Results from our model calculations let us expect that the concentrations (averaged over 24 hours) need not tend to a stationary limit. Rather, beyond an instability they may fluctuate strongly on a scale of a few days.

To get an insight into the possible solutions of the reaction equations we suggest several procedures: i) averaging over times of 24 hours to remove the explicit time dependence, ii) introduction of Groebner bases to obtain *all* fixed points of the system, iii) investigation of the fixed points to trace the instabilities, bifurcations and the attractors, iv) application of the center manifold theorem to obtain the relevant dimension.

To test the feasibility and the results that can be inferred from them we applied all these procedures to our simple model of 6 pollutants. For this model we find: i) averaging is a very good approximation, ii) the application of Groebner bases does not present a problem. We find 12 fixed point but only one of them has positive components. This particular fixed point can become unstable and a Hopf bifurcation occurs beyond which the pollutant concentrations vary strongly within a few days. We present arguments as to why no other attractors are to be expected in the investigated parameter regime. iv) Already in the simplest approximation of the center manifolds, the tangent space approximation, we can show that the relevant dimension is much smaller than the dimension of the reaction equations.

The ease and relative simplicity with which we could apply the suggested procedures to our model let us believe that the same procedures can be applied to a much more realistic model. Work in this direction is in progress.

Acknowledgments The author thanks J. Bene, G. Eilenberger, C. Grebogi, D. Poppe, Ch. Schuette, G. Szabó and T. Tél for useful discussions. He thanks B. Steffen for pointing his attention to the formalism of Groebner bases.

A Appendix

In this appendix we show: real attractors that are not fixed points, have no basin of attraction in the complex space.

Let us first assume that the attractor has at least one positive Lyapunov exponent. Let us assume furthermore that a complex trajectory $\mathbf{z}(t)$ has approached the attractor at time t_0 up to a small distance $\epsilon/2$. Then there is a point on the attractor $\mathbf{x}(t_0)$ with

$$| \delta\mathbf{z}(t_0) | = | \mathbf{z}(t_0) - \mathbf{x}(t_0) | < \epsilon \tag{17}$$

and we can compute $\delta\mathbf{z}(t)$ from the linearized differential equation

$$\dot{\delta\mathbf{z}} = \frac{\partial}{\partial\mathbf{x}}\mathbf{f}(\mathbf{x}(t) \cdot \delta\mathbf{z} + \text{higher order terms} \tag{18}$$

Now $\mathbf{f}$ is real and therefore we get for the imaginary part $\delta\mathbf{y}$

$$\dot{\delta\mathbf{y}} = \frac{\partial}{\partial\mathbf{x}}\mathbf{f}(\mathbf{x}(t) \cdot \delta\mathbf{y} + \text{higher order terms} \tag{19}$$

But we know that for nearly all initial conditions $| \delta\mathbf{y} |$ will increase $\propto e^{\lambda_{max}t}$ where λ_{max} is the largest Lyapunov exponent. Thus there is no complex basin of attraction.

If the attractor is a stable periodic orbit in real space then the largest Lyapunov exponent is zero. Let the period of the orbit be T. Integration over $\Delta t = T$ yields the Poincaré map. Let the point $\mathbf{x}_0$ be on the periodic orbit and let us take new coordinates with origin at $\mathbf{x}_0$ and the first vector of the coordinate system be the tangent vector of the orbit at $\mathbf{x}_0$. Let us denote the new coordinates $\mathbf{u}$ with components u_i and expand the map around the origin. Then we obtain

$$u_1^{(n+1)} = u_1^{(n)} + b_2 \cdot u_1^{(n)} \cdot u_2^{(n)} + \dots$$
$$u_2^{(n+1)} = \dots$$
$$\dot{}$$
$$u_N^{(n+1)} = \dots$$

Note that in the $u_1^{(n+1)}$ component there is no contribution $[u_1^{(n)}]^m$ with $m > 1$. We know that

$$| u_i^{(n+1)} | = \alpha | u_i^{(n)} |, \ \alpha < 1, \ i > 1;$$

if the coordinate system is chosen appropriately. Performing analytic continuation we get for the essential part

$$z_1^{(n+1)} = z_1^{(n)} + ... \tag{20}$$

Therefore, if $\Im\{z_1^0\} \neq 0$ then $\lim_{n\to\infty} \Im\{z_1^{(n)}\} \neq 0$ and consequently there does not exist a basin of attraction in the complex space.

References

1. Hass H. (1991): Mitteilungen aus dem Institut für Geophysik und Meteorologie der Universität zu Köln, editor Ebel A. et.al., No. 83, Cologne, also refer to: http://www.uni-koeln.de/math-nat-fak/geomet/eurad
2. Memmesheimer M. and Bock H.J. (1995): Mitteilungen aus dem Institut für Geophysik und Meteorologie der Universität zu Köln, editor Ebel A. et.al., No. 105, Cologne
3. see e.g. the KAMM/DRAIS model, described in: Adrian G. and Fiedler F. (1991): Beitr. Phys. Atmosph. **64**, 27
4. Stockwell W.R. (1990): J. Geophys. Res. **95**, 16334
5. Kuhn M et.al. (1997): to be published in Atmos. Env.
6. Elbern H., Schmidt H. and Ebel A. (1997): To appear in J. Geophys. Res.
7. Poppe D. and Lustfeld H. (1996): J. Geophys. Res. **101**, 14373
8. Krol M. and Poppe D. (1997): to be published in J. Atmos. Chem.
9. Lustfeld H. (1997): IFF Ferienkurs, **28**, F4.1 , Forschungszentrum Jülich
10. Davenport J.H. et.al. (1988) : Computer Algebra, Academic Press, London
11. Becker T. and Weispfennig V. (1993): Gröbner bases - a computational approach to commutative algebra, Springer Verlag, Berlin
12. Guckenheimer J. and Holmes P. (1983): nonlinear oscillations, dynamical systems, and bifurcations of vector fields, Springer Verlag, Berlin
13. Haken H. (1983): synergetics, an introduction, Springer Verlag, Berlin
14. Verhulst F. (1996): nonlinear differential equations and dynamical systems, Springer Verlag, Berlin
15. Rand R.H. and Armbruster D. (1987): perturbation methods, bifurcation theory and computer algebra, Springer Verlag, Berlin
16. Nicolis G. (1995): introduction to nonlinear science, Cambridge university press, Cambridge
17. Iooss G. and Joseph D.D. (1990): elementary stability and bifurcation theory, Springer Verlag, Berlin
18. Newhouse S.E., Ruelle D. and Takens F. (1978): Comm. Math. Phys. **64**, 35
19. Deuflhard P. and Bornemann F (1994): numerische Mathematik II, DeGruyter Verlag, Berlin
20. Redfern D. (1994): Maple Handbook (MAPLE V RELEASE 3), Springer Verlag, Berlin

Hydrodynamic Singularities

Jens Eggers

Universität Gesamthochschule Essen, Fachbereich Physik,
45117 Essen, Germany

1 Introduction

The equations of hydrodynamics are nonlinear partial differential equations, so they include the possibility of forming singularities in finite time. This means that hydrodynamic fields become infinite or at least non-smooth at points, lines, or even fractal objects. This mathematical possibility is the price one has to pay for the enormous conceptual simplification a continuum theory brings for the description of a many-particle system. Near singularities, the microscopic structure re-emerges, as the flow changes over arbitrarily small distances. Eventually, the singularity is cut off by some microscopic length scale such as the distance between molecules.

The most fundamental question is whether the microscopic structure becomes relevant for features of the flow much larger than the microscopic ones. If this is the case, the continuum description is no longer self-consistent, but has to be supplemented by microscopic information. There are some well-known cases where singularities are artifacts of neglecting diffusive effects like viscosity in the hydrodynamic equations, and there is no more smoothing on small scales. Examples are the singularities widely believed to be produced by the three-dimensional Euler equation (Majda (1991)) or those of Hele-Shaw flows at zero surface tension (Dai, Kadanoff, and Zhou (1991)). There does not seem to be a direct correspondence between the intricate structure of spatial singularities produced by these equations and real flows. For a realistic description of experiments, some tiny amount of viscosity or surface tension needs to be added.

The reason hydrodynamic singularities have nevertheless attracted considerable attention in recent years is the observation that certain singularities have direct physical significance and are not a consequence of inadequacies of the equations. In particular, free-surface flows exhibit a rich variety of experimentally observable singularities, which are responsible for phenomena like the breakup of jets, coalescence of drops, and bubble entrainment.

2 Physical singularities

We attempt to divide these physical singularities into two categories.

2.1 Dynamical singularities

These are singularities which are confined to a point in time. Usually, they are associated with topological transitions like the breakup of a piece of fluid into two pieces or the joining of two pieces into one. For example, the breakup of a viscous jet of fluid is driven by surface tension, which tries to reduce the surface area by diminishing the radius of the jet (Eggers (1997a)). Inertial forces constrain the motion to become more and more localized, since smaller and smaller amounts of fluid have to move. This causes the jet to break at a point in finite time. Only the smoothing effect of viscosity prevents infinite gradients from occurring before the local radius goes to zero.

As the local radius of a fluid thread becomes smaller and smaller during pinching, it inevitably reaches a microscopic scale ℓ_{micro} where the equations cease to be valid. For thread diameters between 10 and 100 nm, short-ranged van der Waals forces come into play, and for even smaller diameters the concept of a sharp interface will certainly loose its meaning. Moreover, a stability analysis (Brenner, Shi, and Nagel (1994)) shows that the pinching thread is very sensitive to thermal fluctuations. This makes even threads of micron size unstable, and leads to a new structure of nested singularities, driven by microscopic fluctuations. After the thread has dissolved, new surfaces form on either side, whose rapid retraction is again governed by the Navier-Stokes equation, but with a new topology. It would appear as if the continuation to the new Navier-Stokes problem should necessarily include the microscopic length ℓ_{micro} at which the thread broke. This is however not the case (Eggers (1997a)), as long as one is looking at scales much larger than ℓ_{micro}. Namely, the final stages occur on very small spatial and temporal scales, and do not affect the flow at a finite distance away from breakup. Thus the outer part of the solution can be used as a boundary condition for the new problem after breakup. A closer analysis reveals that this is sufficient to determine the new solution completely. This means that the dynamics very quickly "forget" the microscopic details of breaking, thus making a consistent hydrodynamic theory of the topological transition possible.

2.2 Persistent singularities

The other important category of singularities are those which exist for a period of time, being either stationary, or moving about in space like the classical example of a shock wave. At finite viscosity, a shock wave is not a true singularity, but maintains a finite width determined by the ratio of the viscosity and the shock strength. However, the width δ of the shock wave is typically of the same order as the mean free path of the gas it is moving in. What is important is thus the fact that the solution remains consistent as δ goes to zero. Indeed, the dissipation inside the shock remains finite in this limit, so on scales much larger than δ the flow field is the same as if δ were zero.

In the realm of free-surface flows, a beautiful example of a stationary singularity has been discovered recently on the surface of the viscous flow between two counter-rotating cylinders (Joseph et al. (1991)). As seen in Fig.1, two counter-rotating cylinders are submerged in a container filled with a very viscous fluid. The relative strength of viscous forces and surface tension is measured by the capillary number

$$Ca = \frac{\eta \Omega r_c}{\gamma},$$ (1)

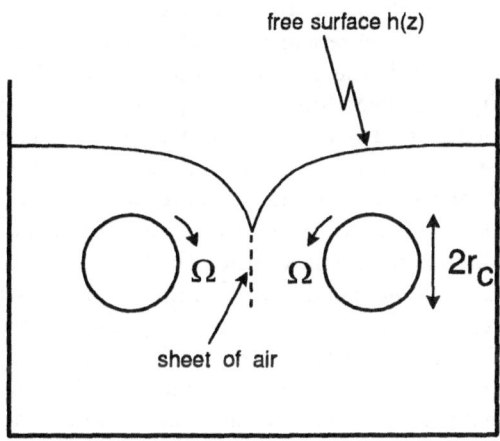

Fig. 1. Sketch of the two-roller apparatus. At a critical capillary number, a pointed cusp appears. At a second, much higher capillary number, a sheet emanates from the cusp.

where γ is the coefficient of surface tension, η the viscosity, and Ω and r_c are the cylinders' rotation speed and radius, respectively. At a critical capillary number, the surface appears to form a cusp, as indicated in Fig.1. Since the flow remains two-dimensional, this corresponds to a line of singularities of the surface. Assuming that it is a true cusp, the original authors analyzed the flow using a solution due to Richardson (1968). This local solution leads to a logarithmic divergence of the dissipation, and thus cannot be consistent with continuum theory or a finite driving power. It was therefore suggested that the divergence is regularized by some microscopic scale. Since the singularity is very weak, the dissipation at the tip would still be small for realistic values of the microscopic length. However, this explanation for eliminating the logarithmic divergence would mean that there is a dependence of the macroscopic flow on a microscopic parameter.

Faced with this possibility, the problem was reanalyzed by Jeong and Moffatt (1992), who solved the Stokes equation exactly, making the simplifying assumption that the two rollers can be represented by a single dipole. The remarkable result of their calculation is that the radius of curvature R at the tip is in fact finite, but exponentially small in the capillary number:

$$R = R_0 \exp\left[-32\pi Ca\right]. \tag{2}$$

For realistic values of the capillary number, this gives radii of curvature far below any physical scale, but is still able to regularize the logarithmic singularity. Thus one finds a finite value of the energy dissipation, making the macroscopic flow independent of the microscopic parameters of the fluid.

For practical purposes, the theoretical value of R is far too small to be realistic. Rather, it most likely is the gas above the fluid which will set the value of R, and this physical effect has been neglected so far. Because of the no-slip boundary condition, gas will be forced into the narrow channel formed by the cusp. A simple calculation based on lubrication theory shows that for $R = 0$ this will lead to a diverging pressure at the tip of the cusp. Thus the gas will force the channel to widen to a finite radius, at which the gas pressure is comparable to the pressure inside the fluid.

It is worth noting that the independence from microscopic parameters is by no means self-evident. A famous counterexample is that of a moving contact line, which occurs for example when a circular drop is allowed to spread on the surface of a table (de Gennes (1985), Brenner and Bertozzi (1993)). Using kinematic arguments alone, one shows that there will be a logarithmic singularity of the energy dissipation at the moving contact line. There is a vast literature on this problem, dealing either with possible mechanisms for a microscopic cutoff, or with a consistent mathematical description of the resulting macroscopic dynamics. The important point to note is that continuum mechanics alone cannot resolve the problem in a self-consistent fashion. It would predict that the spreading of the drop is stalled, contradicting observation.

3 Scaling

The central assumption behind the description of singularities is that of locality. Their spatial and temporal scale becomes arbitrarily small, so that the dynamics should be removed from the large-scale features of the flow. However, consistency between the singular and the large scale dynamics has to be assured by matching conditions between the inner and the outer problems.

A second, closely related assumption is that of self-similarity of the singular flow, which seems a natural concept for a class of problems which lack a typical length-scale. In the case of time-dependent singularities it means that the interface shapes at different times can be mapped on one another by an appropriate rescaling of the axes. For example, the surface profile of the pinch singularity when a fluid thread breaks is (Eggers (1997a))

$$h(z,t) = \frac{\gamma}{\eta}|t_0 - t|\phi\left(\frac{\rho^{1/2}(z - z_0)}{(\eta|t_0 - t|)^{1/2}}\right),$$ (3)

where t_0 and z_0 are the temporal and spatial position where the fluid breaks. Remarkably, the scaling function ϕ is universal, independent of the type of fluid or of initial conditions. For the free surface cusp of Fig.1, the shape of the interface has the scaling form

$$h(z) = h_0 + R^{1/2}f\left(\frac{z}{R^{3/4}}\right),$$ (4)

where $(0, h_0)$ is the position of the cusps' tip.

Naturally, it is of particular interest to compute the scaling exponents. No general understanding of what selects a particular set of exponents exists. Usually, local solutions like (3) or (4) are not exact solutions of the equations, but only balance certain terms that are asymptotically dominant. In the case of the pinch singularity these terms belong to surface tension, viscous, and inertial forces. Knowing that, dimensional analysis alone leads to the correct power laws. However, there are cases like the pinching of a very viscous thread (Papageorgiou (1995)), where the exponents are fixed to irrational values by other consistency requirements.

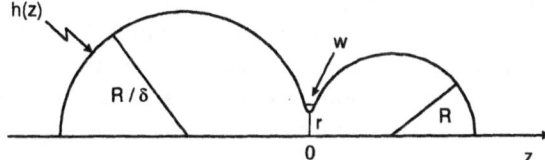

Fig. 2. Coalescence of two drops of radius R and R/δ. Asymptotically, the width w of the bridge between the drops is much smaller than the minimum radius r.

What could possibly keep the local motion from behaving in a self-similar fashion? Two examples for broken self-similarity come from the *coalescence* of two drops. When two drops meet a point, surface tension will try to merge them into one drop, so the minimum radius r of the bridge between the drops will increase, (Fig.2). As long as the bridge is very small, inertial effects can be neglected and the problem is initially governed by the Stokes equation. For geometrical reasons the width of the bridge w is much smaller than the minimum radius r, so dimensionally

$$r(t) = \frac{\gamma}{\eta} t \tag{5}$$

is the only possible power law time dependence of the radius. However, a closer analysis reveals (Eggers, Lister, and Stone (1997)) that (5) has to be corrected by a logarithmic factor $\log(r/R)$, where R is the initial radius of a drop. The reason for this change of the time dependence lies in the long-ranged character of the Stokes equation, which couples very disparate length scales. Hence the width of the bridge cannot be neglected, but enters as a logarithmic factor $\log(w/r) \sim \log(r/R)$.

Another mechanism for broken self-similarity is observed for the same problem, but for much smaller initial size of the drops. In that case the surface-tension-driven motion will first occur on the surface alone, rather than being able to drive a flow in the interior. The equation of motion for this surface diffusion was first given by Mullins (1965). Based on simple scaling arguments, Mullins (1965) gave

$$r(t) \sim t^{1/6} \tag{6}$$

as the evolution of the radius, but this result could not be corroborated by his own numerical simulations. To understand this failure, one needs to take a closer look at the dynamics near the rising bridge (Eggers (1997b)). In contrast to the viscous flow problem, as the gap between the two spheres fills, a bulge of material forms just above the minimum. Eventually, it grows enough for both sides to touch, forming a void inside the material. Of course, at the point of touching a new singularity occurs and the topology of the problem has changed. The naive assumption of a single continuous evolution, underlying (6), is thus incorrect. Self-similarity can at best exist in a discrete sense.

4 Birth of new structures

If one drives the two-roller apparatus of Fig.1 much harder than necessary for the formation of a cusp, a second transition occurs, above which a thin sheet of air emanates from the cusp, and is drawn continuously into the fluid. This sheet is stable in time, but undergoes a three-dimensional instability at its lower end, where it decays into a curtain of tiny droplets. This provides a general mechanism for the entrainment of bubbles into a flow. The existence of the sheet was pointed out by Moffatt (1994), and confirmed in a series of qualitative experiments (Eggers and Smock (1996)), using a silicone oil 10000 times as viscous as water. At sufficiently high driving, the air forced into the cusp experiences a strong enough downward pull to form a stable sheet. A preliminary calculation suggests that the thickness of the sheet scales like

$$\delta \approx \left(\frac{\eta_{air}}{\eta_{fluid}}\right)^{1/2} r_c \qquad (7)$$

This prediction is based on the assumption that there is a return flow in the sheet, which produces very high gradients. So when the sheet becomes very thin the inner flow is able to balance the high sheer stresses produced by the viscous flow. Both surface tension and gravity are not taken into account.

Fig. 3. Tip-streaming in a Couette device, showing a drop of water in silicone oil 1000 times as viscous. To initiate tip-streaming, 200 ppm of surfactant has been added. The image height is 0.5 mm. Photograph courtesy H. Leonhard.

Similar phenomena have been observed for a variety of other stationary singularities. An example, known as "tip-streaming" (Taylor (1934),De Bruijn (1993)), is shown in Fig.3. One sees a drop of water in a shear flow of a very viscous fluid, which produces a cone-shaped singularity at both ends of the drop. Under circumstances that are not well understood, a jet emanates from the tip, and eventually decays into drops due to the Rayleigh capillary instability. This is the precise analogue of the sheet in the two-roller apparatus, but the dimension of the singularity and of the resulting structure are lowered by one. A second example of a zero-dimensional singularity giving rise to a one-dimensional structure is that of a dielectric drop in a strong electric field, where a local cone-shaped solution exists (Taylor (1964)). This "Taylor cone" is never stable, but either oscillates between a rounded and a pointed state, or stabilizes itself by ejecting a jet from its tip. Again, little is known about the conditions under which the jet forms,but the striking similarities between different systems suggest a unifying explanation for the emergence of these structures.

References

Majda A. J. (1991): Vorticity, Turbulence, and Acoustics in Fluid Flow.
SIAM Review **33**, 349–388

Dai W., Kadanoff L. P., Zhou S (1991): Interface dynamics and the motion of singularities. Phys. Rev. A **43**, 6672–6682

Eggers J. (1997a): Nonlinear dynamics and breakup of free-surface flows.
Rev. Mod. Phys. **69**, 865–929

Brenner M. P., Shi X. D., Nagel S. R. (1994): Iterated Instabilities during Droplet Fission. Phys. Rev. Lett. **73**, 3391–3394

Joseph D. D., Nelson J., Renardy M., Renardy Y. (1991): Two-dimensional cusped interfaces. J. Fluid Mech. **223**, 383–409

Richardson S. (1968): Two dimensional bubbles in a slow flow.
J. Fluid Mech. **33**, 475–493

Jeong J.-T., Moffatt H., K. (1992): Free-surface cusps associated with a flow at low Reynolds number. J. Fluid Mech. **241**, 1–22

de Gennes P. G. (1985): Wetting: Statics and Dynamics. Rev. Mod. Phys. **57**, 827–863

Brenner M., Bertozzi A. (1993): Spreading of droplets on a solid surface.
Phys. Rev. Lett. **71**, 593–596

Papageorgiou D. T. (1995): On the breakup of viscous liquid threads.
Phys. Fluids **7**, 1529–1544

Eggers E., Lister J., Stone H. (1997): Coalescence of liquid drops.
manuscript in preparation

Mullins W. W., (1965): Morphological changes of a surface of revolution due to capillarity-induced surface diffusion. J. Appl. Phys. **36**, 1826–1835

Eggers J., (1997b): Coalescence of drops by surface diffusion. unpublished

Moffatt H. K. (1994): private communication

Eggers J., Smock M. (1996): The two-roller apparatus at very high driving.
unpublished experiments

Taylor G. I. (1934): The formation of emulsions in definable fields of flow.
Proc. Roy. Soc. London A **146**, 501–523

De Bruijn R. A.(1993): Tipstreaming of drops in simple shear flows.
Chem. Eng. Sci. **48**, 277–284

Taylor G. I.(1964): Disentigration of water drops in an electric field.
Proc. Roy. Soc. London A **280**, 383–397

Disordered Structures
Analyzed by the Theory of Markov Processes

R. Friedrich[1], Th. Galla[2], A. Naert[3], J. Peinke[2], and Th. Schimmel[4]

[1] Institut für Theoretische Physik, Universität Stuttgart, D-70550 Stuttgart
[2] Experimentalphysik II, Universität Bayreuth, D-95440 Bayreuth
[3] R.I.E.C., Tohoku University, 2-1-1 Katahira, Aobaku, Sendai 980-77, Japan
[4] Angewandte Physik, Universität Karlsruhe, D-76128 Karlsruhe

Abstract. A new application of the theory of Markov processes to the characterization of fractals is presented. We show under which condition distinct stochastic processes of a cascade lead to multifractal scaling behavior. We apply our method to analyze the spatial disorder of the velocity field of fully developed turbulence and the roughness of a gold film of granular-like structure. The results obtained from these experimental data are compared with results from synthetic data of a fractal basin boundary.

keywords: multifractal, multiaffine, small scale turbulence, intermittency, surface roughness

1 Introduction

There is an actual interest to characterize disordered structures, S, exhibiting scaling behavior in terms of fractality or affinity (cf. [1]). In general, the quantification of fractality is based on the introduction of a measurable quantity $Q(l)$ depending on a selected length scale l. A suitably defined average of the q-th power of the quantity $Q(l)$ is then investigated as a function of l. An important question is whether there exists multiscaling behavior in the limiting case $l \to 0$:

$$< (Q(l))^q > \sim l^{\zeta_q} \qquad . \qquad (1)$$

$< \dots >$ denotes an appropriate average over S. If such a scaling behavior exists, exponents ζ_q are deduced which serve as a characterization of the statistical properties of the disordered structure S. If ζ_q is a nonlinear function of q, we say that S is a multifractal or has multiaffine scaling properties [1]. For synthetically constructed complex structures like the Cantor-set or the Sierpinski-gasket (just to mention two well-known examples), the exponents ζ_q can be calculated explicitly. Problems commonly arise in the investigation of natural structures, since the scaling indices usually can not be determined from experimental data with sufficient accuracy.

A different commonly applied method for the characterization of the statistical properties of the disordered structure S is the evaluation of correlations. Let $s(x)$ denote a functional value of S at the location x (x can be thought of as space variable but may also be considered as a time variable). Then $< s^{q_1}(x)s^{q_2}(x+l_1)\ldots s^{q_n}(x+l_{n-1}) >$ is the general form of an n-point correlation. The knowledge of all n-point correlations can be considered as a complete statistical characterization of S.

The multiscaling analysis and the correlation analysis can be linked in the following way. Let us consider the definition $Q(l) := s(x+l) - s(x)$. Here $Q(l)$ is denoted as an increment. In this case the two-point-correlations $< s^{q_1}(x)s^{q_2}(x+l) >$ are directly related to the q-th order moment $< (Q(l))^q >$, with $q_1 + q_2 = q$. We remind the reader that one denotes the moments $< (Q(l))^q >$ of the increments as the q-th order structure function.

2 Evolution Equation for the Probability Distribution

In the present paper we introduce a novel method to analyze disordered structures in a more general way. Furthermore, we shall show that the both approaches mentioned above are incorporated in our present theory. Starting from (1), we know that

$$< (Q(l))^q > = \int (Q(l))^q \, P(Q(l), l) \, dQ(l). \tag{2}$$

Instead of evaluating the scaling behavior of (1) we shall consider the l-dependence of the probability density distribution $P(Q(l), l)$ of the quantity $Q(l)$.

The essential point of our work is that we are able to devise a procedure which allows us to derive an evolution equation for the probability distribution $P(Q(l), l)$ with respect to the size parameter l directly from experimental data. The first step of our approach consists in investigating how the quantity $Q(l)$ at one fixed point x changes with l. This can be easily achieved with the above defined construction of increments. (Also other constructions of $Q(l)$ may be used, like the midpoint construction of increments: $Q(l) := s(x + l/2) - s(x - l/2)$, or l-dependent coefficients of a wavelet transformation at x.) From the data we can evaluate the n-dimensional probability density distributions $p(Q(l_1), l_1; Q(l_2), l_2; \ldots; Q(l_n), l_n)$ or, respectively, the conditional probability distributions $p(Q(l_n), l_n; \ldots; Q(l_r), l_r | Q(l_{r-1}), l_{r-1}; \ldots; Q(l_1), l_1)$. Here, we use the convention $l_{i+1} < l_i$. Next we ask whether these n-dimensional distributions can be expressed by two-dimensional conditional probability distributions. In other words, we ask whether the underlying stochastic dynamics of $Q(l)$ as a function of l corresponds to a Markov process [2]. In order to show evidence for an underlying Markov-process one should show that

$$p(Q(l_n), l_n; \ldots; Q(l_r), l_r | Q(l_{r-1}), l_{r-1}; \ldots; Q(l_1), l_1)$$
$$= p(Q(l_n), l_n; \ldots; Q(l_r), l_r | Q(l_{r-1}), l_{r-1}). \tag{3}$$

This is evidently an impossible task for arbitrary n and r. However, as is well-known, a necessary condition for a process to be Markovian is the validity of the Chapman-Kolmogorov equation [2]:

$$p(Q(l_3), l_3|Q(l_1), l_1) = \int p(Q(l_3), l_3|Q(l_2), l_2) \, p(Q(l_2), l_2|Q(l_1), l_1) \, dQ(l_2).$$
(4)

The validity of this equations yields strong hints that the stochastic process is Markovian.

From the Chapman-Kolmogorov equation an evolution equation for the conditional probabilities can be deduced, which is known as the Kramers-Moyal expansion

$$-\frac{d}{dl} P(Q(l), l) = \sum_{k=1}^{\infty} [-\frac{\partial}{\partial Q(l)}]^k D^{(k)}(Q(l), l) \, P(Q(l), l).$$
(5)

Here, the Kramers-Moyal coefficients $D^{(k)}(Q(l), l)$ are defined by the following conditional moments:

$$M^{(k)}(Q(l), l, \Delta l)$$
$$= \frac{1}{\Delta l} \int dQ(l - \Delta l) \, (Q(l - \Delta l) - Q(l))^k \, p(Q(l - \Delta l), l - \Delta l|Q(l), l),$$
(6)

$$D^{(k)}(Q(l), l) = \lim_{\Delta l \to 0} \frac{1}{n!} M^{(k)}(Q(l), l, \Delta l).$$
(7)

An important feature of the present approach is based on the fact that these Kramer-Moyal coefficients can actually be determined directly from the given data, as we shall indicate below.

3 Markov-Process and Multifractality

In the following we shall discuss implications based on well-known properties of Markov-processes [2], which, to our knowledge have not yet been put into connection with multifractality. (We would like to mention that the presented procedure may be compared with the application of the Frobenius Peron operator for the determination of the natural measure of chaotic attractors [1].)

1. If $D^{(4)} = 0$, Pawula's theorem implies that only the drift term $D^{(1)}$ and the diffusion term $D^{(2)}$ are non-zero. The Kramers-Moyal expansion (5) reduces to the Fokker-Planck equation

$$-\frac{d}{dl} P(Q(l), l) = -\frac{\partial}{\partial Q(l)} D^{(1)}(Q(l), l) P(Q(l), l)$$
$$+ [\frac{\partial}{\partial Q(l)}]^2 D^{(2)}(Q(l), l) P(Q(l), l).$$
(8)

Note, the first minus sign of d/dl is due to the direction of the process starting at large scales and going to small scales.

2. From the equations (5) and (8) the evolution equation for a single event $Q(l)$ at point x in form of a Langevin equation can be derived [2]:

$$-\frac{d}{dl}Q(l) = g(Q(l), l) + h(Q(l), l)\, \Gamma(l). \tag{9}$$

The function g describes the deterministic evolution of $Q(l)$, whereas h takes into account the stochastic fluctuations. In the case of the Fokker-Planck equation the functions g and h are given by $D^{(1)}$ and $D^{(2)}$, where $h \propto \sqrt{D^{(2)}}$. The noise term Γ has the properties of Gaussian white noise.

3. By multiplying equation (5) with Q^q and successively integrating over $Q(l)$, equations for the moments $< (Q(l))^q >$ are obtained [3]:

$$-\frac{d}{dl} < (Q(l))^q > = \sum_{n=1}^{q-1} \frac{q!}{(q-n)!} < D^{(n)} Q^{q-n} >. \tag{10}$$

4. If the Kramers-Moyal coefficient are

$$D^{(n)} = \frac{d_n Q^n}{l}, \tag{11}$$

where d_n are constants, the scaling behavior of equation (1) is guaranteed and we obtain from (10)

$$\frac{l}{< Q^q >} \frac{d}{dl} < Q^q > = -\sum_{n=1}^{q-1} \frac{q!}{(q-n)!} d_n = \zeta_q. \tag{12}$$

For the simple case of a Fokker-Planck equation we see now that $D^{(1)} = -d_1 Q/l$ and $D^{(2)} = d_2 Q^2/l$ implies multiscaling with $\zeta_q = d_1 q - d_2 q(q-1)$. The quadratic Q-dependence of $D^{(2)}$ corresponds to a purely multiplicative noise process, as follows from the discussion of point 2 above. For purely additive noise, i.e. $D^{(2)} = d_2(l)$ and $D^{(1)} = d_1 Q/l$, scaling behavior may be obtained if $d_2 = ca/(l^{1+2d_1})$, with c and a constants. Now $\zeta_q = a - d_1 q$.

5. For the case of the validity of a Fokker-Planck equation the stationary solution

$$P(Q(l), l) = \frac{const}{D^{(2)}} exp(2 \int_{Q_0}^{Q(l)} \frac{D^{(1)}(y, l)}{D^{(2)}(y, l)} dy) \tag{13}$$

is known, where Q_0 is a constant. Taking the condition for scaling behavior of point four (equation (11)), that the drift term is linear and the diffusion term is quadratic in Q, we find that the probability density distribution $P(Q(l), l)$

evaluated by (13) is not normalizable. The stationary distribution is simply the δ-function $\delta(Q(l))$. Including a small additive noise term yields a distribution which for large values of $Q(l)$ has a power-law behavior in $Q(l)$, which is a characteristic feature of a Lévy-distribution. In this case the moments $< Q^q >$ diverge for higher q values and thus the derivation of equation (10) becomes questionable [4]. An argument that the above mentioned results on $< Q^q >$ still hold is that the power law tails of the distribution at infinity are only valid for the stationary solution. For an initial probability distribution at large scales l it would take an infinite cascade (i.e. development with the evolution equation) to create such wings at infinity for Q. Therefore, for any natural structure S, where typically scaling behavior is only found in a finite interval of scales l, we expect that these divergences do not affect our results discussed here.

6. We want to point out that, having shown the Markovian properties, any n-point statistics can be derived from the 2-point statistics $p(Q(l_2), l_2|Q(l_1), l_1)$. In the case of the applicability of the Fokker-Planck equation the n-point statistics can be explicitly given by the knowledge of the two Kramer-Moyal coefficients $D^{(1)}$ and $D^{(2)}$ [2].

4 Application to Experiments

Let us discuss some applications and verifications of the method presented above:

4.1 Velocity Statistics of Turbulence

For the case of fully developed turbulence a major challenge consists in explaining the statistics of velocity increments $V(l)$ (cf. [5]). Usually, the increments of the velocity component in the direction of the distance are measured. (More precisely one should consider in the 3-dimensional space a distance vector, but due to isotropy the direction of this distance vector does not play a role.) The data evaluated here are on one hand from a free jet experiment running with dry air or alternatively with water, which has been built up in Bayreuth, and, on the other hand, from a low temperature turbulence experiment [10]. The presented results are based on the analysis of more than 10^7 measured data points. A segment of a typical time series is shown in Fig.1.

As a testing ground of the Markovian property we have verified the Chapman-Kolmogorov equation, see (4). To illustrate the validity we present in Fig. 2 for a selected triple (l_1, l_2, l_3) the conditional probabilities for three different values of $V(l_1)$. One time the conditional probabilities were evaluated directly from the data set (left hand side of eq. (4)), the other time we have evaluated the conditional probabilities by integration of two conditional probabilities (right hand side of eq. (4)). Taking into account that each bin

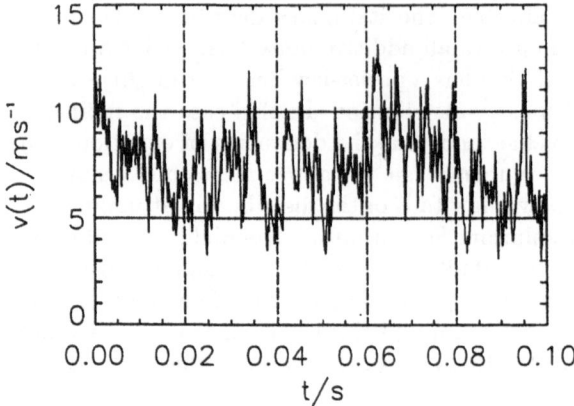

Fig. 1. Part of a typical time series of the velocity measured in the center of a gaseous free jet. The measurement was done with a hot wire in the distance of 30 diameters of the nozzle and a sampling rate of $120kHz$. Based on the Taylor hypothesis of frozen turbulence a time step of $1ms$ corresponds to a distance of $9mm$.

of these probabilities have at least an intrinsic error of the order $\sqrt{N_i}/N$ (N_i is the number of events found for the i-th bin and N is the number of all events for the conditional probability) we have quantified the validity of the Chapman-Kolmogorov equation [6]. Furthermore, recently the Markovian property for the turbulent velocity field has even been proven along eq. (3) in the case of double conditioned probabilities [7].

A quite important result is that apparently the Markovian property based on double conditioned probabilities only holds if the separation of the l_i values is larger than a certain value: L_{mar}. This value is of the order of the Taylor microscale and does not change throughout the cascade. This indicates that smaller structures are correlated, in a statistical sense they have memory. It is a quite interesting question whether this length scale is connected with coherent structures (which then play an analogous role as the elementary molecular interactions in the case of diffusion processes). The fact that the quantity L_{mar} does not depend on the location in the cascade, shows that it is important to incorporate these effects in a the modeling of turbulence. Therefore, shell models for turbulence based on processes with step sizes like a^n (usually $1/2^n$) are questionable.

Based on the conditional probabilities the statistics of the velocity increments can be characterized by Kramers-Moyal coefficients. Recently we have shown experimental evidence that the fourth Kramers-Moyal coefficient vanishes. This suggests that a Fokker-Planck equation describes the increment statistics of turbulence [6]. In Fig. 3 we present the numerical results from experimental data of the first two Kramers-Moyal coefficients. Both coefficients have been evaluated for several scales covering the whole inertial range. (The inertial range in turbulence denotes the range of length scales where scaling

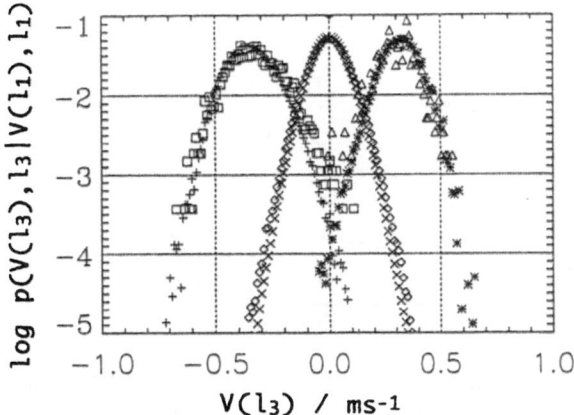

Fig. 2. Verification of the Chapman-Kolmogorov equation (4) for velocity increments. Shown are three sets of conditional probabilities $p(V(l_3), l_3 | V(l_1), l_1)$ for $V(l_1) = -0.45;\ 0;\ 0.45m/s$ and $l_1 = 64\eta;\ l_2 = 54\eta;\ l_3 = 45\eta$, η being the Kolmogorov viscous scale. The open symbols correspond to directly evaluated conditional probabilities, whereas the crosses and stars conditional probabilities represent which were obtained from the integration over $V(l_2)$. The Reynolds number was 27000.

behavior is expected to occur. This range is bounded by a large length scale L dominated by the boundary conditions, like the size of the flow, and the viscous or Kolmogorov length scale η, where dissipation dominates the flow.) We find that $D^{(1)} = \gamma V(l)/l$ and $D^{(2)} = a(l) + b(l)(V(l))^2$. Here, γ turns out to be approximately $-1/3$. If we take the diffusion term $D^{(2)}$ as a small perturbation, we see that $\gamma = -1/3$ leads in a first order approximation to the scaling exponents $\zeta_q = q/3$. This is the well-known Kolmogorov 1941 result. It is easily seen that an additive term in $D^{(2)}$ (like $a(l)$) violates a scaling behavior of the structure functions $< (V(l))^q >$ even if $b(l) = b/l$. From the functional behavior of $D^{(2)}$ on $V(l)$ we can see two features: For small values of $|\ V(l)\ |$ the additive constant dominates, thus we have the approximate case of additive noise. As a consequence the probability density $P(V(l), l)$ has a Gaussian shape at the center, see Fig. 4. For large values of $V(l)$ the quadratic term of $D^{(2)}$ dominates. Here, based on the corresponding Langevin equation (9), we can speak of multiplicative noise acting along the cascade and leading to the pronounced wings in $P(V(l), l)$, i.e., causing the phenomenon of intermittency.

4.2 Energy Cascade of Turbulence

Next, we report on further new findings concerning the statistics of the energy dissipation on different length scales. Following the suggestion of Kolmogorov and Obukhov we evaluate the energy dissipation at scale l according to [8]

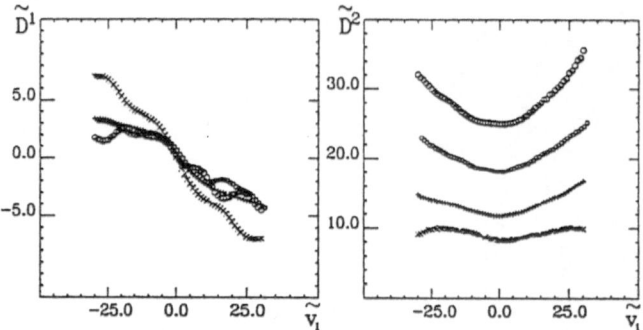

Fig. 3. Scaled drift and diffusion coefficients $\tilde{D}^{(1)} = D^{(1)}/l$, $\tilde{D}^{(2)} = D^{(2)}/l$ and $\tilde{v}_i = V(l_i)(l_i/L_{ref})^{1/3}$ for $l_i = 20, 54, 124, 224\eta$ from down to top of $\tilde{D}^{(2)}$. Note that the first value of l_i is not any more in the inertial range and thus shows a different behavior.

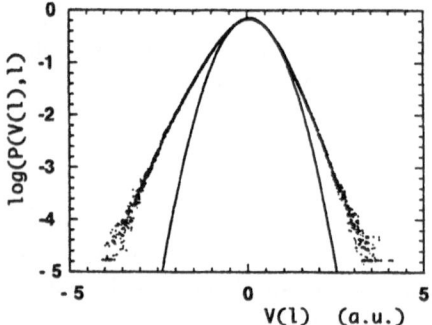

Fig. 4. Exemplary intermittent probability density $P(V(l), l)$ for $l = 56\eta$. The drawn curve corresponds to a pure Gaussian distribution, fitting well the center of the measured probability.

$$\epsilon_l(x) = \frac{15\nu}{l} \int_{x-l/2}^{x+l/2} \left(\frac{dv}{dx'}\right)^2 dx' \quad , \tag{14}$$

where ν denotes the kinematic viscosity. It has been postulated that nongaussian statistics in the velocity increments is due to a lognormal distribution of ϵ_l. We take $log(\epsilon_l)$ as our quantity Q. The variance of $log(\epsilon)$ was assumed to be logarithmic in the length scale. We have already shown that the Chapman-Kolmogorov equation holds [9]. Next we have evaluated the Kramers-Moyal coefficients, and we have indication that the fourth coefficient is zero.

Figure 5 presents the first two Kramers-Moyal coefficients, $D^{(1)}(Q, l)$ and $D^{(2)}(Q, l)$ for different l values of the inertial range. For convenience we have used a logarithmic length scale $\tilde{l} = ln(l/\eta)$, which corresponds to the evaluation of $lD^{(n)}$ in linear scale.

To make the drift terms coincide a l-dependent additive term $F(\tilde{l})$ has been used. The occurrence of this additive term can be justified. In fact it is determined by the constraint of conservation of mean energy.

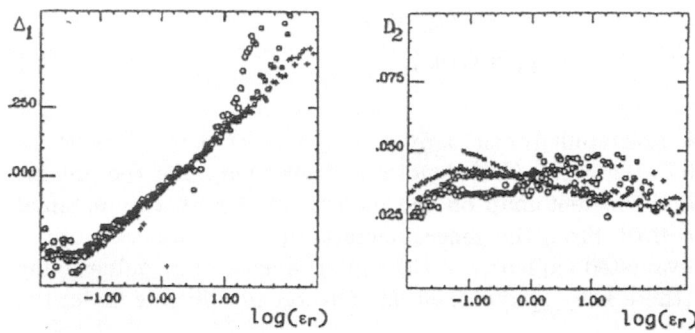

Fig. 5. Drift and diffusion coefficients, $\Delta^{(1)} = D^{(1)} - F(\tilde{l})$ and $D^{(2)}$ for various values of the scale l/η=400, 200, 100, 50, η being the Kolmogorov viscous scale.

In the central region, the drift coefficient is linear in Q and the diffusion coefficient, is constant. Furthermore, the slope of $D^{(1)}$ and the value of $D^{(2)}$ are both scale invariant in the inertial range:

$$D^{(1)}(Q) = \gamma(Q - <Q>) + F(\tilde{l}) \quad ,$$
$$D^{(2)}(Q) = D \quad , \tag{15}$$

with $\gamma \approx 0.2$ and $D \approx 0.03$. Thus, we can summarize that our empirical results show that the statistics of the quantity $log(\epsilon_l)$ is determined by a Fokker-Planck equation with a linear drift and a constant diffusion coefficient. Such a Fokker-Planck equation is denoted as an Ornstein-Uhlenbeck process. The positive slope of $D^{(1)}$ indicates that there is no stationary probability distribution [2]. Comparing this finding with the negative slope of the drift term for the velocity increments, we see that in the case of the energy the statistics starts at large scale with a narrow distribution which becomes broader, whereas the probability distribution of the velocity increments gets narrower as we go down the cascade to smaller scales.

The common argument for the proposed multiscaling (anomalous scaling) of $V(l)$ is that $< (V(l))^q > \propto < \epsilon_l^q > r^{q/3}$, thus multiscaling corrections to the $q/3$-behavior should be due to the ϵ-statistics, displaying scaling behavior of $< \epsilon_l^q >$. Here we present a sensitive methode to investigate experimental data of turbulence whether scaling behavior is present or not. Further results and conclusions are presented in [9].

4.3 Fractal Function

Next we apply our method to other structures. As a first testing ground we have chosen a fractal function, which is the explicit function of a fractal basin boundary [12]. The fractal boundary is generated by the Roessler-equations of a chaotically driven boundary:

$$x_{n+1} = F(x_n) + by_n$$
$$y_{n+1} = G(y_n), \tag{16}$$

where $F(x)$ causes a bistability of the x-dynamics, here $F(x) = x^{1.26}$ with an unstable fixed point at $x = 1.0$. $G(x)$ denotes a chaotic map. For the present case we have chosen the tent map on the interval $[0, 1]$ with the maximal value of 1.0 at $y = 0.76$. From the general equation (16) the function of the boundary can be evaluated explicitly as the y_0-dependence of x_0 values lying on the boundary (these $x_0(y_0)$ values neither converge nor diverge under the iteration with (16):

$$x_0 = \lim_{n \to \infty} F^{-1}(...F^{-1}((1 - bG^n) - G^{n-1})... - bG^0). \tag{17}$$

G^n denotes the n-th iteration of an initial value y_0, thus $G^0 = y_0$. This function is closely related to the Okninsky-Weierstrass function [13]. For the above mentioned functions F and G the numerically evaluated function is shown in Fig. 6. From this function $Q(l) = x_0(y_0 + l) - x_0(y_0)$ is easily evaluated. Eventhough the moments $< (Q(l))^q >$ show good scaling behavior, see Fig. 7, with a nonlinear evolution of the scaling exponents ξ_q in q, we find that the Chapman-Kolmogorov equation is not fulfilled, see Fig. 8. This result could be quantified by means of statistical analysis [14]. Evaluating nevertheless coefficients in analogy to the Kramers-Moyal coefficient, we find that they are strongly oscillating functions of Q and definitely do not follow the equation (11).

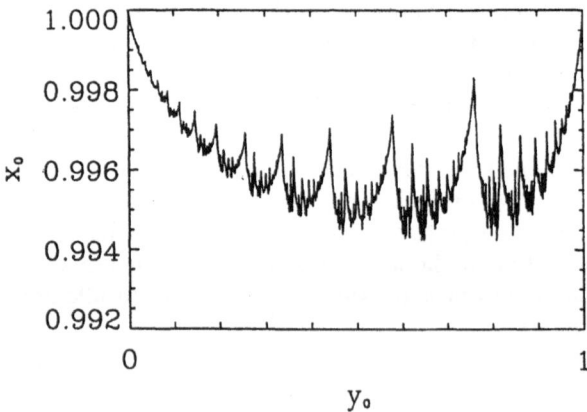

Fig. 6. Graph of the function (17) representing the fractal basin boundary between the basin of attraction of the two attractor at zero and at infinity.

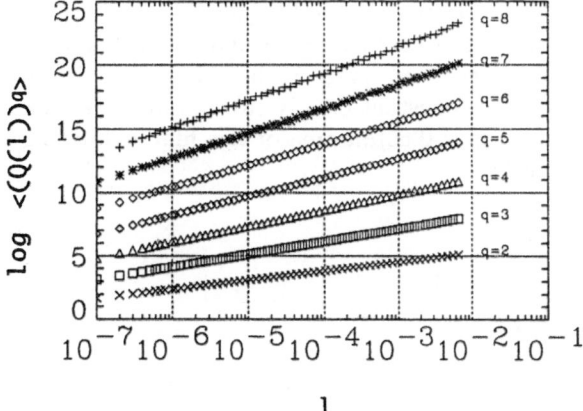

Fig. 7. Double logarithmic presentation of the l-dependence of $< (Q(l))^q >$ for the fractal function. For the uneven moments the absolute value of $Q(l)$ has been taken.

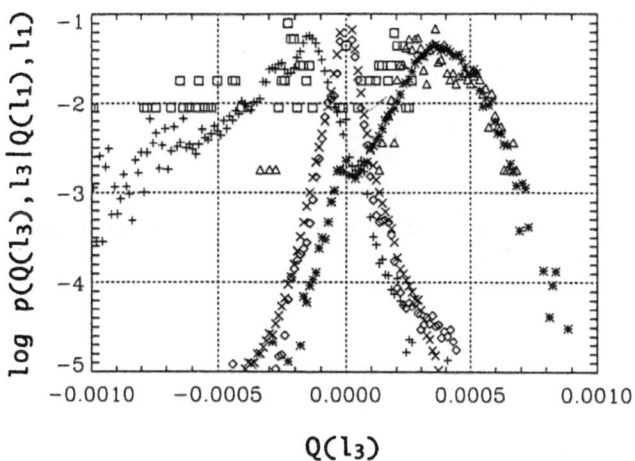

Fig. 8. Verification of the Chapman-Kolmogorov equation (4) for increments of the fractal function. Shown are three sets of conditional probabilities $p(Q(l_3), l_3 | Q(l_1), l_1)$ for $Q(l_1) = -0.510^{-4}$; 0; 0.510^{-4} and $l_1 = 1.5210^{-5}$; $l_2 = 1.2810^{-5}$; $l_1 = 1.0810^{-5}$. The open symbols correspond to directly evaluated conditional probabilities, whereas the crosses and stars conditional probabilities represent which were obtained from the integration over $Q(l_2)$.

4.4 Roughness of a Surface

At last, we want to present results of the analysis of another experimental, disordered system, namely, a rough surface. For this purpose, a gold film was deposited on a glass substrate by thermal evaporation in a vacuum chamber (base pressure approx. 10^{-6} mbar). During the evaporation, the substrate was kept at room temperature. The gold film used in the experiments described below was evaporated at a rate of ca. 0.5 nm per second. The average film thickness of approx. 60 nm was determined with a quartz microbalance.

The resulting films are stable in air and were subsequently examined by atomic force microscopy (AFM) in the contact mode under ambient conditions. The investigations were performed using V-shaped silicon nitride cantilevers with sharpened tips (force constant 0.03 N/m). Figure 9 shows a topographic AFM image of a $0.65\mu m$ x $0.65\mu m$ area of this film, exhibiting granular structure of the gold surface.

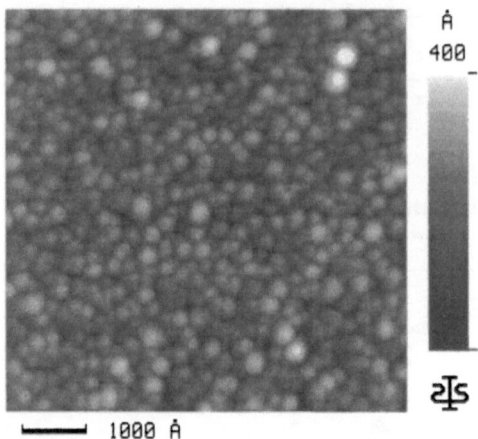

Fig. 9. AFM image of a gold film obtained by thermal evaporation on a glass substrate. Image size: $0.65\mu m$ x $0.65\mu m$.

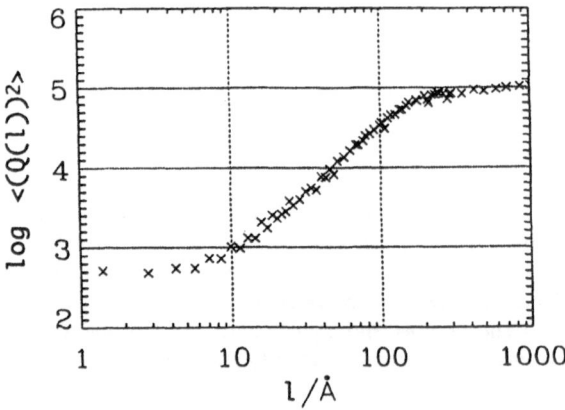

Fig. 10. Double logarithmic presentation of the l-dependence of $< (Q(l))^2 >$ for the surface roughness.

From more than 600 images of scan size, corresponding to more than 10^7 data points, the height- height differences $Q(l) = h(x + l) - h(x)$ have been analyzed. As shown in Fig. 10, good scaling behavior of the moments of Q

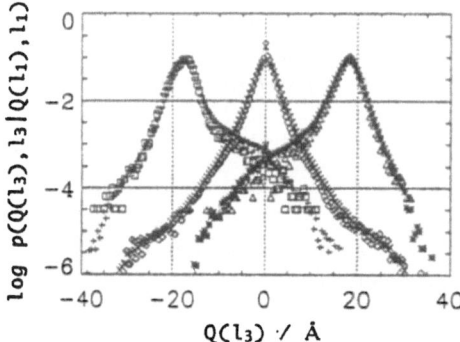

Fig. 11. Verification of the Chapman-Kolmogorov equation (4) for increments of the high of the rough surface. Shown are three sets of conditional probabilities $p(Q(l_3), l_3|Q(l_1), l_1)$ for $Q(l_1) = -21; 0; 21\text{Å}$ and $l_1 = 68, 7\text{Å}; l_2 = 62, 5\text{Å}; l_1 = 56, 2\text{Å}$. The open symbols correspond to directly evaluated conditional probabilities, whereas the crosses and stars conditional probabilities represent which were obtained from the integration over $Q(l_2)$.

is observed for $l > 20\text{Å}$. Furthermore, the Chapman-Kolmogorov equation is well fulfilled, as shown exemplarily in Fig. 11. For the Kramers-Moyal coefficients we found that $D^{(1)}$ is linear, with negative slope, and $D^{(2)}$ is quadratic in Q. In contrast to the velocity data of the turbulent flow we obtain in the case of the surface data that the fourth Kramers-Moyal coefficient does not converge to zero. Approximating $D^{(n)}$ according to expression (11) we see that in the scaling range d_4 is about $d_2/4$. Thus, already for $q = 5$ the influence of d_4 on ξ_4 plays a role. We conclude that the surface is generated by multiplicative noise, but the noise is not Gaussian. As a characteristic length scale where the Markovian properties seem to be violated we find a value of 20Å, a value which is already approaching the atomic scale. Note the lattice constant for Au(111) surface is 2.9Å. For further details on the evaluation of the surface roughness see Ref. [15].

5 Summary

To summarize we have presented a new method to analyze the statistical properties of a disordered structure. We have discussed how our approach can be related to multifractal scaling. This method has now been applied to experimental data, such as velocities of turbulence, energy of turbulence, and the heights of a rough surface. Furthermore, it has been applied to a fractal function. We found different results depending on the quantity we have investigated. From the fractal function we have seen that multifractality is not always based on a Markovian process. In the case of surface roughness and turbulence, multiscaling and intermittency is due to multiplicative noise but the type of noise differs. For the turbulence we found for our experimental data no multiscaling behavior is present. In conclusion, we have presented a method with obviously allows one to obtain further insight into hierarchical processes generating disorder. We are convinced that the presented method is more sensitive to characteristic features of disorder than looking merely at scaling behavior.

Acknowledgments

Helpful discussions with Martin Greiner, Peter Lipa. JP acknowledges financial support by the Deutsche Forschungsgemeinschaft.

References

1. R. Badii and A. Politi, *Complexity* (Cambridge University Press, Cambridge 1997);
   B. Mandelbrot, *The Fractal Geometry of Nature* (Freeman. San Francisco 1982);
   K.J. Falconer, *Fractal Geometry , Mathematical Foundations and Applications* (John Wiley, Chichester 1990);
   J. Feder, *Fractals* (Plenum Press, New York 1988);
   T. Vicsek, *Fractal Growth Phenomena* (World Scientific, Singapore, 1992).
2. P. Hänggi and H. Thomas, Physics Reports **88**, 207 (1982);
   N.G. van Kampen, *Stochastic Processes in Physics and Chemistry* (Elsevier, Amsterdam 1990);
   H. Risken, *The Fokker-Planck Equation* (Springer-Verlag Berlin, 1984).
3. For the integration it is assumed that the probability distributions $P(Q(l), l)$ decay at infinity fast than any power of Q.
4. private communication with Peter Lipa.
5. K.R. Sreenivasan and R.A. Antonia, Annu. Rev. Fluid Mech. **29**, 435 (1997);
   U. Frisch, *Turbulence* (Cambridge University Press, Cambridge 1995).
6. R. Friedrich, J. Peinke, Physica D **102**, 147 (1997);
   R. Friedrich, J. Peinke, Phys. Rev. Lett. **78**, 863 (1997).
7. R. Friedrich, J. Zeller, J. Peinke, preprint submitted to Europhys. Lett.
8. A. M. Obukhov, J. Fluid Mech. 13, 77 (1962);
   A. N. Kolmogorov, J. Fluid Mech. 13, 82 (1962).
9. A. Naert, R. Friedrich, J.Peinke, A stochastic equation for the energy cascade in turbulence, Phys. Rev. **E** (in press).
10. B. Chabaud, A. Naert, J. Peinke, F. Chillà, B. Castaing, B. Hébral, Phys. Rev. Lett., **73**, 3227 (1994).
11. B. Castaing, Y. Gagne, E. Hopfinger, Physica (Amsterdam) 46D, 177, (1990).
12. J. Peinke, M. Klein, A. Kittel, G. Baier, J. Parisi, R. Stoop, J.L. Hudson, and O.E. Roessler, Europhys. Lett. **14**, 615 (1991);
    J. Peinke, M. Klein, A. Kittel, A. Okninsky, J. Parisi, and O.E. Roessler, Physica Scripta **T49**, 672 (1993).
13. A. Okninsky, J. Stat. Phys. **52**, 577 (1988).
14. Th. Galla, *Diplomarbeit*, Experimentalphysik II (Bayreuth 1997).
15. H. Bayer, R. Friedrich, Th. Galla, J. Peinke and Th. Schimmel, preprint.

Transition to Turbulence in Shear Flows

Bruno Eckhardt, Kerstin Marzinzik and Armin Schmiegel

Fachbereich Physik, Philipps Universität Marburg
35032 Marburg, Germany

Abstract. We review recent work on the transition to turbulence in shear flows, in particular plane Couette flow. In the linearized system the non-normality of the linear operator gives rise to non-orthogonal eigenvectors and to a significant amplification of noise. We present a typical result for a Fokker-Planck equation with non-normal relaxation matrix. As the driving becomes stronger, further stationary states are born in a saddle node bifurcation. This is observed in a two degree of freedom phenomenological model, in the a few mode approximation to a simple shear flow and in full numerical studies of plane Couette flow. The stationary states give rise to a fractal border between decaying and turbulent states.

1 Introduction

The transition to turbulence in shear flows differs from other hydrodynamics stability problems, like e.g. the much studied and reasonably well understood case of a fluid heated from below, in many respects. In particular, (i) turbulent states arise at control parameters where the laminar flow is still stable, (ii) whether a transition is induced depends sensitively on the perturbation applied and (iii) the excited turbulent state seems to be transient rather than permanent (for a recent summary of the state of affairs see Grossmann 1996). Results from long time numerical simulations, simple models and ideas from dynamical system theory have sparked a wave of interest in this transition which hopefully will lead to a consistent picture for it. Some of the elements we believe to be essential will be summarized here. The presentation is bottom-up, from the simplest models to simplified flow models to full numerical simulations of plane Couette flow.

Many investigations of linear stability of a flow have focussed on the eigenvalues of the linear operator (e.g. Chandrasekhar, 1961; Drazin and Reid, 1981). However, since the linear operator is not normal, the eigenvectors are not orthogonal and the decay of small perturbations need not be monotonic (Boberg and Brosa 1988, Trefethen *et al.* 1993, Gebhardt and Grossmann 1994). Rather, there can be a transient initial period of growth in which the perturbation is amplified sufficiently to invalidate the linear approximation. Since the amplification is of the order of the Reynolds number Re, perturbations that decay will have to be ever smaller as the Reynolds number increases. This behaviour can be captured in small models with just two degrees of freedom. An example is discussed in section 2.

Much insight into the the transition in Rayleigh-Bénard situations could be gained from few degree of freedom models, such as the Lorenz model (Lorenz

1963) or a generalization for binary fluids (Cross and Hohenberg 1993). On a technical level, the reason these models get by with three or five degrees of freedom lies in the fact that the main nonlinearities contain the temperature and the flow, so that nontrivial couplings can be found with three modes already. In shear flows, the only nonlinearity is the advection term $(\mathbf{u} \cdot \nabla)\mathbf{u}$ and it requires more Fourier modes to obtain non-vanishing contributions. An example is discussed in section 3, where a sinusoidally driven shear flow with periodic no-penetration boundary conditions is reduced to 19 dynamical degrees of freedom (Eckhardt and Mersmann 1997). This model shares with the two degree of freedom model of section 2 the non-normality of the linear part and the occurence of a saddle-node bifurcation. The presence of more degrees of freedom results however in a more complicated structure of the stable and unstable manifolds and a fractal transition region to turbulence. The new stationary states turn out to be all unstable, and the turbulent structures seen are transient.

Turning to a full numerical simulation of plane Couette flow with a spectral scheme and about 962 dynamically active degrees of freedom, we again find many of these features, qualitatively similar but quantitatively modified (Schmiegel and Eckhardt 1997). A new feature in this high-dimensional phase space is that the region influenced by the saddle-node nifurcation is highly anisotropic, so that some initial conditions are not effected until much higher Reynolds numbers. These results are described in section 4.

Finally, in section 5 we summarize our results and compare to experiments.

2   The linear problem: non-normality and noise

In the last years several low-dimensional models for the transition in shear flows. As discussed by Baggett and Trefethen (1997) these models have similar basic features: the linear part is non-normal and the nonlinearity preserves some norm, reminiscent of the energy conservation of the advection term in the Navier-Stokes equation. For illustration we take the simplest model with two degrees of free-

dom,
$$\begin{aligned}
\dot{x} &= -x + Re\,y - y\sqrt{x^2 + y^2} \\
\dot{y} &= \quad -2y \quad + x\sqrt{x^2 + y^2}
\end{aligned} \tag{1}$$

This coincides with the Cornell model (Baggett and Trefethen, 1997) if time and coordinates are scaled appropriately ($\mathbf{x} = Re\,\mathbf{x}_{Cornell}$, $t = t_{Cornell}/Re$). One eigenvector points in the x-direction and the second one has a y-component that decreases with increasing Re. Thus the eigendirections collapse onto the x-axis as Re approaches infinity. Perturbations pointing into the y-direction will first grow in total size before decaying. The nonlinearity mixes the two variables without affecting the total energy.

Viewed as a dynamical system, the analysis of the behaviour of the system starts from an investigation of the stationary points. For eq. (1) there are one, three or five stationary points, depending on Re. With

$$D = \sqrt{8 + 2Re^2 - 2Re\sqrt{Re^2 - 8}} \qquad (2)$$

they are

$$
\begin{aligned}
\mathbf{x}_0 &= (0,0) \\
\mathbf{x}_1 &= \left(\left(2Re - 2\sqrt{Re^2 - 8}\right)\Big/ D, \ \left(Re^2 - 4 - Re\sqrt{Re^2 - 8}\right)\Big/ D \right) \\
\mathbf{x}_2 &= \left(\left(2Re + 2\sqrt{Re^2 - 8}\right)\Big/ D, \ \left(Re^2 - 4 + Re\sqrt{Re^2 - 8}\right)\Big/ D \right) \\
\mathbf{x}_3 &= -\mathbf{x}_1, \qquad \mathbf{x}_4 = -\mathbf{x}_2.
\end{aligned}
\qquad (3)
$$

The solution for the laminar flow, $\mathbf{x}_0$, exists for all Reynolds numbers, while the other ones arise in a saddle node bifurcation at $R = \sqrt{8}$. The stationary points $\mathbf{x}_0, \mathbf{x}_2$ and $\mathbf{x}_4$ are stable, $\mathbf{x}_1$ and $\mathbf{x}_3$ are unstable.

For a transition to the stationary points $\mathbf{x}_2$ or $\mathbf{x}_4$, a perturbation of finite amplitude is necessary to push the system out of the basin of attraction of $\mathbf{x}_0$. The borders of the basin of attraction of $\mathbf{x}_0$, $\mathbf{x}_2$ or $\mathbf{x}_4$ are the stable manifolds of $\mathbf{x}_1$ and $\mathbf{x}_3$. The full phase space structure at $Re = 3$ is shown in Fig. 1. As the Reynolds number increases, the saddles $\mathbf{x}_1$ and $\mathbf{x}_3$ approach the origin like

$$\mathbf{x}_1 \sim \left(\frac{2}{Re}, \frac{2}{Re^2} \right) \quad \text{as} \quad Re \to \infty. \qquad (4)$$

The stable manifolds of these saddles limit the basin of attraction of the laminar state and their closest distance to the origin may be smaller than that of $\mathbf{x}_1$ or $\mathbf{x}_3$. In either case, the maximal size of an isotropic perturbation that decays to the origin will decrease rapidly with increasing Re. On the other hand, suitably chosen perturbations of arbitrary strength will still find their way to the origin if they lie between the two manifolds. The large scale behaviour of these manifolds is that of a spiral so that the decay to the laminar fixed point $\mathbf{x}_0$ will be oscillatory.

So far we have discussed the deterministic dynamics. Adding noise to the systems gives rise to an interesting problem of its own, the solution of a Fokker-Planck equation with non-normal linear relaxation term. Focussing on the linear part, the Langevin equations are

$$
\begin{aligned}
\dot{x} &= -x + Re\, y + \xi_x \\
\dot{y} &= -2y + \xi_y
\end{aligned}
\qquad (5)
$$

with the noise characterized by

$$
\begin{aligned}
\langle \xi_i(t) \rangle &= 0, \\
\langle \xi_i(t)\xi_j(t') \rangle &= \delta_{ij}\delta(t - t')
\end{aligned}
\qquad (6)
$$

where $i, j \in \{x, y\}$. The strength of the noise can be scaled out in the linear model.

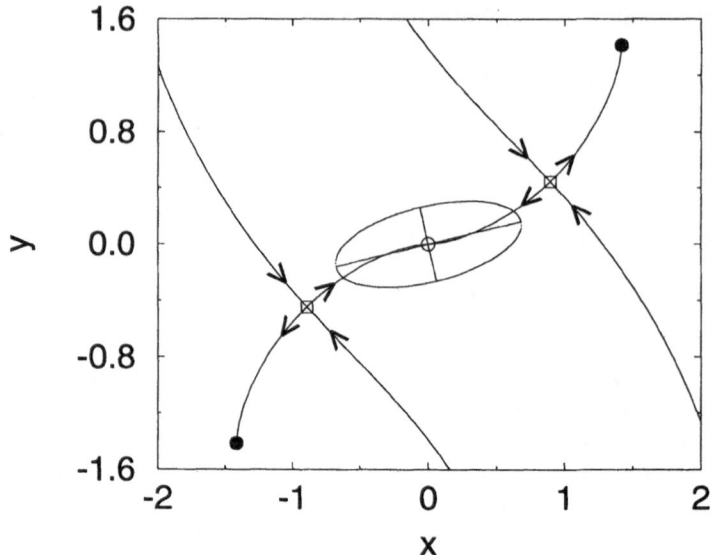

Fig. 1. Stable (circles) and unstable (squares) fixed points of equation (1) at $Re = 3$. The lines through the unstable points are the stable and unstable manifolds of these points. The stable manifolds are the borders between the basin of attraction of the three stable stationary states. The ellipse denotes an equiprobability line of the probability distribution for the linearized model with noise (see equation 5).

The corresponding Fokker-Planck equation for the density $P(x, y, t)$ reads

$$\dot{P} = \partial_x \left[(x - Re\, y)\, P \right] + \partial_y \left[2yP \right] + \Delta P. \tag{7}$$

The stationary solution of this equation is Gaussian (Risken, 1984),

$$P(x) \sim \exp[-\mathbf{x}^{\mathbf{T}} Q \mathbf{x}], \tag{8}$$

with a symmetric matrix Q. The relationship between this matrix and the linear flow matrix is somewhat complicated and its main axes typically do not coincide with the eigenvectors of the linear problem. However, as the constant density contour in Fig. 1 shows, the invariant density does stretch out into the cone between the eigenvectors. The eigenvalues of Q are

$$\lambda_{\pm} = \frac{3}{4} \pm \frac{3}{4\,(Re^2 + 9)} \sqrt{(Re^2 + 3)^2 + 4Re^2}\,; \tag{9}$$

they show a tendency to form narrow, elongated ellipses for large Re.

Combined with the previous analysis for the location of the fixed points and their stable manifolds, this analysis can shed light on the time distribution for the transition to the other fixed point. For stronger noise or increasing parameter Re more of the stationary density P will overlap with the basin of attraction of the second fixed point, thus inducing a transition to that state. Formulated as a first passage time problem, one can calculate the time it takes for the transition to be induced after start up of the system, and this may be experimentally accessible (Marzinzik and Eckhardt, in preparation).

3   A few mode model for a shear flow

The main shortcomings of the two degree of freedom model of the previous section are that it has too few degrees of freedom to show any complicated time-dependence or chaos (as a precursor to spatio-temporal turbulence) and that it cannot be derived from the Navier-Stokes equation in any systematic way. Following the example of Saltzmann (1961) and Lorenz (1963) we now present an example of a few degree of freedom model for a shear flow that is based on a Fourier representation of the Navier-Stokes equation. It differs from the plane Couette flow presented in the next section in the boundary conditions and the driving. However, both models share a number of qualitative features.

We assume that the velocity field can be expanded in a Fourier series,

$$\mathbf{u}(\mathbf{x}, t) = \sum_{\mathbf{k}} \mathbf{u_k}(t) e^{i\mathbf{k}\cdot\mathbf{x}}. \tag{10}$$

Incompressibility requires $\mathbf{u_k} \cdot \mathbf{k} = 0$. The Navier-Stokes equation with a volume force $\mathbf{f}$ reads

$$\dot{\mathbf{u}} + (\mathbf{u} \cdot \nabla)\mathbf{u} = -\nabla p + \frac{1}{Re}\Delta\mathbf{u} + \mathbf{f}. \tag{11}$$

It has a nonlinearity in the advection part $(\mathbf{u} \cdot \nabla)\mathbf{u}$ only, and the Fourier modes have to selected such that they contain non-trivial contributions from this term. A simple analysis shows that one has to include $\mathbf{k}$-vectors that add up to a triangle. However, the simplest case, with six vectors in a plane forming a regular hexagon, only shows trivial dynamics. A model with a deformed hexagon and two modes driven can mimic the dynamics of the two dimensional model of section 2 (B. Eckhardt, unpublished). Thus more vectors are needed. The model used in Eckhardt and Mersmann (1997) has 18 wave vectors (and thus 36 degrees of freedom), but by partially exploiting symmetries, the number of degrees of freedom is subsequently reduced to 19. Further reductions are conceivable, but perhaps at the cost of losing some features of the dynamics.

Thus while we end up with a few degree of freedom system, the set of equations is too complicated to be displayed here and was calculated with the help of a algebraic manipulation program (see Eckhardt and Mersmann 1997). However, it is amenable to detailed and extensive numerical investigations.

For low Reynolds numbers, there is a single stable stationary state that corresponds to a flow following the driving, hence referred to as the laminar state. This laminar flow is stable for all Reynolds numbers. At a Reynolds number of about 190 new stationary states are born in saddle node bifurcations. As is typical for saddle node bifurcations, they come in pairs of states where one has one unstable direction more than the other. In addition, these states come in groups related to each other by symmetry transformations. In the present example, the saddle-node bifurcation takes place in 8 regions in phase space simultaneously!

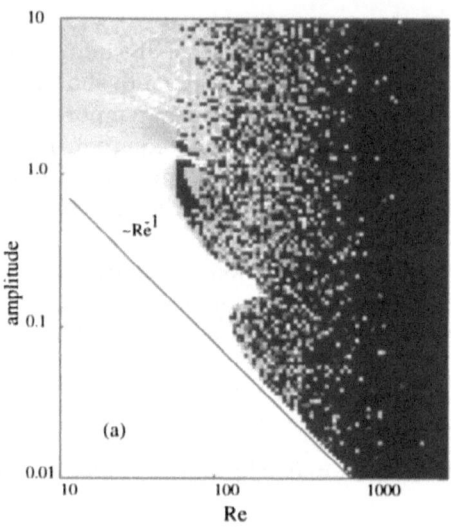

Fig. 2. Life time of perturbations as a function of amplitude and Reynolds number Re for a model shear flow (Eckhardt and Mersmann 1997). Black boxes indicate life times larger than the integration time, light boxes indicate short life times. The boxes stand for a single trajectory started on a rectangular grid in this parameter plane. Magnifications are shown in Fig. 3.

Moreover, *all* new stationary states are saddles with stable and unstable manifolds. The dynamics of perturbations outside the domain of attraction to the laminar profile thus will be much more complicated. Phase space trajectories might be attracted along the stable manifold of some saddle and might leave along some unstable direction. Further away, another saddle might dominate, attract along its stable direction and push out along an unstable one and so forth. As there are no stable excited states to relax to, this can go on forever. However, there is a small chance that the trajectory will escape this tangle and find its way to the laminar profile, in which case one would speak of a transient turbulent excitation.

One way to test this picture is to map out life times of perturbations as a function of amplitude and Reynbolds number as shown in Fig. 2. For small amplitude or Reynolds number perturbations decay smoothly on a short time scale. For $Re > Re_{cr} \approx 190$ there is a growing domain in initial amplitude where the life time changes abruptly with varying amplitude. Magnifications of the interval by 10^7 still show rather rapid variations (Fig. 3). It is thus fair to expect these fluctuations to continue on all scales. This type of behaviour is reminiscent of that in irregular scattering, e.g. of vortex pairs (Eckhardt and Aref 1988, Eckhardt 1988).

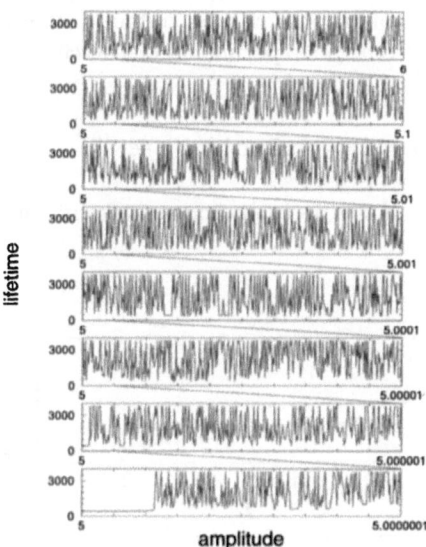

Fig. 3. Successive magnifications of the life time of perturbations as a function of amplitude for fixed Reynolds number (Eckhardt and Mersmann 1997).

4 Plane Couette flow

We finally turn to full numerical simulations of plane Couette flow, a shear flow between two parallel walls, a distance h apart and moving with velocity $2U_0$ relative to each other. The Reynolds number is defined as $Re = U_0 h/2\nu$ with ν the kinematical viscosity. The coordinate system is oriented such that the x-axis is along the flow direction, the y-axis along the neutral direction and the z-axis along the shear direction, perpendicular to the plates.

In our numerical model we approximate **u** by a finite set of Fourier modes in the x, y-plane with periodic box-lengths 2π and normalized Legendre polynominals in z-direction,

$$\mathbf{u} = \sum_{k_x,k_y,p} \tilde{\mathbf{u}}_{k_x,k_y,p} e^{i(k_x x + k_y y)} L_p(z) . \tag{12}$$

Boundary conditions and incompressibility are taken into account with a Lagrange formalism of the first kind (a slight modification of the usual Galerkin approach). All in all we have 962 independent dynamical degrees of freedom.

As initial conditions we take localized vortex rings which can be oriented in various directions. Here we take a ring in the x-z-plane with axis along the y-direction. Such initial conditions may be induced experimentally by a wire spanned across the flow as in the experiments of Dauchaut and Daviaud (1995) The initial flow is

$$\mathbf{u}(t=0) = \operatorname{curl}\operatorname{curl}\delta(x,y)\,e^{-10z^2}\,\mathbf{e}_y . \tag{13}$$

As a measure of the size of the perturbation we take its energy, $E = \int d^3x\,|\mathbf{u} - \mathbf{u}_0|^2$.

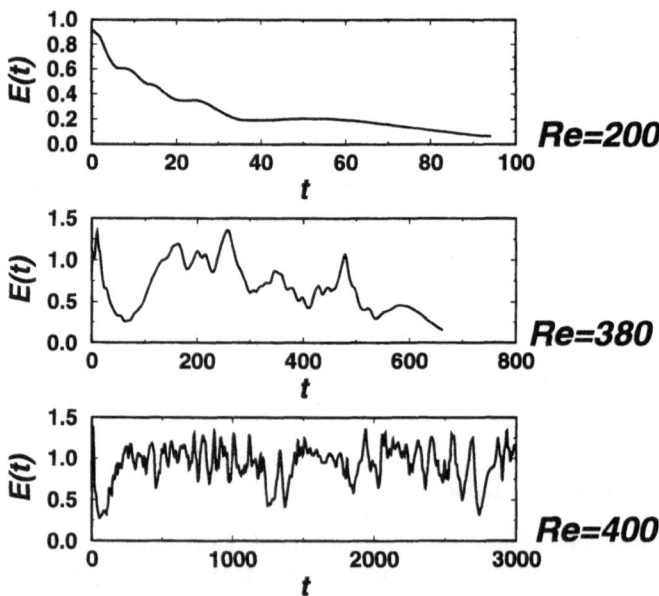

Fig. 4. Time evolution of the energy of a disturbance (13) for three different Reynolds numbers. Note the different scales of the time axis.

In Fig. 4 we show the time evolution of the same perturbation for different Reynolds numbers. For $Re = 200$ the energy content decays until about $t = 40$, after which one notes a weak increase and then the final decay to zero. A closer inspection of the energy distribution shows that at $t \approx 40$ the energy content of the modes with flow component in the z-direction has essentially decayed. The system then enters the linear regime and the subsequent hump is the linear amplification due to the non-normality.

For the larger Reynolds number $Re = 380$ there is an initial decay of the energy followed by a built up of large oscillations for times up to about $t = 550$. This is followed by a hump and a final decrease to zero. In the dynamical system picture used before, this intermediate long lived state reflects the reshuffling between different stationary states before the system can finally relax to the laminar state.

For the even higher Reynolds number $Re = 400$ one notes a more complicated time evolution extending for more than five times the life time of the perturbation at $Re = 380$, without any sign of decay. This corresponds to a fully turbulent state.

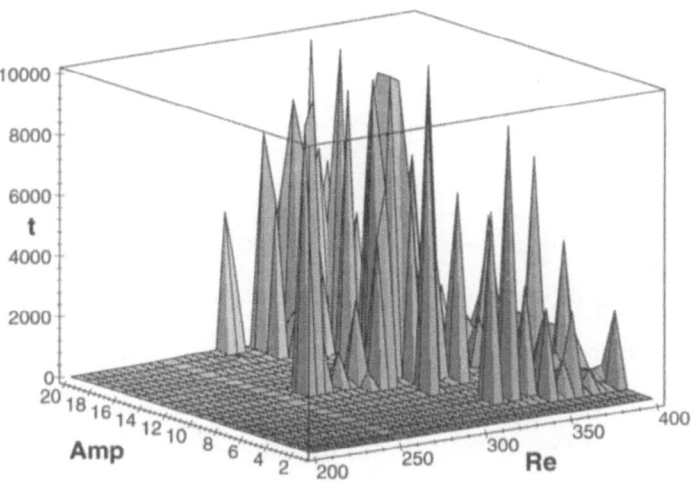

Fig. 5. Landscape plot of life time versus amplitude and Reynolds number for a perturbation of type (13). The numerical cut-off in the computational time was $T_{\max} = 10000$. For the experiments of (Daviaud, Hegseth and Bergé, 1992) on water and a Reynolds number of about $Re = 400$ this corresponds to about 10 min.

A complete map of life time versus Reynolds number and amplitude (Fig. 5) again shows the fractal and non-smooth behaviour familiar from Fig. 2. There is a sensitive dependence on the size of the perturbation and on the Reynolds number (for further details see (Schmiegel and Eckhardt, 1997)).

Thus far the behaviour of the model of section 3 and the full numerical simulations has been rather similar. The novel feature connected to the high dimensionality of the full simulation is that the onset of long lived perturbations need not be close to the saddle-node bifurcation. The bifurcation occurs for our aspect ratio at about $Re = 267$, whereas the first long lived states appear near $Re \approx 270$. If the initial conditions are changed to rings with axis perpendicular to the flow direction (Schmiegel and Eckhardt, 1997), the first long lived states appear as late as $Re \approx 350$! Thus the stable and unstable manifolds of the saddles are curled up in parts of this high-dimensional space and do not explore it isotropically.

We close by showing a stationary state at $Re = 267$ in Fig. 6. It consists of a local vortex in the direction of the flow, similar to the structures found by Nagata (1990) and Clever and Busse (1997). The other state born in that bifurcation has a slightly different modulation in the flow direction and the opposite orientation of the vorticity.

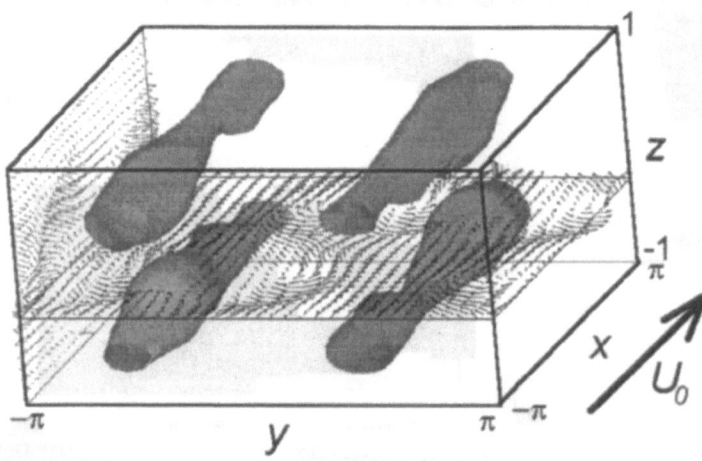

Fig. 6. Steady state at $Re = 267$. The shaded structures are surfaces of the same kinetic energy and indicate the tubular structure of the stationary state.

5 Summary

Before turning to a comparison with experiment, let us summarize the main features of the models studied. The simple two degree of freedom model of section 2 showed a transition to a nontrivial stationary point for sufficiently large perturbations as the control parameter exceeded a critical value. For large values of the parameter, the basin of attraction of the laminar state became smaller. In addition to these features the model shear flow of section 3 showed new stationary states that were saddles rather than attractors. As a consequence the dynamics beyond the basin of attraction of the laminar state becomes complicated and the life time distribution of perturbations shows fractal features. Finally, the simulation of the full plane Couette flow in section 4 revealed significant anisotropies in phase space, with some finite amplitude perturbations becoming unstable at Reynolds numbers much higher than the critical one for the onset of bifurcations.

Many of these findings are in qualitative agreement with experiments. For instance, the experiments of (Ahydin and Leutheusser, 1979) use initial perturbations similar to a vortex ring with axis perpendicular to the flow and report a critical Reynolds number of $Re \approx 280$. The experiments of (Tillmark and Alfredsson, 1992) on the other hand, have a vortex ring essentially in the z-direction and report a critical Reynolds number of about 380. That the first is lower than the second is in agreement with our findings. Small quantitative differences may be attributed to dependencies on aspect ratio, which we did not yet investigate.

The most spectacular prediction of the present calculations is the fractal behaviour of the life time distribution. This is difficult to study experimentally, but it is clear that the wide variation of life times means that experiments at a fixed Reynolds number, even with a very good control on the size and shape of the initial perturbation will show large fluctuations in the life time of the perturbation. The large fluctuations observed by (Daviaud, Hegseth and Bergé, 1992) may be taken as evidence for that.

Further evidence for this fractal stability border as well as a hint for the universality of it may be derived from experiments of (Darbyshire and Mullin, 1995) on pipe flow. They monitored perturbations of well defined amplitude at different Reynolds number for some finite time. In an amplitude vs. Reynolds number plane there will be mostly decaying states for small a and/or Re and turbulent, long lived ones for large a and Re. The intermediate transition region shows turbulent states next to decaying ones and no sharp border. Adapted to our calculations, their life time picture corresponds to a cut through the landscape of Fig. 2 or Fig. 5 at some time T, with initial conditions below T identified as decaying and above T as turbulent. One then ends up with a diagram close to theirs.

The discussion here has been narrowed to a single observable, the life time of perturbations. Further insights can presumably be gained from an analysis of the time series as well as the spatial modulation of the flow field. Some aspects of this are currently under investigation.

References

Ahydin, M. and Leutheusser, H.J. (1979): Novel experimental facility for the study of plane Couette flow. *Rev. Sci. Instrum.* **50**, 1362-1366

Baggett, J.S. and Trefethen, L.N. (1997): Low-dimensional models of subcritical transition to turbulence, *Phys. Fluids* **9**, 1043-1053

Boberg, L. and Brosa, U. (1988): Onset of turbulence in a pipe, *Z. Naturforsch.* **43a**, 697

Chandrasekhar, S. (1961): *Hydrodynamic and hydromagnetic stability*, Oxford University Press

Clever, R.M. and Busse, F.H. (1997): Tertiary and quarternary solutions in plane Couette flow, *J. Fluid Mech.* **344**, 137

Cross, M.C. and Hohenberg, P.C. (1993): Pattern formation outside equilibrium, *Rev. Mod. Phys.* **65**, 851

Darbyshire, A.G. and Mullin, T. (1995): Transition to turbulence in constant-mass-flux pipe flow, *J. Fluid Mech.* **289**, 83-114

Dauchot, O. and Daviaud F. (1995): Streamwise vortices in plane Couette flow, *Phys. Fluid.* **7**, 901-903

Daviaud F., Hegseth J. and Bergé P. (1992): Subcritical transition to turbulence in plane Couette flow, *Phys. Rev. Lett.* **69**, 2511-2514

Drazin, P.G. and Reid, W.H. (1981): *Hydrodynamic stability*, Cambridge University Press

Eckhardt, B. (1988): Irregular scattering, *Physica D* **33**, 89

Eckhardt, B. and Aref, H. (1988): Integrable and chaotic motion of four vortices: II collision dynamics of vortx pairs, *Phil. Trans. R. Soc. London A* **326**, 655

Eckhardt, B. and Mersmann, A. (1997): Transition to turbulence in a shear flow, preprint

Grossmann, S. (1996): in *Nonlinear physics of complex systems*, J. Parisi, S.C. Müller and W. Zimmermann, Springer, Berlin, pp. 10

Gebhardt, T. and Grossmann, S. (1994): Chaos transition despite linear stability, *Phys. Rev. E* **50**, 3705

Lorenz, E.N. (1963): Determinisitic nonperiodic flow, *J. Atmos. Sci.* **23**, 130

Lundbladh, A. and Johansson A.V. (1991): Direct simulation of turbulent spots in plane Couette flow, *J. Fluid Mech.* **229**, 499-516

Nagata, M. (1990): Three-dimensional finite-amplitude solutions in plane Couette flow: bifurcation from infinity, *J. Fluid Mech.* **217**, 519-527

Risken, H. (1984): *The Fokker-Planck Equation. Methods of Solution and Applications*, Springer

Saltzmann, B. (1961): Finite amplitude free convection as an initial value problem I, *J. Atmos. Sci.* **19**, 329

Schmiegel, A. and Eckhardt, B. (1997): Fractal stability border in plane Couette flow, preprint

Tillmark, N. and Alfredsson, P.H. (1992): Experiments on transition in plane Couette flow, *J. Fluid. Mech* **235**, 89-102

Trefethen, L.N., Trefethen, A., Reddy, S. and Driscoll, T. (1993): Hydrodynamic stability without eigenvalues, *Science* **261**, 578

Deformation of Charge Density Waves in Quasi–One–Dimensional Semiconductors Visualized by Scanning Electron Microscopy

Georg Heinz[1], Vadim Ya. Pokrovskii[2], Matthias Goldbach[1], Achim Kittel[1], and Jürgen Parisi[1]

[1] Faculty of Physics, Department of Energy and Semiconductor Research, University of Oldenburg, D–26111 Oldenburg, Germany
[2] Institute of Radioengineering and Electronics, Russian Academy of Science, Moscow, Russia

Abstract. It is well known that charge density wave systems exhibit a variety of phenomena related to the existence of metastable states: Electric measurements reveal hystereses in both their thermal and electric behavior as well as different kinds of memory effects. Although these integral methods provide strong evidence for metastability, they can not uncover the spatial structure of the charge density wave. As predicted by the theory of Fukuyama, Lee, and Rice, the charge density wave should organize in domains obeying a characteristic length scale. We have performed investigations on a TaS_3 sample with a low–temperature scanning electron microscope, using the electron beam as weak local heat source and recording the thermopower as a function of the beam position. We demonstrate that the structures observed can be related to metastable deformations of the charge density wave.

1   Introduction

Since the beginning of the seventies, many different compounds have been synthesized that exhibit a strong anisotropy in their electric conductivity (Wudl (1975), Zahradnik et al. (1971)). Already in the thirties, Peierls predicted that one–dimensional metals are unstable against a static lattice distortion with the wave vector equal to twice the Fermi vector ($q = 2k_F$) (Peierls (1955)). Due to the interaction between the charge carriers and the phonon system, an energy gap opens around the Fermi level, giving rise to a transition from metallic to semiconducting behavior just at the Peierls temperature T_P. In some of these chain–like crystals, the charge carriers condense into a macroscopic quantum state, the so–called charge density wave (CDW). It consists of a static lattice distortion with the wave vector $q = 2k_F$ that is accompanied by a periodic modulation of the density of the charge carriers. Under the influence of an external electric field larger than a threshold value E_T, the CDW can overcome the pinning forces of the impurities that are always present in a real system, leading to a nonlinear current–voltage characteristic. Another evidence for the collective motion of the condensate is the appearance of current and voltage oscillations (narrow band noise) above the threshold electric field.

Other phenomena like hystereses in the thermal and electric behavior and those summarized under the expression "memory effects" could be explained with the existence of many metastable states of the CDW (Dumas and Schlenker (1983)): Because of its inner degrees of freedom, the CDW is deformed in the potential of the randomly distributed pinning centers, the configuration of the deformations being dependent of the electric and thermal history of the sample. In the following, we image the spatial structure of those metastable states via low–temperature scanning electron microscopy.

2   The Phase–Slip Model

Let us first have a look at the consequences of deformations of the CDW, following the phase–slip model described by Pokrovskii and Zaitsev-Zotov (1989). Under the assumption that the length scale of the deformations is much larger than the average distances between the impurities, the influence of the pinning centers can be expressed via the threshold electric field E_T (in the literature discussed as weak pinning model). Due to the fact that any deformation of the CDW is accompanied by a separation of charges, the deformations also lead to changes in the local chemical potential ζ.

Taking note of the fact that the CDW begins to creep as soon as the internal electric field exceeds the threshold field E_T, the spatial derivative of the chemical potential $d\zeta/dx$ corresponding to the local electric field can not exceed E_T. This means that the relation

$$\left|\frac{d\zeta}{dx}\right| \leq e_0 E_T \tag{1}$$

is valid all over the sample (see Fig. 1(a)), e_0 being the electron charge. Further allowing that the CDW will brake up (phase–slip act) as soon as the internal stress exceeds a critical value, a second condition for the deviation of the chemical potential from its equilibrium value, $\delta\zeta = \zeta - \zeta_{eq}$, can be derived, namely, $|\delta\zeta| \leq \zeta_{cr}$.

Figure 1(a) gives a one-dimensional scheme for the chemical potential extending along the sample before (solid) and after (dotted) a phase–slip act, i. e., the addition or removal of one period of the CDW. A simple calculation shows that the length, over which the CDW rearranges, is given by

$$L_{2\pi} = \sqrt{\frac{4\pi}{e_0 E_T}\frac{d\zeta}{dq}}, \tag{2}$$

where q describes the wave length of the CDW. Equation (2) corresponds to the length of a Fukuyama–Lee–Rice domain, and an estimation ends up with values in the range of $40\mu m$ for real systems (Gruener (1994)).

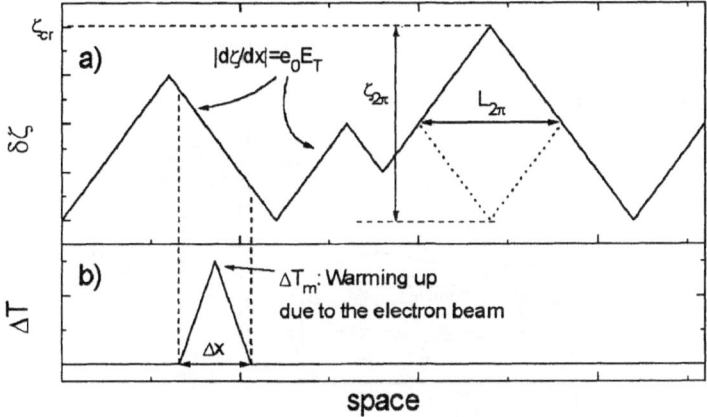

Fig. 1. *(a): The phase–slip act. The variation of the chemical potential is plotted as a function of space. The CDW rearranges in a region of the coherence length. In (b) the variation of the temperature is plotted as a function of space. The local heating due to the electron beam is assumed to be of triangular shape.*

3 The Thermoelectric Effect

The above considerations have shown that a deformation of the CDW leads to a spatial variation of the chemical potential. It is well known that not only a difference in temperature at the contacts (Seebeck effect) produces a voltage drop along the sample. There are also thermoelectric effects that arise from a nonuniform distribution of temperature along the sample, even though we have the same temperature at the contacts. It holds when the chemical potential along the sample is inhomogeneous. The resulting thermopower S for a p–type conductor, like the investigated TaS$_3$ samples, is given by

$$S = \frac{\Delta - \zeta}{e_0 T},\qquad(3)$$

where Δ denotes the energy gap (Artemenko et al. (1996)). The resulting thermal voltage is given by the integral

$$V_{th} = \int S dT$$
$$= \int_0^L S(x)\frac{dT}{dx}dx\qquad(4)$$

where L is the length of the sample, and the thermopower S being dependent on the spatial coordinate x.

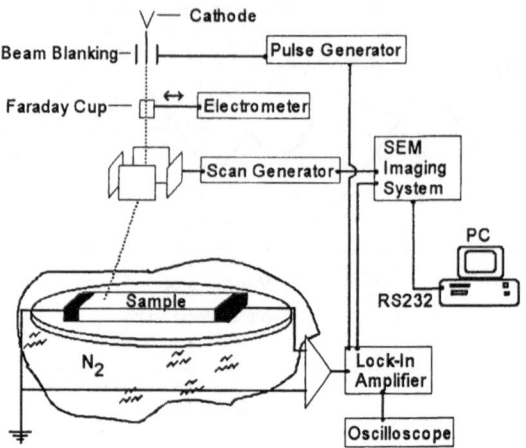

Fig. 2. *Scheme of the experimental set-up for the investigations with the low–temperature scanning electron microscope. The thermovoltage across the sample is measured as a function of the coordinates of the electron beam. The continuous electron beam is chopped by the beam–blanking unit with the frequency f_B, allowing the use of a lock–in amplifier for the detection of small signals. The beam current I_B at the position of the sample can be determined with a removable Faraday cup.*

If the maximum local heating of the electron beam is ΔT_m and the diameter of the spot is Δx (as depicted in Fig. 1(b)), the thermal voltage reads

$$V_{th} = \frac{\Delta x \, \Delta T_m}{2} \frac{dS}{dx}$$
$$= \frac{\Delta x \, \Delta T_m}{2 e_0 T} \cdot \frac{d\zeta}{dx}, \tag{5}$$

being proportional to the local variation of the chemical potential and, thus, to the deformation of the CDW.

4 Experimental Details

Figure 2 shows the experimental set-up for thermovoltage measurements in the scanning electron microscope. A more detailed treatment can be found elsewhere (Heinz (1997)). A cryostage has been constructed that allows to cool the sample holder with liquid nitrogen. The drift of the temperature, an important aspect for these measurements, was less than $20 \; mK$ during 15 minutes. To achieve a good thermal contact to the nitrogen medium, the sample was fixed on a sapphire disc. The thermal contact to the substrate has been proven to be homogeneous all along the sample surface. The electric

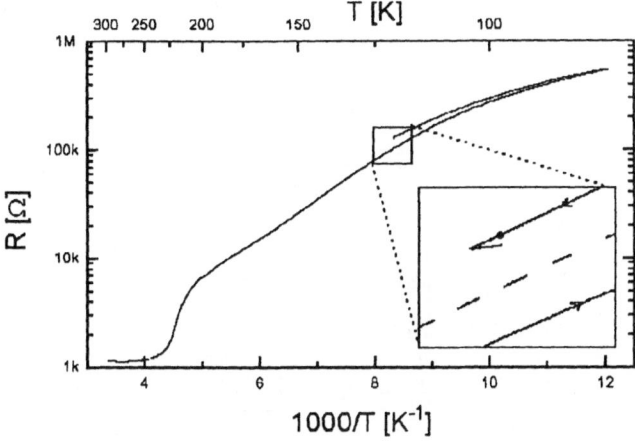

Fig. 3. *The resistance of a TaS₃ sample as a function of the inverse temperature. The inset shows the influence of the electron beam on the resistance of the sample. Starting from the dot, the heating due to the beam ($\Delta T \approx 0.5\,K$) leads only to a minute change in R.*

contacts were made of indium. The dimensions of the TaS$_3$ sample investigated are 1250x4x1 μm^3. One end of the sample was grounded, while the other one was floating. To improve the signal to noise ratio, a high–impedance low–noise preamplifier has been used to measure the voltage drop along the sample. The output of the preamplifier was led to a lock–in amplifier, using the beam–blanking frequency (f_B) as reference signal.

To prevent destruction of the metastable states in the CDW, the power of the beam was reduced as much as possible. The acceleration voltage V_{HV} was chosen to be only 5 kV, and the beam current I_B was typically 1 nA. For a further reduction of the local heating, the electron beam was defocussed to a diameter of 1.5 μm. The heating of the electron beam could be determined directly by measuring its influence on the resistance of our sample. Since the thermal behavior of the resistance is known from electric measurements (as shown in Fig. 3), we can calculate the local change in temperature leading to values of about 0.5 K (with $V_{HV} = 5\,kV$ and $I_B = 1\,nA$). As depicted in the inset of Fig. 3, such a small change in temperature leads only to a weak perturbation of the metastable states; the resistance does not reach its equilibrium value (dashed line in the middle of the hysteresis loop).

5 Experimental Results

Figure 4 displays two images where the thermal voltage induced by the electron beam was used to define the brightness of each pixel. Two measurements

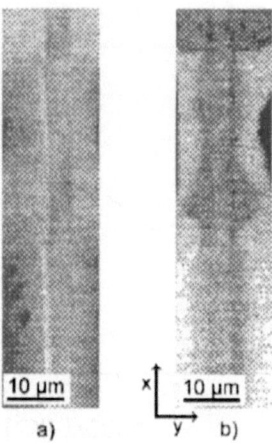

Fig. 4. *Spatial distribution of the thermovoltage in a TaS₃ sample obtained by scanning electron microscopy. In (a) the upper contact was grounded, while the lower one was floating. In (b) the connections were exchanged. Bright regions correspond to large voltages, dark regions to small voltages
(T = 155 K, V_{HV} = 5 kV, I_B = 1 nA, f_B = 490 Hz).*

with exchanged connection to ground were performed, in order to eliminate
the contribution of the injected current (via the electron beam) to the voltage drop along the sample. Because the width of the sample investigated
is only 4 μm and the beam was defocussed up to a diameter of 1.5 μm, the
data corresponding to the sample in Fig. 4 was averaged along the y–axis.
Thus, some kind of a one–dimensional sample will be assumed in the following
considerations.

The elimination procedure is as follows: The electron beam (beam current I_B at position x) will induce a voltage drop in the part of the sample
between x and grounding, $V_B(x) = R(x)I_B$. With the total resistance of the
sample, R_S, and the length of the sample, L, we can write $R(x) = R_S x/L$.
By changing the connection to ground, we obtain two data sets of the form

$$V_1(x) = V_{th}(x) + R(x)I_B \tag{6}$$
$$V_2(x) = V_{th}(x) - R(L - x)I_B, \tag{7}$$

where V_{th} is the signal of interest, namely, the voltage due to thermoelectric effects. Both data sets are plotted in Fig. 5 (curves (b) and (d)). The
difference between the data sets gives the (nearly constant) injection contribution $V_1 - V_2 = R_S I_B$ (curve (a)). Upon knowing the resistance of the
sample, we can determine the injection current I_B that corresponds to the
number of electrons which penetrate into the sample. The sum of both data
sets is plotted in curve (c), the dashed curve being the net thermal voltage

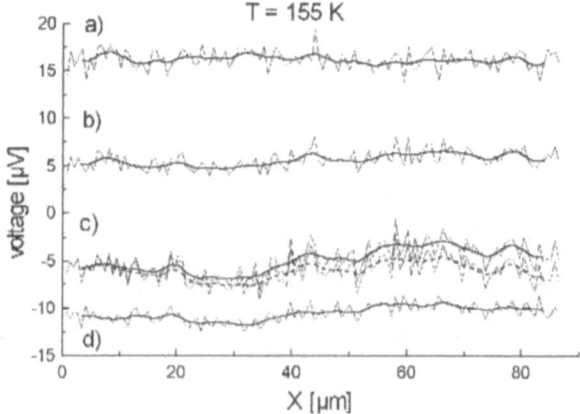

Fig. 5. *The data sets (b) and (d) represent the data of Fig. 4(a) and (b), respectively, averaged over the sample width along the y direction. As guide for the eye, the original data and a filtered version are plotted. For the determination of the injection contribution of the electron beam, the difference of (b) and (d) is plotted in (a). In (c) the solid line is the sum of (b) and (d) and the dashed curve represents the thermal voltage after elimination of the injection contribution.*

after elimination of the injection contribution.

Following (5), in Fig. 6 we calculate the inner electric field (corresponding to $d\zeta/dx$) from three sets of pictures, the first measured at $T = 147\,K$, the second at $T = 155\,K$, and the third at the same temperature, yet after applying a voltage pulse of amplitude $V > V_T$ (V_T means the threshold voltage according to the threshold electric field E_T). As expected from the phase–slip model, the maximum value of the variations in Fig. 6(a) is of the same order as the threshold electric field $E_T \approx 2\,V/cm$, extracted from current–voltage characteristics at the corresponding temperatures. Moreover, the length scale of these variations ranging from about 10 to 20 μm is in good agreement with the theoretical prediction (Gruener (1994)) of about 40 μm.

Even though the sample has drifted approximately 10 μm to the left during change and stabilization of the temperature, the main structures are visible in all three curves. The temperature rise from 147 K to 155 K did not change the signal appreciably. However, application of a voltage pulse with an amplitude exceeding the threshold value led to a decrease of the variations in $d\zeta/dx$. The explanation could be as follows: At electric fields larger than the threshold value E_T, the CDW begins to slide. During movement, the internal stress can relax such that the CDW gets a uniform periodicity along the whole sample. After switching off the electric field, the CDW again has to rearrange (i. e., to deform) into the potential of the impurities, in order to minimize its potential energy. As a consequence of the relatively large

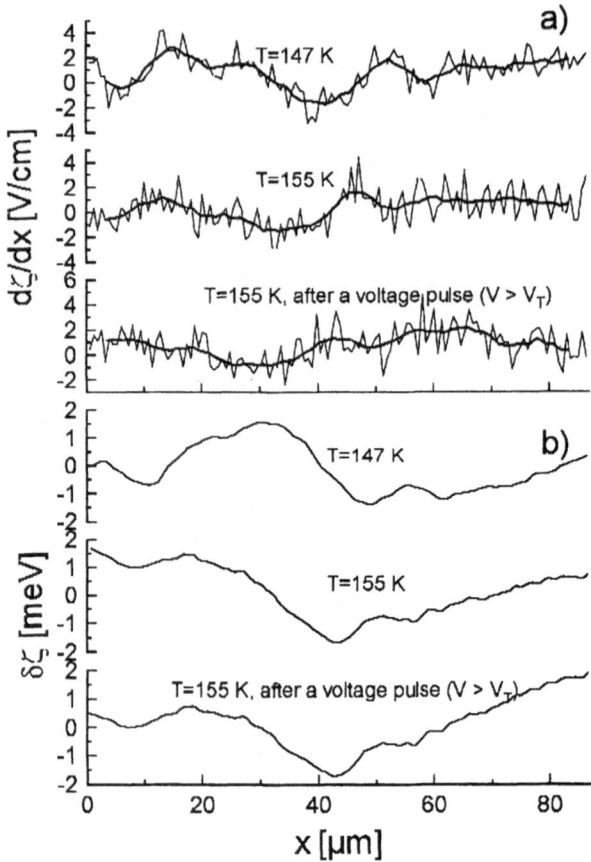

Fig. 6. *In (a) dζ/dx (corresponding to the inner electric field times e_0) is plotted as a function of the x-position for the temperatures $T = 147\,K, T = 155\,K$, and $T = 155\,K$ after applying a voltage pulse with an amplitude larger than the threshold voltage $(V > V_T)$. The integration of the data of (a) gives the spatial variation of the chemical potential δζ, as shown in (b).*

relaxation times in these systems, the deformations of the CDW in the lower curve of Fig. 6(a) are not fully developed, ending up with lower inner electric fields.

The integration of $d\zeta/dx$, finally, leads to the local variations of the chemical potential δζ, plotted in Fig. 6(b). Of course, we can not determine the absolute value of the chemical potential, because the integration produces an undefined constant. But our measurements clearly reveal variations in the chemical potential of about $3\,meV$ that well coincide with values reported by other groups (Artemenko et al. (1996)).

6 Conclusion

Our investigations have shown that a low–temperature scanning electron microscope is a powerful tool for the detection of small changes in the spatial distribution of the chemical potential. Still variations of only $3\,meV$ due to the deformation of the CDW could be detected. The agreement between the results obtained from the pictures gained with the scanning electron microscope and the values obtained from electric measurements turns out to be unexpectedly good. In accordance with the phase–slip model, the inner electric field does not exceed the threshold value E_T. Moreover, the measured changes in the chemical potential ($3\,meV$) also agree well with model calculations (Artemenko et al. (1996)). Unfortunately, we could not detect an abrupt change in the signal (after variation of the temperature and/or application of a voltage pulse) that could be interpreted as a phase–slip act (addition or removal of one period of the CDW). One reason may be the fact that the diameter of the samples investigated was much larger than the transverse coherence length, i. e., the width of the Fukuyama–Lee–Rice domains. Such a restriction strongly complicates the detection of a phase–slip act in only one domain, since the measurement indeed averages over all neighboring domains (inside the defocussed beam). Future investigations on much thinner samples should give an answer to these questions.

Acknowledgment

We thank M. J. Bünner and J. Voit for stimulating discussions and Yu. I. Latyshev, Ya. S. Savitskaya, and V. V. Frolov for furnishing the samples. This work was partially supported by the Deutsche Forschungsgemeinschaft, the Russian Foundation for Basic Research (grant 95-02-05392), and MNTP "Physics of Solid State Nanostructures" (grant No 97-1052).

References

Artemenko S. N., Pokrovskii V. Ya., and Zaitsev–Zotov S. V. (1996): Sov. Phys. JETP **83**, 590.

Dumas J. and Schlenker C. (1983): Solid State Commun. **45**, 885.

Gruener G. (1994): *Density Waves in Solids* (Addison-Wesley, Reading).

Heinz G. (1997): *Räumliche Strukturbildung in quasi–eindimensionalen Leitern*, PhD Thesis (Oldenburg).

Peierls R. E. (1930): Ann. Phys. Leipzig **4**, 121; Peierls R. E. (1955) *Quantum Theory of Solids* (Oxford University Press, New York).

Pokrovskii V. Ya. and Zaitsev-Zotov S. V. (1989): Synth. Metals **32**, 321.

Wudl F. (1975): J. Am. Chem. Soc. **97**, 1962.

Zahradnik R., Carsky P., Huening S., Kiesslich G., and Scheutzow D. (1971): Int. J. Sulphur Chem. C **6**, 109.

Probing Nonlinear Carrier Transport in Semi– and Superconductors via Low–Temperature Scanning Laser Microscopy

A. Kittel[1,2], P. Stagge[1], W. Friedmann[1], and J. Parisi[2]

[1] Physical Institute, University of Bayreuth, D–95440 Bayreuth, Germany
[2] Faculty of Physics, University of Oldenburg, D–26111 Oldenburg, Germany

Abstract. In order to be able to visualize spatial pattern formation during low–temperature carrier transport processes in solid state physics, we demonstrate a new concept of a scanning–laser microscope that is integrated inside an encapsulated cryogenic sample stage. All the components of the microscope, i.e., a laser diode, the deflection unit, and the focussing optics, are cooled to the temperature of the sample under investigation. Furthermore, unwanted excitation of the sample, like excitation due to background radiation, is suppressed. Our technical solution meets the challenging experimental claims of a perfect shielding and low–temperature applicability further enables the sample to be exposed to strong magnetic fields. First results on imaging current inhomogeneities in both a semi– and a superconductor experiment will be outlined.

1   Introduction

In recent years, the investigation of complex charge carrier transport in solid-state physics has come into the center of interest of many scientists. The local inhomogeneity of the current density is an interesting example, because it is correlated with the density of dissipation and/or represents the key quantity to prove theoretical predictions. Such kind of phenomena can be observed in various super– and semiconducting transport processes of different dimensionality (i.e., one dimension: charge density waves; two dimensions: Josephson and quantum Hall effect; three dimensions: flux flow and impact ionization breakdown, etc. (Schlenker (1989), Grüner (1994), Huebener (1988), Mayer et al. (1988), Peinke et al. (1992))). We point out that most of the experimental systems exhibiting those phenomena are extremely sensitive against any kind of excitation. As a consequence, experiments have to be carried out at low temperatures and without any background radiation. We take the sensitivity against radiation, in order to perturb the sample under investigation locally via a focussed laser beam. The change of the integral current through the biased sample related to the laser perturbation at a distinct location on the sample surface is used to get information about the physical quantity of interest (say, the current density) subsisting in the perturbed volume. By the help of a scanning–laser microscope, a complete two–dimensional image of the current density distribution of the sample can be obtained. But due to the experimental requirements, a particular construction of the microscope is demanded that contains an overall shielding against external irradiation, i.e., a noncompromising solution for technical realization of the microscope. So far, conventional low–temperature scanning–laser microscopes (Wilson et al. (1984)) are constructed with a laser and the scanning unit located *outside* an optical cryostat (Brandl et al. (1990), Itkis et al. (1986)). Such kind of solution does not allow for the investigation of those radiation–sensitive systems mentioned above, because it is definitely not possible to avoid that infrared illumination of the sample due to the ubiquitous background radiation penetrates the cryostat on the same way the laser beam is coupled to the sample. Note that either the window of the optical cryostat or the end of the optical fibre at room temperature are nothing but a black body radiator at $300\,K$ temperature. Obviously, we would have severe shielding problems with the well–established scanning–laser microscope, as a consequence of the inevitable lack left for penetration of the beam probe.

The only possible solution of this problem is to incorporate the laser diode, the deflection unit, and the focussing optics *inside* an encapsulated sample stage and to immerse the whole apparatus into the cryostat, in order to cool down all components. There remains a further experimental demand, namely, the possibility either to shield unwanted magnetic fields from the sample (in case of investigating SQUID's) or to apply strong magnetic fields to the sample (in case of investigating quantum Hall systems). Since the compact scanning–laser microscope is located in close vicinity to the sample, all components of the microscope are subjected to and, thus, influenced by the strong magnetic field. This fact has to be taken into account when selecting the possible realization of detail problems. It is worth mentioning that the area of application of our microscope is by far not limited to carrier transport experiments. There are also investigations of fluorescence, reflection or other physical quantities which have to be carried out at low temperatures, the absence of background radiation, and/or strong magnetic fields.

2   The apparatus in detail

In the following, the construction of the low–temperature scanning–laser microscope (LTSLM) is introduced and discussed. To capture the different components of the LTSLM in a systematic way, the order of treatment is defined by the sequence the laser beam passes the individual components.

The LTSLM is fixed inside an evacuated metal cylinder (that is, a tube sealed by a top and a bottom flange all made from stainless steel with $200\,mm$ height and $50\,mm$ diameter) immersed into a conventional bath cryostat. All components are mounted on a base plate made of brass or copper which is arranged along the rotational axis of the cylinder and thermally coupled to the cooling liquid (e.g., LHe or LN_2) via the bottom flange. The latter also carries the sample. Moreover, the plate serves as cooling bath capable to fix the temperature of the electrical wiring of the laser diode, the scanning unit, the focussing unit, the temperature sensor, and the sample.

The laser diode we are using is a *GaAs* semiconductor laser which emits light of $670\,nm$ wavelength at room temperature. Certainly, we can also use any other semiconductor laser. Due to the fact that a semiconductor laser is a device where the majority carriers are responsible for the functioning of the device, it even works at low temperature. Since the electric losses determined by recombination processes of the charge carriers (without light emission) are decreasing with temperature, the threshold current value I_{th} for the laser mode diminishes accordingly (in our case for a $1\,mW$ laser made by Hitachi, I_{th} typically is $40\,mA$ at room temperature and $7\,mA$ at temperatures of liquid helium). Another effect that comes into play when lowering the temperature is the decrease of the wavelength of the emitted light.

The light is emitted from the diode with an aperture of about 30°. It then passes a collimator optics which focusses the light to infinity having a diameter of $3\,mm$ and an angular error of less than a few arcseconds. Further, we have inserted two polarizing films into the beam path to enable a coarse tuning of the light intensity and to polarize the laser light for polarization–direction–sensitive measurements.

Afterwards, the beam passes the scanning unit. During construction, we have spent a large amount of attention on that part of the LTSLM for the following reasons. First, large deflection angles (of about 5°) are necessary for a satisfying scanning area (of about $5 \times 5mm^2$). Second, the overall scanning unit has to be extremely compact. Third, it has to work within an extended range of different temperatures. Fourth, no magnetoelectric actors can be used under (possible) application of strong magnetic fields. Optical deflection based on, e.g., the Kerr effect does not promise any alternative, because high voltage is necessary, the elements are too large, and/or the deflection angle is by far too small. Therefore, we have selected a rotatable resting mirror with a bimorphous piezoelectric actor (furtheron called "piezo"). Such a solution makes it possible to have a compact arrangement embracing the demanded features and a satisfying deflection angle. Applying an electric field to the piezo element leads to a bending of the piezo. To transpose the bending into a rotation of the mirror, a refined construction is necessary, because only the lever ratio defines the deflection angle. The problem remains to optimize the deflection angle versus the mechanical hysteresis, as a consequence of a finite precision of the bearing of the deflecting mirrors. We use watch–maker bearings, in order to get a maximum of precision. But, nevertheless even these bearings release some mechanical play. In fact, the residual play limits the reachable deflection angle by increasing the lever ratio. The coupling between the piezo and the mirror is managed via a bar made of fibre–strengthened epoxide resin. The latter has two kink positions, to get a length compensation between the mirror and the piezo, because both elements have different centers of rotation. The kink position is defined by parts of the bar at which the fibres are not filled with epoxide resin. It takes a lot of sure instinct and some experience to prepare bars that are mobile enough, even at low temperatures, but without unwanted mechanical play.

The next component in the light path is the focussing unit which has to focus the parallel laser beam onto the sample surface with the smallest possible diameter of the Airy disc. The diameter determines the extension of the area where the sample is perturbed by the laser beam and, therefore, the spatial resolution of the scanning–laser microscope. In case of a linear or a well–known reaction of the measured signal on the intensity of perturbation, it is possible to improve the resolution by computing the deconvolution of the measured signal with the response function. But in general, the latter represents a complex and unknown function such that the resolution limit is immediately defined by the perturbed area. For all that, we use a "lens of best shape" which is optimized for the demand of focussing a parallel beam. As a possible improvement, one also can select an objective, like it is used for standard optical microscopes with a shorter focus length. Herewith, the resolution of the microscope can be improved by a factor of four, that is, we end up with a diameter of the Airy disc of about $3\,\mu m$. It is necessary to alter the position of the lens to compensate for the changes in distance between the lens and the sample surface conditional on the temperature variation. For realization, we have constructed a piezoelectric motor that follows a stick–slip principle and even works under vacuum and low–temperature conditions. The motor enables one to move up and down the lens over a few millimeters. It is completely made of stainless steel and consists of three parts, namely, a support for the three bimorphous piezoelectric actors at one end, an outer part of the thread glued to the piezos at the other end, and the inner part of the thread which carries the focussing lens just representing the rotor. One advantage of our solution compared to a screw drive based adjustment is that no mechanical feedthrough from outside the vacuum chamber to the lens support is necessary. Therefore, a compact construction can be achieved.

The sample itself is mounted on a sapphire window, for the sake of having an electrically insulating substrate with a high thermal conductivity. The sapphire substrate is sealed with indium and fixed by a flange. The back plane of the sapphire is in direct contact with the cooling liquid. The temperature can be reduced by lowering the pressure of the cooling liquid. For the case of the medium liquid helium, temperatures down to $1.6\,K$ can be reached. On the other hand, heating and temperature stabilization can be performed by the help of a modified sample cooling stage. Hereto, we complete the apparatus with a heating wire and a temperature sensor, both fixed at a flange made of copper. Temperatures ranging from $1.6\,K$ to $350\,K$ are now accessible.

Figure 1 (a) gives the scheme of the LTSLM. The base plate (1) at the top contains a set of drill holes (2) that enable thermal coupling of the elec-

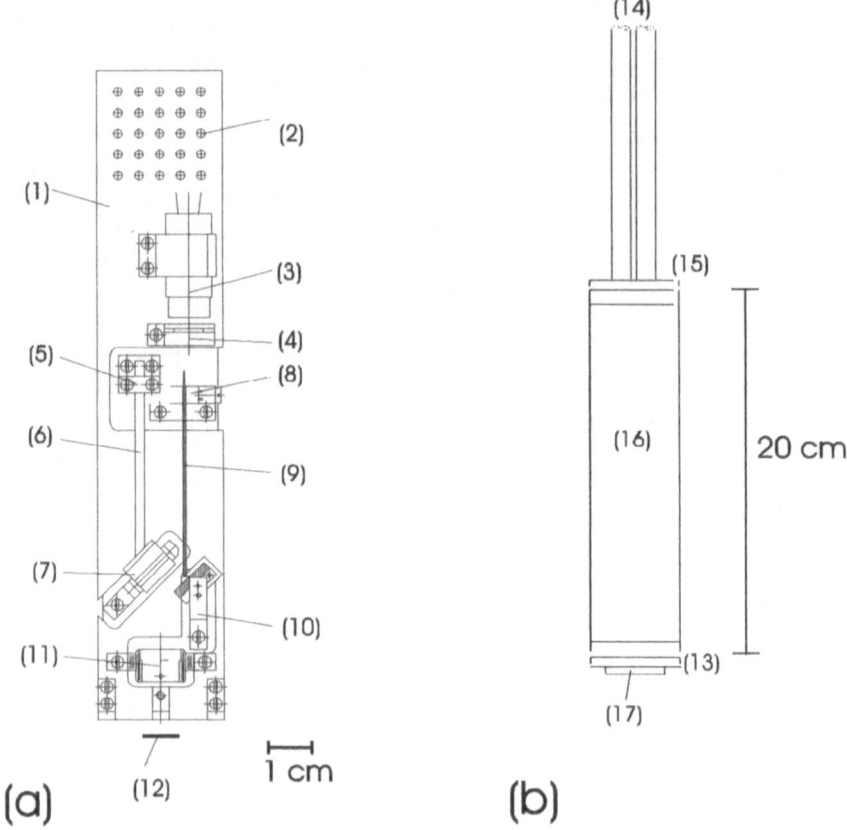

Fig. 1. (a) Scheme of the scanning–laser microscope: (1) base plate, (2) drill holes for thermal coupling of the electric wiring, (3) laser diode with collimator optics, (4) polarization filter, (5) fixing of the x–direction piezo, (6) x–direction piezo, (7) x–direction mirror, (8) fixing of the y–direction piezo, (9) y–direction piezo, (10) y–direction mirror, (11) focussing lens, (12) sample position. (b) Scheme of the lower part of the microscope holder: (13) bottom flange, (14) tubes for fixing the can in the cryostage, (15) top flange, (16) center cylinder, (17) window for the substrate. The length scales are indicated.

tric wiring. Below, we have arranged the laser diode including the collimator optics (3) together with a polarization filter (4) and the deflection unit. The latter embraces the fixing (5), the bimorphous piezoelectric element (6), and the mirror (7) for beam deflection in the x–direction as well as the corresponding fixing (8), piezo (9), and mirror (10) in the y–direction. At the bottom of the base plate, the focussing lens (11) completes the "flying spot" part of the scanning–laser microscope. The position of the sample to be investigated is marked by (12). Figure 1 (b) outlines the lower part of the microscope holder with the vacuum can. Inside the can, the microscope shown in Fig. 1 (a) is mounted on the bottom flange (13). Three tubes (14) made of stainless steel fix the can in the cryostage. Moreover, they serve as a connection to the vacuum pump (not shown here) and guide the electric wiring for the supply of the laser diode, the pin diode, the piezo, and the sample. The top flange (15) is welded to the three tubes and sealed with indium to the center cylinder (16) of the vacuum can, which in turn is sealed to the bottom flange (13). The bottom flange possesses a window (17) for the substrate carrying the sample. To stabilize and tune the temperature, the bottom flange can be substituted by a flange with an integrated heater element combined with a resistance temperature device or a thermocouple.

Finally, the operating principle of the LTSLM is briefly addressed. The basic idea behind the imaging process starts from focussing the laser beam on the surface of the sample to be investigated. As a result of the interaction, one can obtain local information on such physical quantities like the reflection, the fluorescence, or the change in electric conductivity. For measuring the reflection or fluorescence of the sample, we have properly arranged a pin diode next to the focussing lens. By scanning the laser spot over the whole surface, the quantity of interest can be mapped as a function of the beam position. Upon transcribing to a gray scale representation, we also take the brightness–modulated image to illustrate the local distribution of the system response.

3 Characterization of the apparatus

In order to demonstrate the functioning of our microscope, we first want to present some pictures simply taken by recording the reflected laser light via the pin diode at different operation conditions. Figure 2 (a) exhibits a scanned image of a negative film with different test patterns taken at liquid nitrogen temperature of $77\,K$ without any magnetic field applied. The length of the square of the chequer–board–like structure is about $84\,\mu m$. The scanning region chosen spans an area of $1\,mm$ times $0.32\,mm$. The details visible as small speckles (already present as pattern distortion on the film) are reproducible by rescanning the area. The image is scanned from the lower left to the upper right corner. The fast movement of the mirror is directed along the vertical direction of the figure. At the lower part of the figure, a stripe–like structure can be seen. It results from a damped oscillation excited by the back movement of the mirror after a line scan. For demonstrating the reproducibility of the scanning process, two successive line scans are plotted in Fig. 2 (b). The spatial resolution limit determined from the measured line scans is about $14\,\mu m$, in correspondence to the theoretical value (about $12\,\mu m$) estimated from the geometrical characteristics.

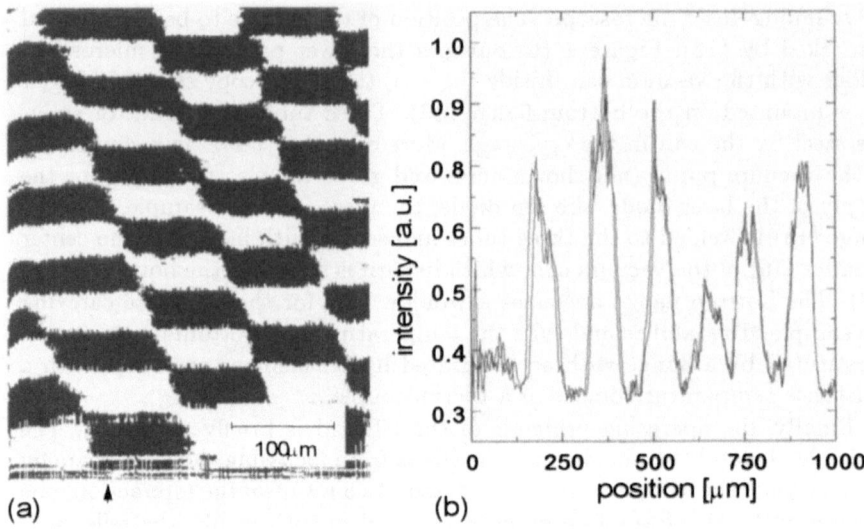

Fig. 2. (a) Picture of a chequer–board–like pattern taken with the scanning–laser microscope at liquid nitrogen temperature by detecting the scattered light. The length scale is indicated at the bottom right. (b) Two neighboring line scans across the picture shown in part (a) made at the position marked by an arrow (scan direction from bottom to top).

Next, we check the operation of the microscope when exposed to strong magnetic fields. Figure 3 exhibits a reflection picture, gained in the same way as mentioned in Fig. 2, yet now taken at room temperature. While part (a) is recorded without applying a magnetic field, part (b) accordingly captures the identical clipping area in the presence of an external magnetic field of about $5\,T$. Comparison of Fig. 3 (a) and (b) gives rise to the conclusion that the functioning of our microscope continues when subject to strong magnetic fields. The only apparent influence of the magnetic field leads to a slight shift of the scanning area.

4 Imaging transport phenomena in semi- and superconductor physics

The main application field of the LTSLM appears to aim at spatially (and also temporally) resolved imaging of electric properties (e.g., electric conductivity, photo current, reverse current, etc.) in a variety of solid state systems. To measure inhomogeneities of such quantities, we take advantage of some

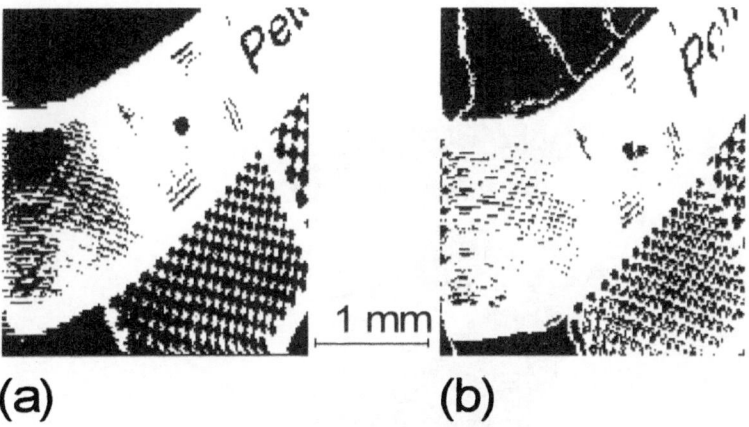

(a)   **(b)**

Fig. 3. Pictures of a test pattern on a black–and–white negative taken with the scanning–laser microscope at room temperature by detecting the scattered light. The length scale is indicated in between the two pictures. The picture of part (a) was taken without an external magnetic field, (b) with an external magnetic field of about 5 T oriented perpendicular to the sample surface and parallel to the largest edge of the base plate.

standard methods already known from scanning–electron microscopy. For example, these are the electron–beam–induced current or voltage (EBIC or EBIV, respectively) and the β–conductivity. Noise reduction techniques like lock–in amplification can easily be implemented by chopping the laser beam. Due to the fast processes inside the laser diode leading to the laser mode, relatively high chopping frequencies are reachable (some tens of MHz).

For illustration, Fig. 4 shows a laser–beam–induced voltage (LBIV) picture of the quasi–one–dimensional conductor blue bronze ($K_{0.3}MnO_4$) cooled down to $77\,K$. Such kind of material is capable to condensate to a (in principle, phase–coherent) charge density wave ground state below the characteristic Peierls temperature (at about $180\,K$ for the present case) (Schlenker (1989), Grüner (1994)). The sample of dimension $550 \times 180 \times 15\mu m^3$ contains ohmic contacts made of indium that surround the opposite ends of the sample in the form of a cap. Such contact configuration enables a current flow into the direction of the longest edges which coincides with the crystallographic b–axis. The sample is glued on a sapphire disc (diameter of $20\,mm$), the largest surface of which is facing the laser beam. During the measurement, the sample was biased by a constant current source of $70\mu A$. We have detected the change of the voltage drop across the sample in direct response to switching on and off the laser diode, the beam of which is focussed to a certain position on the sample surface. As already stated above, the two–dimensional picture results from scanning the laser beam over the sample

356

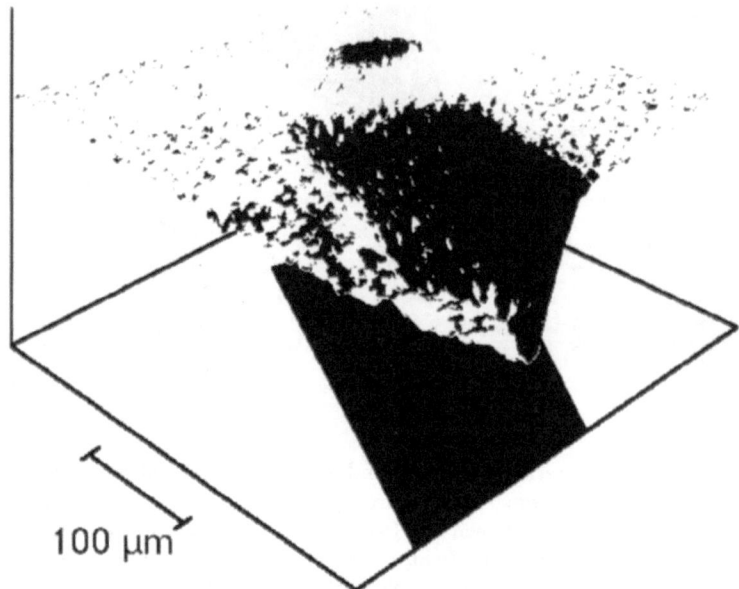

Fig. 4. Picture of the change in the voltage drop induced by a laser beam switched
on and off during scanning over the sample surface. The sample made of blue bronze
exhibits charge density wave conductivity under a constant current bias. The largest
sample surface can be scanned by the laser focus. During the scanning process, the
electric response, i.e., a change in the voltage drop across the sample, is measured
and plotted along the vertical axis in the picture. Note that the longest edges of
the sample are directed nearly along the diagonal of the scanning area pointing
at the rear corner. The extension of the scanning area is larger than that of the
sample surface. For illustration, the area covered by the sample is marked by a gray
rectangle and the contacts by black areas. In the vicinity of the rear contact, an
increase of the voltage drop can be observed. Near the opposite contact, a decrease
of the voltage drop can be measured. (Parameters: temperature $77\,K$, bias current
$70\,\mu A$, sample dimensions $550 \times 180 \times 15 \mu m^3$).

area of interest and monitoring the local system response (i.e., the change of
the voltage drop) as a function of the beam coordinate. After having checked
the relaxation time of the sample due to an abrupt change in laser intensity,
we choose the chopping frequency (i.e., the frequency of switching on and off
the laser beam) to be $748\,Hz$. The change in the voltage drop is measured
via a lock–in amplifier (locked to the chopping of the laser diode) and plotted

in Fig. 4. Due to the local heating of the sample via the deposited laser light, the chemical potential is shifted and, thus, the voltage drop across the sample changes depending on the local constitution of the electronic properties of the sample. We point out that similar investigations using a conventional scanning–laser microscope have been carried out by other groups, in order to detect phase–slip phenomena in these materials (Itkis et al. (1986)).

Another highly interesting field of application is the investigation of superconducting films. As discussed in Gross (1994), a scanning microscope — the low–temperature scanning–electron microscope (LTSEM) and, of course, the LTSLM — can be used to measure the distribution of the critical current density and the distribution of the critical temperature, i.e., weak conducting areas are detectable. The advantages of the LTSLM compared to the LTSEM are as follows: (i) it needs, in principle, no vacuum, (ii) high magnetic fields are applicable, (iii) the scannable area is of a larger extension, and (iv) it is less expensive.

As an example for the investigation of such systems, we have tested a $BiO_{1.5}$–SrO–CaO–CuO multilayer with the dimensions $800 \times 700 \times 0.3 \mu m^3$. The sample was biased under constant current conditions with different current values. The temperature was kept constant at 87 K. For the electrical measurements, we used a four–terminal set–up to reduce the influence of the interfaces. The resulting gray–scale pictures are shown in Fig. 5. Figure 5 (a) displays a picture of the surface where the PIN diode mentioned above was used to measure the reflected light. Bright and dark areas correspond to regions of high and low reflectivity, respectively. The indium–made contact areas are bright and the dark areas are connected to the surface of the superconducting film. For convenience, the borders of the sample are marked with a white rectangle. In Fig. 5 (b) – (d), the measured change of the voltage drop across the sample during switching the laser beam on and off is coded in the gray scale. Here, bright areas are connected with areas of the sample which are sensitive against heating up due to the laser spot. That means that at these areas the superconductivity throughout the sample cross–section breaks down and becomes resistive. There, the superconducting current is quenched into the sensitive regions, resulting in high current densities. The extension where the sample reacts to the heating of the laser increases from Fig. 5 (b) to Fig. 5 (d), i.e., more and more parts of the sample are close to their local critical current density. For further details of the measurement process, see Gross (1994). The results can be understood with the knowledge that the substrate (MgO) on which the sample is processed is cleaved (Vengalis (1994)), resulting in steps on the surface. The steps weaken the superconducting film over a wide part of the cross–section. Only a small part serves as a superconducting connection.

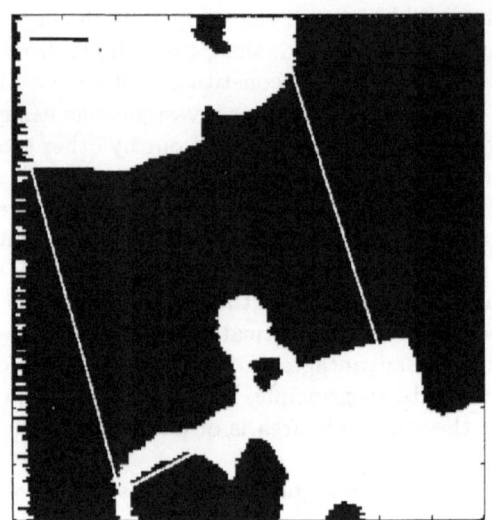

(a)

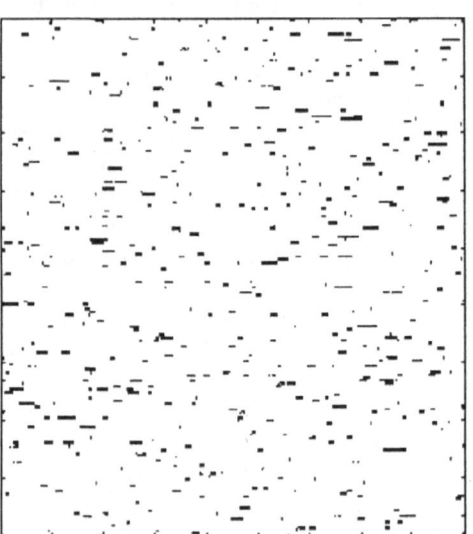

(b)

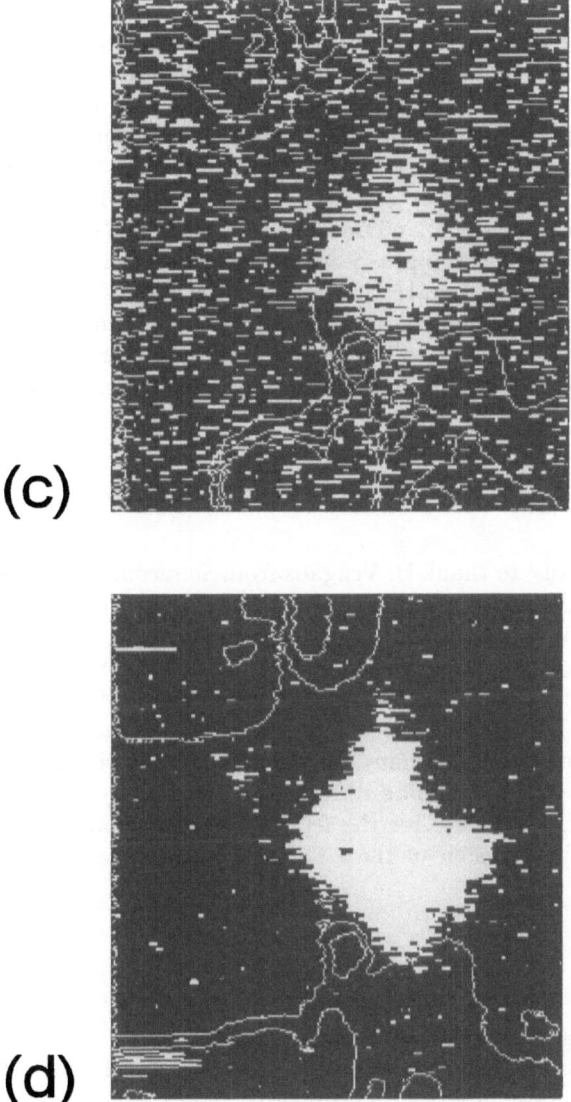

(c)

(d)

Fig. 5. (a) Picture of the surface of the superconducting film detected via the PIN diode. (b) – (d) Pictures of the change of the voltage drop induced by a laser beam switched on and off during scanning obtained for different values of the bias current. (Parameters: bias current $10\,\mu A$ (b), $170\,\mu A$ (c), and $6\,mA$ (d), temperature $87\,K$, sample dimensions $800 \times 700 \times 0.3 \mu m^3$).

Our future activities in applying the present LTSLM technique to various questions of charge carrier transport will certainly embrace different super– and semiconducting physical systems. For example, one can think of superconducting bulk materials (Dimitrenko et al. (1993)), interfaces (Dimitrenko et al. (1992)), and devices (Josephson junctions (Chang et al. (1985)), SQUIDS (Sivakov et al. (1994)) etc.), as well as of one– and two–dimensional conductors (quantum Hall effect (Merz et al. (1993)) etc.), semiconductor materials (Mayer et al. (1988)), and devices (solar cells, FET, etc.). Hereto, further improvements concerning a decrease of the spatial resolution limit (ending up at about $1\,\mu m$), a correction of the optical errors with the help of the computer used for the control of the microscope, and an enlargement of the scan area are possible. Due to its noncompromising overall construction, our microscope will have the potential to provide a challenging and powerful tool for looking at interesting new physics.

Acknowledgments

The authors would like to thank B. Vengalis from Semiconductor Physics Institute, Vilnius, Lithuania, for putting generously the superconducting films, Siemens AG Redwitz, Germany, the piezo–electric elements, watchmaker and jeweller Hacker Bayreuth, Germany, the bearings of the mirrors, and the company H.O.T. Nürnberg, Germany, for coating the thread of the focussing unit free of charge. Furthermore, we acknowledge Dr. Thomala from Richard Bergner GmbH Schwabach, Germany, for the assistance during the construction of the thread of the focussing unit. One of us (A.K.) wants to thank A. Willmann and H.-G. Wener for the fruitful collaboration during the development of the first version of the LTSLM and T. Nissel for continuous support.

References

Brandl A. and Prettl W., Festkörperprobleme **30**, 371 (1990).
Chang Jhi-Jiun, Ho C.H., and Scalapino D.J., Phys. Rev. B **31**, 5826 (1985).
Dimitrenko I.M., Zhuravel A.P., and Sivakov A.G., Sov. J. Low. Temp. Phys. **18**, 676 (1992).
Dimitrenko I.M., Grip P.A., Sivakov A.G., Turutanov O.G., and Zhuravel A.P., Sov. J. Low. Temp. Phys. **19**, 259 (1993).
Gross R. and Kölle D., Rep. Prog. Phys. **57**, 651 (1994).
Grüner G., *Density Waves in Solids* (Addison–Wesley, Reading, 1994).
Huebener R.P., in *Advances in Electronics and Electron Physics*, Vol. 70, ed. P.W. Hawkes (Academic Press, New York, 1988) p. 1.
Itkis M.E., Nad F.Ya., and Pokrovskii V.Ya., Sov. Phys. JETP **63**, 177 (1986).
Mayer K.M., Parisi J., Peinke J., and Huebener R.P., Physica D **32**, 306 (1988).
Merz R., Keilmann F., Haug R.J., and Ploog K., Phys. Rev. Lett. **70**, 651 (1993).

Peinke J., Parisi J., Rössler O.E., and Stoop R., *Encounter with Chaos* (Springer, Berlin, 1992).

Schlenker C. (ed.), *Low–Dimensional Electronic Properties of Molybdenum Bronzes and Oxides* (Kluwer, Dordrecht, 1989).

Sivakov A.G., Zhuravel A.P., Turatanov O.G., Dimitrenko I.M., Hilgenkamp J.W.M., Brons G.C.S., Flokstra J., and Rogalla H., Physica C **232**, 93 (1994).

Vengalis B., Deksnys A., Jukna A., Lisauskas V., Jasutis V., Balevicius S., and Flodström A.S., Physica C **235**, 699 (1994).

Wilson T. and Sheppard C.J.R., *Theory and Practice of Scanning Optical Microscopy* (Academic Press, New York, 1984).

Nonlinear Spatio-temporal Emission Dynamics of Broad Area Laser Diodes

Ingo Fischer[1], Ortwin Hess[2], and Wolfgang Elsäßer[1]

[1] Darmstadt University of Technology, Institute of Applied Physics, Schloßgarten-
str. 7, D-64289 Darmstadt, Germany
[2] DLR, Institute of Technical Physics, Pfaffenwaldring 38-40, D-70569 Stuttgart,
Germany

Abstract. We present the combination of microscopic model calculations and direct experimental evidence for ultrafast nonlinear spatio-temporal dynamics of a
laser system. In particular, we have studied the nearfield dynamics of a broad area
semiconductor laser on picosecond timescales. The mechanisms inducing the complex behavior can be identified as self-focusing, diffraction, and hole burning. We
find the onset of dynamical filamentation within the turn-on dynamics of such a
laser and demonstrate that this is a non-transient behavior.

1 Motivation

Spontaneous spatio-temporal pattern formation has found rising interest in
the field of nonlinear optical systems within the last decade [1, 2, 3, 4] The
motivation to study optical systems in this context is twofold. On the one
hand it has turned out that many transverse optical systems exhibit complex spatio-temporal dynamics covering a great variety of phenomena, partly
showing close analogies to hydrodynamical or chemical systems. The particular advantages of optical systems are that various different geometries with
different number of spatial dimensions can be easily realized, with different
nonlinearities and good parameter control. Besides, optical systems allow
easy access to near- and farfield, exhibit unique peculiarities such as light polarization properties and show dynamical time scales covering ~ 12 orders of
magnitude, ranging from seconds for photorefractive media down to picosec-
onds for semiconductor lasers. On the other hand the aim in laser design to
move towards higher intensities is associated with the stronger influence of
optical nonlinearities and to the realization of cavities with large apertures.
Due to that, spontaneous pattern formation is often found to disturb laser
applications and the detailed study of these phenomena will help to extend
the knowledge about general properties of optical systems that have been
disregarded for a long time.

In this contribution we present an optical system that exhibits complex
spatio-temporal dynamics with one transverse spatial dimension and on ultrafast timescales: the broad area laser diode and its nearfield emission properties. Broad area laser diodes are designed as semiconductor lasers with high

output power and therefore have large technological relevance. However, the increase in output power which is achieved by enlarging the transverse width of the active area from $w \sim 1\mu m$ for a usual narrow stripe laser to typical sizes of $w \sim 50\ \mu m$ to $500\ \mu m$ results in poor beam quality and coherence properties. This is related to the formation of transverse structures in the nearfield of the laser and the breakdown of the diffraction limitation in the corresponding farfield direction[6, 7, 8]. Recent experiments in combination with model calculations have given evidence for underlying strikingly complex spatio-temporal dynamics in the optical field [9, 10, 11, 12].

2   Modelling

Our simulations are based on the space-dependent Maxwell-Bloch equations for semiconductor lasers [5] which consist of a system of microscopic material equations self-consistently coupled to the system of partial differential equations for the counterpropagating complex optical fields $E^+(x, z, t)$ and $E^-(x, z, t)$, the polarizations $P^+(x, z, t)$ and $P^-(x, z, t)$ as well as for the macroscopic charge carrier density $N(x, z, t)$

$$\pm \frac{\partial}{\partial z} E^\pm + \frac{n_l}{c} \frac{\partial}{\partial t} E^\pm = \frac{i}{2} \frac{1}{k_z} \frac{\partial^2}{\partial x^2} E^\pm - \left(\frac{\alpha}{2} + i\eta \right) E^\pm + \frac{i}{2} \frac{\Gamma}{n_l^2 \epsilon_0 L} P^\pm \quad (1)$$

$$\frac{\partial}{\partial t} N = D_f \left(\frac{\partial^2}{\partial x^2} + \frac{\partial^2}{\partial z^2} \right) N + \Lambda - \gamma_{nr} N - W + G, \quad (2)$$

with the macroscopic generation rate

$$G = -\chi'' \frac{\varepsilon_0}{2\hbar} \left(|E^+|^2 + |E^-|^2 \right) + \left\{ \frac{(-i)}{2\hbar} [E^+ P^{+*} + E^- P^{-*}] + \text{c.c.} \right\}. \quad (3)$$

The macroscopic polarization variables are at every spatial point self-consistently linked to the microscopic polarization functions by summation over all momentum states $P^\pm = \frac{2}{V} \sum_{\mathbf{k}} d_{cv}(k)\, p_{\mathbf{r}}^\pm(k)$, where $d_{cv}(k)$ is the optical dipole matrix element; isotropic Fermi-surfaces are assumed, i.e. only $k = |\mathbf{k}|$ is considered. The dynamics of the *microscopic* polarization functions $p_{\mathbf{r}}^\pm(k)$ along with those of the distributions of electrons $f_{\mathbf{r}}^e(k)$ and holes $f_{\mathbf{r}}^h(k)$, where the index $\mathbf{r} = (x, z)$ indicates their parametrical dependence on space, is governed by the microscopic system of material equations presented and discussed in [5, 10]. In the macroscopic wave equations (1), passive waveguiding properties are represented by $\eta = \eta(x)$ and the confinement factor $\Gamma = \Gamma(x)$ [5]. The (linear) refractive index of the semiconductor material is denoted by n_l, c is the speed of light, $L = 200\mu m$ the total (longitudinal) length of the laser cavity and ϵ_0 the permittivity of free space. The wave number in (longitudinal) propagation direction is given by $k_z = n_l k_0$, with k_0 being the unperturbed wave number in the vacuum. In (1) and in the generation rate (3), the linear part of the polarization is conveniently included in the linear absorption term α of the semiconductor and in the imaginary part of the susceptibility [11], $\chi'' = \alpha n_l^2/2\pi\lambda$, respectively, with λ being the wavelength of the laser. Diffusion of charge carriers is represented in (2) via the diffusion coefficient D_f, where the influence of spatial variations of N on D_f is disregarded. The rates of nonradiative and spontaneous recombination are given by γ_{nr} and W, respectively [5]. The transversely dependent injection of charge carriers along the contact stripe is represented by the pumping term $\Lambda(x) = j(x)\eta_i/ed$, where the current density $j(x)$ is uniform along z and transversely only different from zero from $x_0 - w/2 \leq x \leq x_0 + w/2$, x_0, being the center of the laser stripe, $w = 100\mu m$ its width and $\eta_i = 0.8$ the fraction of carriers which actually reaches the active region; e is the electron charge and $d = 0.15\mu m$ the (vertical) thickness of the layer. The *longitudinal boundary conditions* are given by $E^+(x, z = 0, t) = -\sqrt{R_1}\, E^-(x, z = 0, t)$ and $E^-(x, z = L, t) = -\sqrt{R_2}\, E^+(x, z = L, t)$. They represent reflection of the optical fields at the facet mirrors at $z = 0$ and $z = L$ of the laser structure. The *transverse boundary conditions* $\partial E^\pm/\partial x = -\alpha_w E^\pm$ and $\partial N/\partial x = -\tilde{v}_{sr}N$ at $x = +w/2$ as well as $\partial E^\pm/\partial x = +\alpha_w E^\pm$ and $\partial N/\partial x = +\tilde{v}_{sr}N$ at $x = -w/2$ account for the strong absorption (constant α_w) of the optical fields outside the pumping stripe and surface recombination effects of the charge carriers.

3   Experimental Methods

Before directly comparing our numerical results with corresponding experimental ones we present a short outline of our experiments. The measurements of the turn-on dynamics in the nearfield of the broad area laser (BAL) have been performed using a GaAs/GaAlAs laser (SDL-2430-C) emitting at $\lambda = 814nm$. Its active region incorporating a quantum well active layer structure has a transverse width of $w = 100\,\mu m$. The threshold current of the laser is $I_{th} = 290\,mA$.

The measurement of the spatio-temporal nearfield dynamics requires the detection of single events with ps-time resolution and μm spatial resolution. For this reason we make use of a single-shot streak camera technique [9]. The experimental setup is schematically depicted in Fig. 1. A magnified image of the nearfield of the light intensity emitted at the output facet is obtained using a microscope objective which projects the output facet of the laser onto the input slit of a Hamamatsu streak camera C1587 with a single sweep unit M1952. This allows to measure single-shot spatio-temporal traces with a temporal resolution down to $\sim 10ps$. The main pulse generator serves as current source for the broad area laser and as trigger source for the streak camera. The broad area laser is driven with rectangular pulses of $25\,ns$ length with a rise time of about $3\,ns$. The repetition rate of the pulses is 100 Hz which ensures that junction heating can be minimized which otherwise would additionally contribute to thermal lensing effects. The amplitude of the pulses corresponds to an injection current of ~ 2 times the threshold current. The

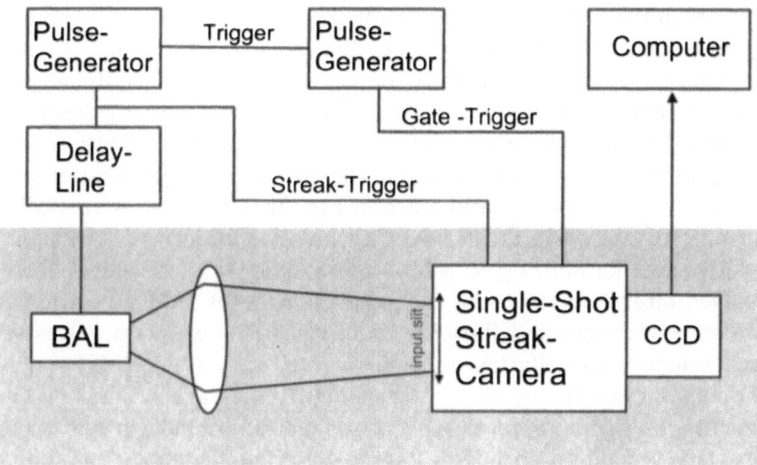

Fig. 1. Experimental setup for the spatio-temporally resolved nearfield measurements of a broad area semiconductor laser

streak camera is triggered by one single, arbitrarily chosen pulse delayed with respect to the current pulse, allowing to choose the temporal position of the spatio-temporal nearfield traces. The traces which are visualized on the phosphor screen of the streak tube are read out by a cooled high-sensitivity CCD-camera which is linked to a computer via a frame grabbber card where the traces can be stored.

For these driving conditions the laser emits several transverse (and longitudial) modes corresponding to a spectral width of less than 2 nm which is obtained from time-integrated measurements by an optical spectrum analyzer. The time-integrated nearfield shows a nearly periodic spatial modulation of the intensity, consisting of different transverse modes, which is a well known behavior called multi-filamentation, fundamental solitons or higher nonlinear modes. [15, 7, 8, 13, 14]. Under these operation conditions complex spatio-temporal behavior can be observed, as we will demonstrate in the following.

4 Experiments and Modelling Results

In the following we shall present and discuss the spatio-temporal intensity traces, extracted from the numerical modelling and directly compare them to those obtained in the measurements. In the first part we concentrate on the first $1.6ns$ after applying the current pulse to the laser. Fig. 2a depicts a numerically obtained nearfield trace, Fig. 2b an experimental streak trace for $I \sim 2 \cdot I_{th}$. The x-axis represents the position on the output facet of the BAL. The emitted intensity is linearly coded via greyscales such that light shading corresponds to high intensities and dark shading to low intensities. The temporal evolution is depicted in the y-direction. The time window depicted here has a length of 1.6ns. The modelled trace in Fig. 2a exhibits a spatially nearly homogeneous relaxation oscillation as response of the semiconductor laser to the current pulse. Starting already with the second relaxation oscillation the pulse breaks up into seperate spots ending up in an irregular spatio-temporal emission dynamics. The experimental trace in Fig. 2b also shows a spatially nearly homogeneous relaxation oscillation. The emission intensity at the right edge of the active area for $80\mu m \leq x \leq 100\mu m$ is a little bit lower which, however, has evolved after some ns to be the same as at other lateral positions. In correspondence with the modelling the intensity profile breaks up already after this first relaxation oscillation into irregular spatio-temporal dynamics where in the experiment the symmetry breaking occurs earlier than in the modelling results. Futhermore, after the first relaxation oscillation peak the emission intensity is modulated in the lateral direction showing 7 maxima. This modulation is present for the following temporal evolution of the intensity. This nearly static modulation is well-known as multi-filamentation or higher nonlinear modes [6, 7, 8]. This is different compared to the modelled behavior which is not surprising, because the waveguiding properties have not been adopted. We want to concentrate here on the dynamical changes within the lateral beam profile. In qualitative agreement with the modelling we find an irregular spatio-temporal behavior starting directly with the relaxation oscillations. This is not a transient behavior as we will demonstrate in the following.

Therefore, we have concentrated on a time window starting 8ns after turning on the injection current. In Fig. 3, once again, in a) a spatio-temporal trace obtained from the numerical modeling and in b) an experimental trace are depicted. Here, the length of the time window is 1.5 ns. The selected time window gives a representative image of the spatio-temporal dynamics in the post-relaxation oscillation period. In the complex spatio-temporal dynamics we observe irregular pulsing behavior and a striking phenomenon we call migrating optical filaments. The migrating filaments manifest themselves in spots of high intensity which seem to change their position of emission within a few hundred nanoseconds from one edge of the laser to the opposite one at the edges seemingly being reflected and migrating back. In the numerical results shown in Fig. 3a it takes about $\Delta t \simeq 500\,ps$ for an optical filament to migrate over the width of the active zone. The spatial width of such a filament is $s \simeq 10$ to $20\,\mu m$. In Fig. 3b we have depicted a typical experimental trace obtained by the single-shot streak camera corresponding to the situation in Figure 3a. In our experiments, the migrating filaments can indeed be clearly recognized bouncing back and forth.

From the modelling we can identify the interplay of self-focussing, diffraction and spatial hole burning which again depends on spatial carrier diffusion as the relevant physical mechanisms. In a simple picture this can be

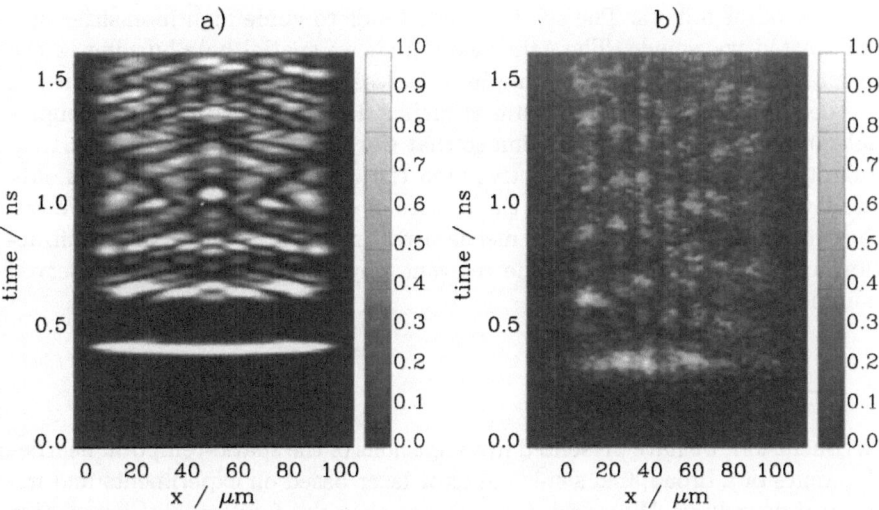

Fig. 2. Spatio-temporal nearfield traces of the emitted light intensity from 0..1.6 ns after turning the laser on. Traces obtained from: (a) numerical modelling, (b) experiments

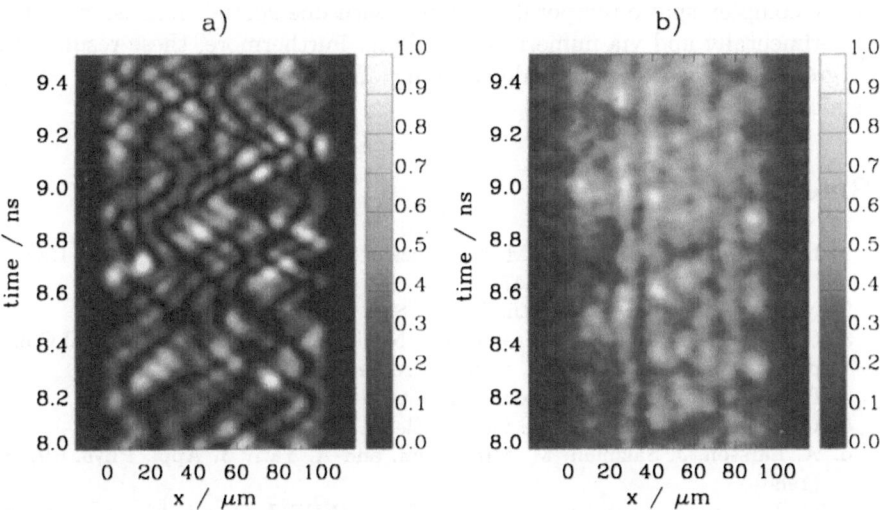

Fig. 3. Spatio-temporal nearfield traces of the emitted light intensity from 8.0..9.5 ns after turning the laser on. Traces obtained from: (a) numerical modelling, (b) experiments

understood as follows. The self-focussing tends to guide high intensities in a self-created waveguide. The gain medium shows spatial hole burning in the regions of high intensity resulting in a decrease of the optical gain. Thus, in the neighbouring regions the gain is higher. In addition, diffraction couples light into this neighbouring region so that the spot of high intensity starts to migrate. At the edges of the active area the coupling via diffraction is only effective to one side, leading to the change of direction of migration. We can thus identify a combination of mechanisms involving self-focussing, diffraction and carrier diffusion as the relevant physical effects for this observed nonlinear dynamical behavior.

5 Conclusion

In conclusion, we have presented investigations of the spatio-temporal nearfield dynamics of a broad area semiconductor laser based on experiments and numerical modelling. We could demonstrate that the fundamental interaction between light and the active semiconductor medium, i.e. the balance between diffraction, self-focussing, hole burning and carrier diffusion leads to spontaneous complex spatio-temporal dynamics, which occurs directly after turning on the injection current and prevails for later times. Thus, broad area semiconductor lasers have been demonstrated to be very interesting devices to study complex spatio-temporal dynamics with one spatial dimension, both, experimentally and via numerical modelling. Furthermore, these results are of great relevance for the understanding and improvement of high power semiconductor lasers.

References

1. L. A. Lugiato, F. Prati, L. M. Narducci, P. Ru, J. R. Tredicce, D. K. Bandy (1988) Phys. Rev. **A 53**, 3847
2. N. B. Abraham and W. Firth, J. Opt. Soc. Am. B. **7**, 951 (1990).
3. R. G. Harrison, J. S. Uppal (editors), "Nonlinear Dynamics and Spatial Complexity in Optical Systems", SUSSP Proceedings **41** (1992).
4. L. A. Lugiato, Chaos, Solitions & Fractals, **4**, 1251 (1994).
5. O. Hess and T. Kuhn, Prog. Quant. Electr., **20**, 85 (1996).
6. A. Larsson, J. Salzman, M. Mittelstein, and A. Yariv J. Appl. Phys. **60**, 66 (1986).
7. C. J. Chang-Hasnain, E. Kapon, E. Colas, IEEE J. Quant. Electr., **QE-26**, 1713 (1990).
8. R. J. Lang, A. G. Larsson, J. G. Cody, IEEE J. Quant. Electr., **QE-27**, 312 (1991).
9. I.Fischer, O.Hess, W.Elsäßer, E.Göbel, Europhys.Lett. **35**, 579 (1996)
10. O. Hess, Chaos, Solitons & Fractals **4**, 1597 (1994).
11. O. Hess, S. W. Koch, and J. V. Moloney, IEEE J. Quant. Electr., **QE-31**, 35 (1995).

12. J. Martin-Regalado, S. Balle, N.B. Abraham, IEE Proc.-Optoelectron. **143**, 17 (1996)
13. D. W. Mc Laughlin, J. V. Moloney, A. C. Newell, Phys. Rev. Lett., **51**, 75 (1983).
14. G. Khitrova, H. M. Gibbs, Y. Kawamura, H. Iwamura, T.Ikegami Phys. Rev. Lett., **70**, 920 (1993).
15. D. Mehuys, R. J. Lang, M. Mittelstein, J. Salzman, A. Yariv, IEEE J. Quant. Electr., **QE-23**, 1909 (1987).
16. A. H. Paxton, G. C. Dente J. Appl. Phys., **70**, 2921 (1991).

Gravitational Slowing Down of Clocks Implies Proportional Size Increase

Otto E. Rössler[1], Heinrich Kuypers[1], and Jürgen Parisi[2]

[1] Division of Theoretical Chemistry, University of Tübingen, Auf der Morgenstelle 8, D-72076 Tübingen, Germany
[2] Faculty of Physics, Department of Energy and Semiconductor Research, University of Oldenburg, D-26111 Oldenburg, Germany

Abstract. In searching for a potentially nonunique mapping between regions of high and low redshift, something simpler was found instead: A proportional size increase in a high-redshift region. Unlike special-relativistic effects, where slowed-down (departing) objects shrink in size, a size increase occurs in a gravitational potential. As the horizon of a black hole is approached, the rest of the universe when watched from there shrinks in size to zero. More general implications have yet to be derived.

Nonlinear dynamics and nonlinear space-time theory, conceived by Poincaré and Einstein, respectively, have a potential for fusion. A very simple phenomenon, the relativistic slowing-down of clocks in a gravitational potential, discovered by Einstein in 1907 [1], seemed worth scrutinizing under this aspect. It seemed to hold the promise of a "nonlinear mapping" being generated by a light ray that is sent back and forth between a high- and a low-redshift region. A nonunique relationship of the type of a "folded over!" logistic map was hoped to be implicit. This is not the case. Instead, an unexpected quite different effect was found.

The effect is illustrated with the aid of Fig. 1. Two horizontal lines, marked t_1 and t_2, are shown. They describe the flow of time valid for two clocks that live in two different gravitational potentials, the upper ("outer") clock being unchanged in its time course, the lower ("inner") clock being slowed down by a certain factor. In between the two lines, there is a network of light rays, sent back and forth by means of mirrors between the locations of the two clocks. The diagram looks like an anagram of the two capital letters "W" and "M". Hence its name.

The lower clock is slowed down by a factor of 4 in the picture. Think of the temporal wave length of a Cesium atom's light (or a constant multiple of it), represented by the two symbolic sine waves. One sees that the light which arrives upstairs, emitted at the beginning and the end of a full wave downstairs, is indeed "redshifted" compared to the temporal behavior of a Cesium atom upstairs. Similarly, the light which arrives downstairs from the

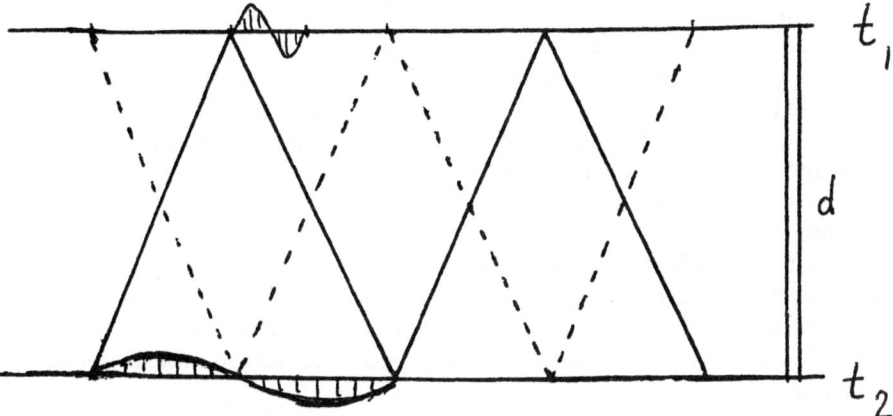

Fig. 1. "WM-diagram": Bidirectionally unique connections formed by light signals between a non-redshifted clock (upper line) and a redshifted clock (lower line) in a strong gravitational field. Two sine-shaped, locally equivalent periodic time signals are also indicated. So is the mutual distance (d). The implication - a different value of d on the two levels - is discussed in the text.

beginning and the end of the upper wave is "blueshifted" compared to the temporal behavior of a Cesium atom downstairs.

The duration of a "roundtrip signal" is therefore different when measured by the upper and the lower clock. However, by Einstein's general covariance principle, the velocity of light is the same at both places. Therefore, the distance (d) between the two places is four times smaller when evaluated downstairs than when evaluated upstairs.

This result has unexpected consequences. All objects in a slowed-down region seemingly increase in size by the same factor by which their time course is slowed down. This follows from general covariance, which states that locally, everything has to look unchanged when watched from a free-falling vantage point. Therefore objects like meter sticks of a given size necessarily "look normal" in each of the two frames. That is, light is bound to pass by the same number of atoms (atoms of iron or molecules of lignin, respectively) that make up the stick, in the same amount of time, in the two frames.

This size change of sticks is, by the way, in accord with a prediction made by Boscovich [2]. What are the consequences?

As the Schwarzschild radius of a black hole is approached, z (the local redshift factor by which clocks are slowed down), approaches infinity as is well known [3]. Therefore, the external world is bound to shrink to zero diameter from the vantage point of a local observer who is comoving with the clock in question. Conversely, if the size of the black hole is assumed fixed relative to the external world (as it is according to current belief), the observer must be "wrapped around" the black hole infinitely many times shortly before reaching the Schwarzschild radius. Obviously, the two assumptions made - constant size of the black hole; growing size of the observer - are mutually incompatible.

Another unsolved problem concerns mass and charge. According to general covariance, they ought to be unchanged in each frame. But the size change and the time change, combined, both amount to a kind of "dilution effect". Compared to the outside world, mass and charge may therefore be reduced.

To conclude, a counterintuitive new implication of the equivalence principle (and general relativity) appears to exist. Gravitational redshift - the slowing down of clocks in a gravitational field - appears to go hand in hand with a proportional increase in size. As a consequence, highly dilute black holes (for which tidal forces are negligible [3]) nevertheless cause an unbounded change in size: "flattening". Most of the implications of the new scaling law are open.

Acknowledgments We thank Andreas Dress, Dieter Fröhlich, N. Nescit, Normann Kleiner, Joachim Peinke and Christof Kruelle for discussions. For J.O.R.

References

[1] A. Einstein (1907). Relativitätsprinzip und die aus demselben gezogenen Folgerungen. Jahrbuch der Radioaktivität 4, 411-462.

[2] R.J. Boscovich (1755). On space and time as they are recognized by us (in Latin). For an English translation, see: O.E. Rössler, Boscovich covariance. In: Beyond Belief, Randsomness, Prediction and Explanation in Science (J.L. Casti and A. Karlqvist, eds.), pp. 69-87. CRC Press, Boca Raton 1991.

[3] O.E. Rössler, H. Kuypers, H.H. Diebner and M.S. El Naschie (1997), Almost black holes, an old new paradigm. Chaos, Solitons and Fractals (in press).

Lecture Notes in Physics

For information about Vols. 1–469
please contact your bookseller or Springer-Verlag

New Series m: Monographs